AMINO ACID - NEW INSIGHTS AND ROLES IN PLANT AND ANIMAL

Edited by **Toshiki Asao**
and **Md. Asaduzzaman**

Amino Acid - New Insights and Roles in Plant and Animal
http://dx.doi.org/10.5772/66064
Edited by Toshiki Asao and Md. Asaduzzaman

Contributors

Shengfa Liao, Taiji Wang, Mark Crenshaw, M. Shamimul Hasan, Guoyao Wu, Danilo Vargas Gonçalves Vieira, Fernando Guilherme Guilherme Perazzo Costa, Matheus Ramalho De Lima, José Geraldo De Vargas Júnior, Danilo Teixeira Cavalcante, Talita Pinheiro Bonaparte, Marilú Santos Sousa, Érika Martins Figueireido, Sarah Gomes Pinheiro, Ana Carolina Müller Conti, Mamoru Tanaka, Takeaki Okamoto, Paolo Iadarola, Simona Viglio, Marco Fumagalli, Monica Di Venere, Maddalena Cagnone, Stefan Loic, Xupeng Cao, Song Xue, Xuran Fan, Uener Kolukisaoglu, Juan Suarez, Xiao-Ting Zou, Xinyang Dong, Takuji Ohyama, Norikuni Ohtake, Yuki Ono, Kotaro Tsutsumi, Manabu Ueno, Sayuri Tanabata, Takashi Sato, Yoshihiko Takahashi, Kuni Sueyoshi, Oaikhena Zekeri Esezobor, Amraibure Odia, Vanesa Fernandez, Adina-Elena Segneanu, Silvia Velciov, Sorin Olariu, Florentina Cziple, Daniel Damian, Ioan Grozescu, Fumio Tanaka, Arthit Nueangaudom, Kiattisak Lugsanangarm, Somsak Pianwanit, Sirirat Kokpol, Nadtanet Nunthaboot, Seiji Taniguchi, Haik Chosrowjan, Mario Ordóñez, José Luis Viveros-Ceballos, Ivan Romero-Estudillo

Published by InTech
Janeza Trdine 9, 51000 Rijeka, Croatia

Publishing Process Manager Dajana Pemac
Technical Editor SPi Global
Cover InTech Design team

First published June, 2017
Printed in Croatia
Legal deposit, Croatia: National and University Library in Zagreb

Additional hard copies can be obtained from orders@intechopen.com

Amino Acid - New Insights and Roles in Plant and Animal, Edited by Toshiki Asao and Md. Asaduzzaman
p. cm.
Print ISBN 978-953-51-3241-7
Online ISBN 978-953-51-3242-4

For EU product safety concerns:
IN TECH d.o.o., Prolaz Marije Krucifikse Kozulić 3, 51000 Rijeka, Croatia,
info@intechopen.com or visit our website at intechopen.com.

Meet the editors

Dr. Toshiki Asao is a specialist in hydroponic crop production and professor at Department of Agriculture, Faculty of Life and Environmental Science, Shimane University, Japan. Dr. Asao is a native of Kyoto, Japan. His main research focus is the development of hydroponic techniques for vegetables and ornamentals in greenhouses and also the development of specialty vegetables through hydroponics under controlled environment agriculture providing human health benefits beyond basic nutrition. His other research project is studying autotoxicity in vegetables and ornamentals in hydroponics and developing possible control measures. He has published a number of scientific articles, book chapters, and edited books.

Dr. Asaduzzaman is a native of Bangladesh and received his PhD degree majoring in Bioproduction Science from Tottori University, Japan, and currently working at Horticulture Research Centre, Bangladesh Agricultural Research Institute, Bangladesh. His main research focuses are the development of hydroponic techniques for fruits, vegetables and ornamentals in greenhouse and also production of specialty crops under controlled environment agriculture. His other research project includes studying the autotoxicity, a phenomenon of intraspecific allelopathy in vegetables and ornamentals through hydroponics, and developing suitable control measures to overcome. He has published a number of original research articles, book chapters, and edited books. He has been awarded Gold Medal from Bangladesh Agricultural University in 2011 and BAS-TWAS 2016 Prizes for the Young Scientists from Bangladesh.

Contents

Preface

Amino acids are the nitrogenous compound that forms the basic component of all living cells. It has a great role in the growth and development of plants and animals. It can either be applied exogenously on plant leaves or supplemented with animal feed. In addition, it has the positive impact as foliar spray on plants under stress conditions. While in animals, amino acid improves product quality through the influence in different biochemical processes of protein synthesis.

This book provides new aspects of amino acid structure, synthesis reactions, dietary application, and also metabolism in plants. The first few chapters describe the therapeutic uses, antiallergic effects, new aspects in the D-amino acid structure, historical background of desmosines, and stereoselective synthesis of α-aminophosphonic acids. The role of amino acids on plant includes new insights and aspects of D-amino acids, metabolism and transport in soybean, changes during energy storage compounds accumulation, etc. Dietary supplementation of amino acids for Japanese quails, laying hens, and finishing pigs is described in the later part.

Interesting research on amino acids from around the world is brought together to produce this resource for teachers, researchers, and advanced students of biological science.

Publication of this book would have been impossible without the interesting research work of many researchers around the world. Acknowledgment goes to the chapter contributors, who volunteered their valuable time to publish this book.

Dr. Toshiki Asao
Department of Agriculture
Faculty of Life and Environmental Science
Shimane University, Matsue, Japan

Dr. Md. Asaduzzaman
Horticulture Research Centre
Bangladesh Agriculture Research Institute
Gazipur, Bangladesh

Section 1

Role of Amino Acids in General

Chapter 1

Therapeutic Uses of Amino Acids

Amraibure Odia and Oaikhena Zekeri Esezobor

Additional information is available at the end of the chapter

http://dx.doi.org/10.5772/intechopen.68932

Abstract

Amino acids, which are the building blocks of peptides and proteins, are indispensable chemicals needed by the body for optimal metabolism and proper body functioning. Classified as essential, nonessential and conditionally essential, amino acids play vital roles in the body such as in protein synthesis and as precursors in the production of secondary metabolism molecules. Amino acid oxygenases also play vital metabolic roles such as in prevention of diseases; as a result, amino acids and their oxygenases isolated from various organisms are potent candidates in treatment of diseases which include cancers, inflammations, as well as antibacterial agents.

Keywords: amino acids, oxygenase, therapeutic, bacteria, flavoprotein, enzyme

1. Introduction

The use of amino acids in medicine today continues to be explored using clinical research and applications. Amino acids play several roles in the body [1]; they are essential in the synthesis of proteins and precursors in the formation of secondary metabolism molecules [2], and as a result, amino acids are found in all parts of the body [1].

Amino acids are mainly found as L-enantiomers in all forms of life. However, significant amounts of D-amino acids are produced by bacteria, which are the major producers of D-amino acids [3]. In bacteria, D-amino acids are involved in the synthesis and cross-linking of peptidoglycan [4].

In humans, amino acids participate in various physiological processes, such as skeletal muscle function, atrophic conditions, sarcopenia, and cancer. They play key roles in cell signalling, homeostasis, gene expression, synthesis of hormones, phosphorylation of proteins and also possess antioxidant abilities [2, 5]. Amino acids are also key precursors in the synthesis of low molecular weight nitrogenous compounds, which have numerous biological importance. The existence of amino acids and their metabolites, such as glutathione, polyamines, taurine, serotonin and thyroid hormones, in physiological amounts is important for proper body functions [5].

Traditionally, amino acids were classified as essential and nonessential amino acids [5]. However, another class known as conditionally essential amino acids now exists. These classifications are based on whether the body is able to synthesise the amount that it needs for metabolic maintenance [1]. Essential amino acids are those that cannot be synthesised or those that are synthesised inadequately by the body relative to needs and hence must be obtained from diets to meet physiological requirements. Amino acids which the body can synthesise in sufficient amounts to meet the body's maximum requirements are known as nonessential amino acids. Conditionally essential amino acids are those which the body can synthesise in adequate amounts, but under situations of higher utilisation rate, the body obtains them from diets in order to meet optimal requirements [5].

Inadequate intake of amino acids from diets and below optimal synthesis by the body may expose an individual to amino acid deficiency symptoms, such as weight loss, poor growth and development. Because amino acids are not stored in the body for long periods of time and in sufficient amount, meeting maximum daily requirements from diets and/or amino acid supplements is necessary for healthy living [1].

The therapeutic use of amino acids presents a viable and important option for natural medicine. Some of the most prominent areas of therapeutic applications of amino acids are for treatment of brain metabolism and neurotransmission imbalances. Other areas in which amino acids also find key applications are immune function, cardiovascular and gastrointestinal (GI) health [1], treatment of liver diseases, fatigue, skeletal muscle damage, cancer prevention, burn, trauma and sepsis, maple urine disease and diabetes [6].

2. D-Amino acids

Bardaweel [2] investigated the antibacterial activities of some D-amino acids, which include D-alanine, D-lysine, D-serine and D-proline (**Figure 1**), against *Bacillus subtilis, Staphylococcus aureus, Staphylococcus epidermidis, Escherichia coli, Pseudomonas aeruginosa* and *Xanthomonas vesicatoria* and reported that the amino acids exhibited relatively low inhibitory effectiveness against the pathogens. However, D-lysine, followed by D-alanine, was more potent than the other amino acids examined, even though their minimum inhibitory concentration (MIC) values were in the millimolar ranges.

A study by Hochbaum et al. [7] reported that D-amino acids were effective in preventing biofilm (communities of cells held together by a self-produced extracellular matrix typically consisting of protein, exopolysaccharide and often DNA) development in *S. aureus*, which is a leading cause of hospital-acquired infections. The D-isomers that were found to be active in inhibiting biofilm formation were D-phenylalanine (**Figure 2a**) and D-proline and D-tyrosine (**Figure 2b**). Mixture of D-tyrosine, D-proline and D-phenylalanine was more effective in preventing biofilm formation

http://dx.doi.org/10.5772/intechopen.68932

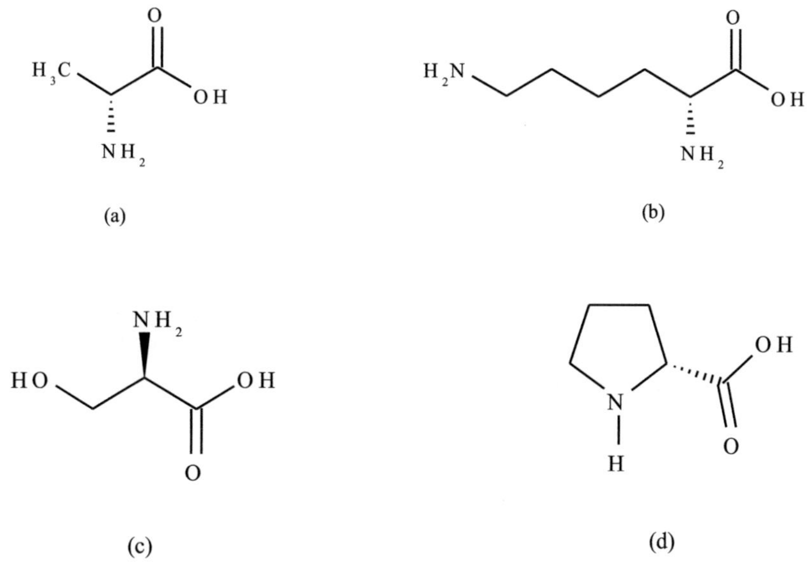

Figure 1. Chemical structures of (a) D-alanine, (b) D-lysine, (c) D-serine and (d) D-proline.

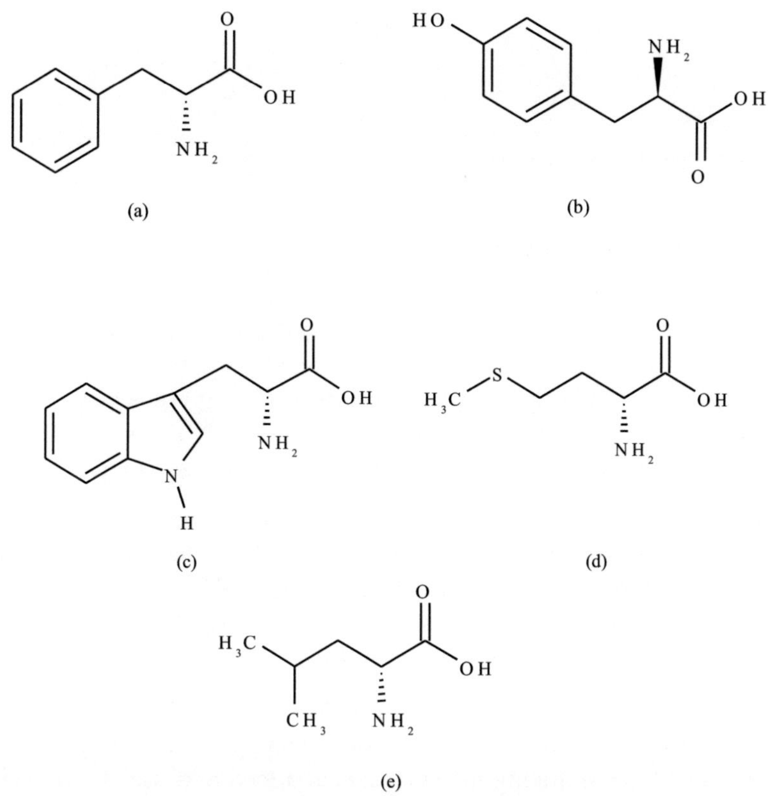

Figure 2. Chemical structures of (a) D-phenylalanine, (b) D-tyrosine, (c) D-tryptophan, (d) D-methionine and (e) D-leucine.

than the mixture of D-tryptophan, D-methionine, D-leucine (**Figure 2c–e**) and D-tyrosine. Earlier, the study by Kolodkin-Gal et al. [8] reported that D-tyrosine, D-leucine, D-tryptophan and D-methionine were active in inhibiting biofilm formation by *B. subtilis*, whereas D-isomers of other amino acids, such as D-phenylalanine, were inert in inhibiting biofilm formation.

The therapeutic potential of D-amino acid oxidase (DAAO) inhibitors, a flavoenzyme that degrades D-amino acids through the process of oxidative deamination, in schizophrenia patients has also been studied [9]. DAAO catalyses the metabolism of D-serine, a known full agonist at the allosteric glycine binding site of the *N*-methyl- D-aspartic acid (NMDA) (**Figure 3**) receptor, which has been reported to improve negative and cognitive symptoms of schizophrenia [10]. As a result, several studies have focused on the design and development of selective DAAO inhibitors, which when administered to schizophrenia patients have been shown to increase the concentrations of D-serine in the blood and the brain [9].

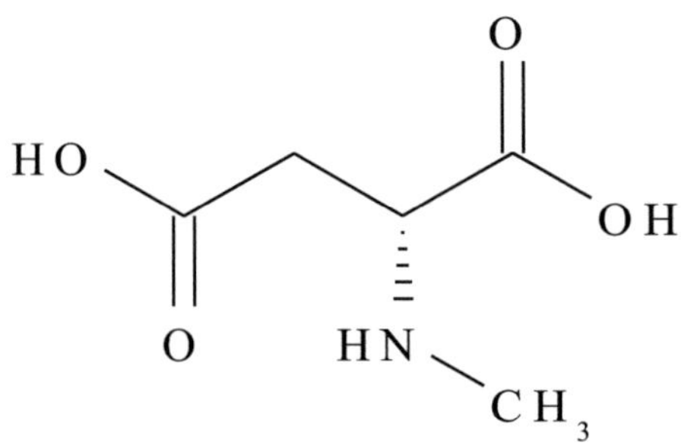

Figure 3. Chemical structure of *N*-methyl-D-aspartic acid.

3. Branched-chain amino acids

Branched-chain amino acids (BCAAs) are essential amino acids required for synthesis of body proteins. BCAAs play vital roles in regulation of protein synthesis and maintenance of glutamate-glutamine levels in the body. BCAAs are oxidised during high-energy-demanding and stressful conditions, and as a result, limit their accessibility in body tissues, which in the long run upsets mechanisms controlling the synthesis of proteins and body glutamate-glutamine pool [6].

The use of BCAA supplements in treatment of diseases is a developing nutritional strategy in disease management. Several studies have reported that when BCAA supplements are administered, patients experience improvements in health, although there are some disease conditions where BCAAs showed no effects. However, increased levels of BCAAs in the body have been observed to be involved in disease pathology [6].

The BCAAs—leucine, isoleucine (**Figure 4**) and valine—are metabolically very active. In the peripheral tissues, they may be oxidised to produce energy or act as anticatabolic factors (particularly leucine) by stimulating the synthesis and lowering the rate of degradation of muscle protein [11]. They are three of nine essential amino acids that are not synthesised by

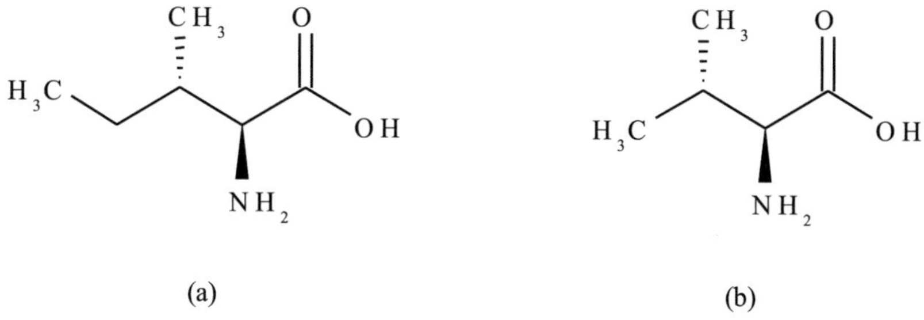

Figure 4. Chemical structures of (a) isoleucine and (b) valine.

the human body and therefore must be obtained from diet. Approximately 35% of indispensable muscle proteins and 40% of total amino acids required by mammals are composed of these BCAAs [6]. The three BCAAs either together or with leucine alone can stimulate protein synthesis and can also inhibit protein degradation depending on the context [12].

Although most of the amino acids are degraded in the liver, BCAAs are primarily broken down in the extrahepatic tissues (muscle, adipose, kidney and brain). Catabolism of these amino acids is initiated by transamination reaction with α-ketoglutarate to form glutamate and branched-chain keto acids (BCKAs). Then, the glutamate is converted to glutamine by the action of the glutamine synthetase enzyme [6]. Glutamine (**Figure 5**), which is derived mainly from skeletal muscles [13], is one of the most abundant amino acids in the body [14]. It is utilised readily by the liver, kidneys, GI tract and the immune system. Glutamine transports nitrogen and carbon inside the organs and plays a vital role in proper immune system function and GI integrity, as well as maintenance of overall amino-acid balance in the body [13, 14].

BCAAs have been considered as potential intervention for repair of damaged muscle tissues and some studies have suggested that BCAA supplementation may improve the repair of re-induced damaged muscle [15]. BCAAs, particularly leucine, have been reported to possess anabolic potential. They stimulate the metabolic pathways that initiate protein synthesis [16] and are involved in the control of protein breakdown (proteolysis) in impaired muscles [17].

The transamination product of BCAAs, α-ketoisocaproate (α-KIC) (**Figure 6**), is known to prevent the enzymatic action of branched-chain α-keto dehydrogenase complex (BCKDH), which increases the oxidation of BCAAs [18]. These anabolic potentials of BCAAs have led

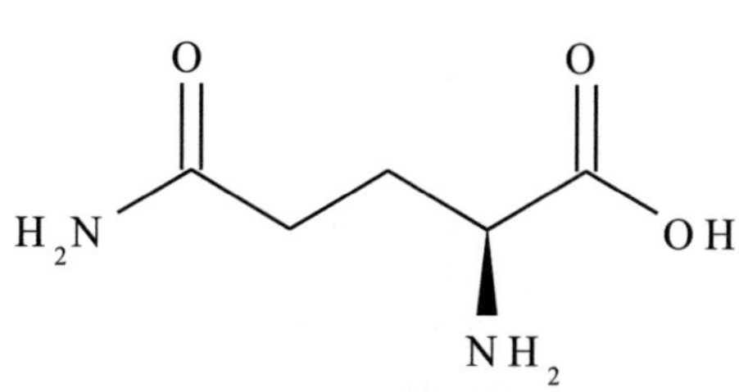

Figure 5: Chemical structure of glutamine.

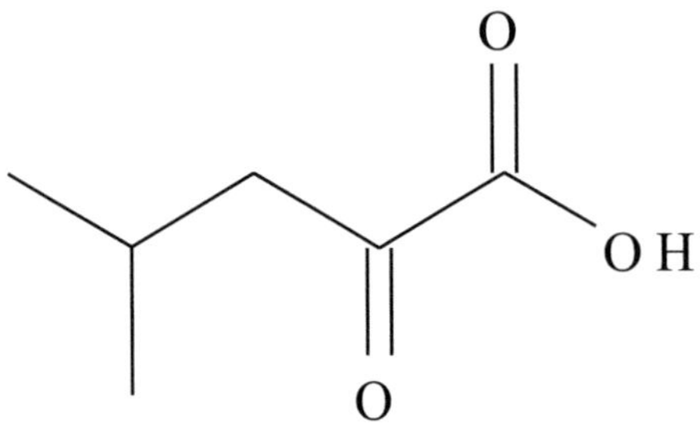

Figure 6. Chemical structure of α-ketoisocaproate.

to suggestions that BCAA supplementation could stimulate repair of impaired muscles by reducing oxidation of proteins, promoting the formation and development of muscle components and improving muscle functioning ability [15].

In a study by Soomro et al. [11], BCAAs were reported to be effective in the management of hepatic encephalopathy. While comparing the recovery and recurrence of hepatic encephalopathy of patients who were on BCAAs given initially intravenously and then orally with those of group without BCAAs, the results showed that those on BCAAs showed early improvements and recovery and subsequently on follow-up visits at 4 months. Improvements were observed in ammonia levels which were initially raised, but however decreased subsequently at 6 days and on 4 months of follow-up. In comparison to patients who did not receive BCAAs, the albumin levels of patients administered with BCAAs also increased from initial reading noticed at 4 months of follow-up.

BCAAs act as energy substrates, substrates for gluconeogenesis and modulators of muscle protein metabolism. These properties make their use in amino acid-enriched solutions theoretically appropriate for the management of the metabolic alterations that occur in sepsis. Forty-five percent branched-chain amino acid-enriched solutions have been suggested to intensify synthesis of proteins in the liver [19] and proteins whose plasma levels are elevated or reduced (acute-phase proteins) during severe illness [20]. Because acute-phase proteins might play vital roles in a septic patient's defence mechanisms against infections, the administration of solutions containing high amounts of BCAAs may increase the likelihood of quick recovery and survival for such patients [19].

4. Amino acid oxidases

Amino acid oxidases (AAOs) are flavoenzymes that catalyse the oxidative deamination of amino acids to α-keto acids with the generation of ammonia and hydrogen peroxide [21], as shown in (**Figure 7**) [22]. Depending on the amino acid isomer used as a substrate, it is possible to differentiate between L-amino acid oxidases and D-amino acid oxidases [23]. However, of particular interest in AAOs are the L-amino acid oxidases (LAAOs) because AAOs are highly specific for L-amino acids, and generally hydrophobic amino acids (such as phenylalanine,

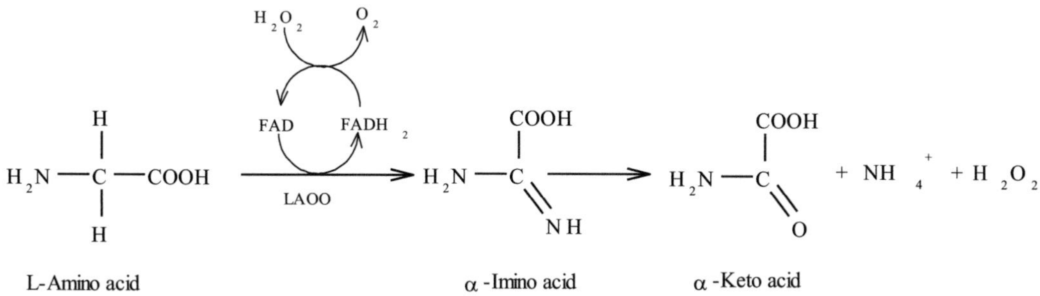

Figure 7. Mechanism of chemical reaction catalysed by L-amino acid oxidases (LAAOs) [22].

tryptophan, tyrosine and leucine) [24] are the best substrates [25]. These flavoenzymes are found in diverse organisms, such as bacteria, fungi, algae, fish, snails as well as venoms of snake families [26].

Several kinds of LAAOs have been isolated and their crystal structure presented (**Figure 8**). Most of the LAAOs isolated and characterised structurally to date are flavoproteins, which exist as dimers. The subunits in the structures are joined by noncovalent bonds with flavin mononucleotide (FMN) or flavin adenine dinucleotide (FAD) [27]. The venoms of many snakes have characteristic yellow colour which has been attributed to the flavin component in LAAOs isolated from the snakes. The flavins have also been reported to contribute to the toxicity of such venoms because of the oxidative stress caused by H_2O_2 production [26].

In a study by Joseph et al. [24], LAAOs isolated from snake venom induced platelet aggregation and cytotoxicity in various cancer cell lines. The enzyme also showed antibacterial activity by inhibiting the growth of Gram-positive *(B. subtilis)* and Gram-negative *(E. coli)* bacteria. Snake venom LAAOs have also been reported to exhibit oedema-inducing, apoptotic-inducing as well as anti-bacterial, anti-coagulant and anti-HIV effects [25].

In a study of the king cobra *(Ophiophagus hannah)* venom L-amino acid oxidase, Lee et al. [32] reported that the heat-stable enzyme exhibited very potent anti-proliferative activity against human breast and lung tumorigenic cells, but not in their non-tumorigenic counterparts. They further reported that after eight weeks of treatment of mice samples with the isolated LAAOs, the enzyme markedly inhibited PC-3 tumours when compared to the control group.

In another study of the heat stable L-amino acid oxidase isolated from the king cobra *(O. hannah)* venom, Phua et al. [33] reported that the LAAO showed antibacterial activity against Gram-positive bacteria, such as *B. subtilis, Bacillus cereus, S. aureus* [including methicillin-resistant *S. aureus* (MRSA)], and *S. epidermidis*. The LAAO also showed antibacterial activity against gram-negative bacteria such as *Salmonella enteridis, P. aeruginosa, Serratia marcescens, Klebsiella pneumoniae, E. coli* and *Enterobacter cloacae*. They further reported that the snake venom showed the highest antibacterial activity against *Staphylococcus* spp. and *E. coli*, even though the inhibition zones increased with increasing concentration of venom in all cases.

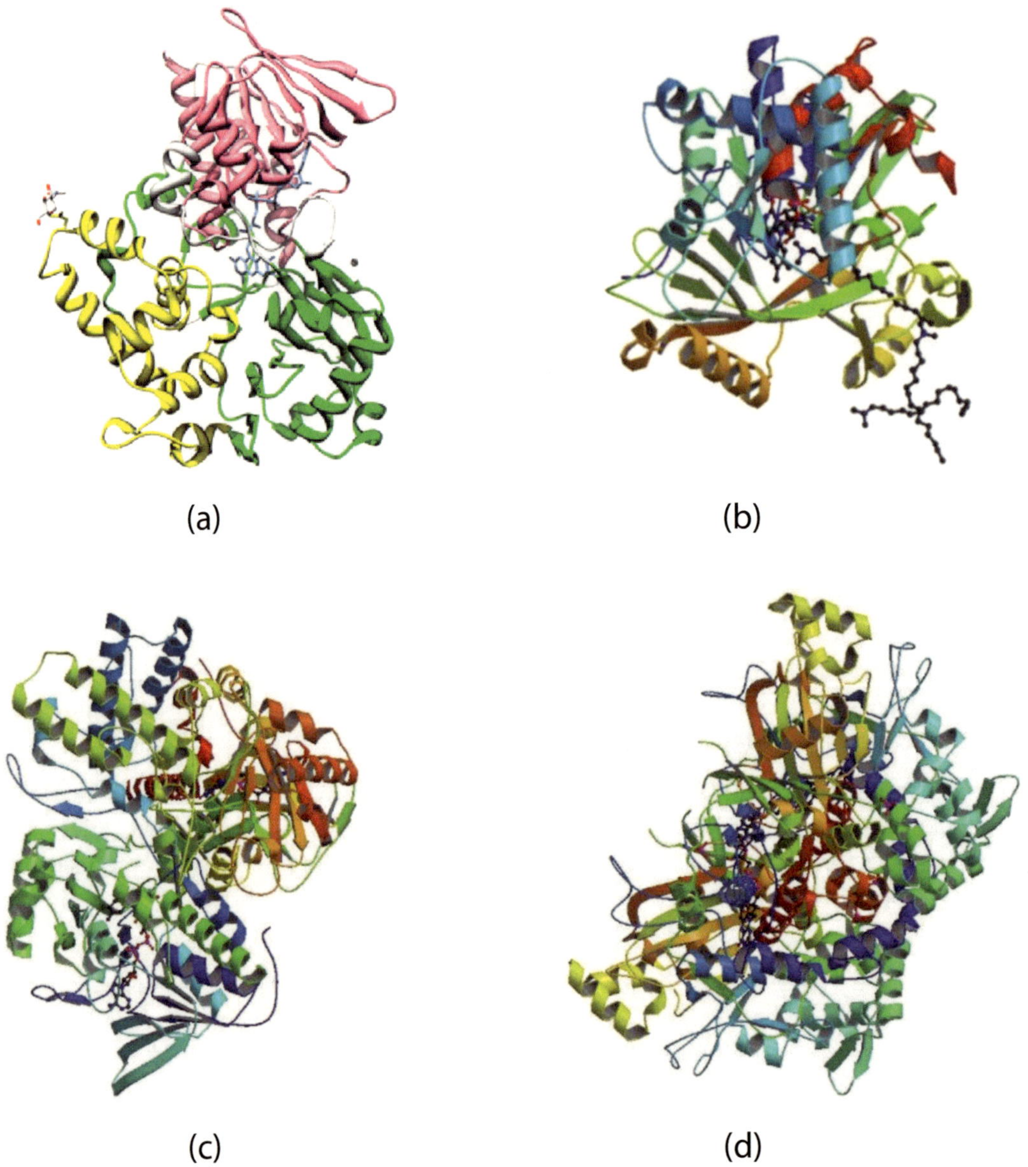

Figure 8. Crystal structures of LAAOs isolated from (a) the venom of *Vipera ammodytes* [28], (b) *Proteus vulgaris* [29], (c) *Rhodococcus opacus* [30] and (d) *Streptomyces* sp. [31].

LAAOs isolated from the venom of *Calloselasma rhodostoma* were also reported by Costa et al. [34] to induce acute inflammatory responses *in vivo*, with recruitment of neutrophils and release of IL-6, IL-1β, LTB4 and PGE2. An *in vitro* study showed IL-6 and IL-1β production by peritoneal macrophages stimulated with LAAOs, which was dependent on the activation of the Toll-like receptors TLR2 and TLR4. They also reported that LAAOs promoted apoptosis of HL-60 and HepG2 tumour cells mediated by the release of hydrogen peroxide and activation of immune cells, resulting in oxidative stress and production of IL-6 and IL-1β that triggered a series of events, such as activation of caspase 8, 9 and 3, and the expression of the pro-apoptotic gene BAX.

L-Lysine α-oxidase (LysOx) isolated from the extracellular growth medium of *Trichoderma cf. aureoviride* was reported by Pokrovsky et al. [35] to have shown considerable cytotoxicity and anti-tumour effects *in vitro* against a panel of murine and human tumour cell lines and *in vivo* on murine tumours and on animals with human tumour xenografts (breast cancer SKBR3, melanoma Bro, colon cancer HCT116 and ovarian adenocarcinoma SCOV3). L-Amino acid oxidase isolated from *Bothrops marajoensis* has also been reported [36] to cause nephrotoxicity in isolated perfused kidney and cytotoxicity in MDCK renal cells.

5. Conclusion

The therapeutic effects of amino acids and amino acid oxygenases present interesting prospects for the use of these chemicals in management of diseases. The future potential of amino acid-based therapeutics in treatment of diseases and the diverse effects of naturally occurring amino acid oxygenase is far reaching.

Author details

Amraibure Odia[1]* and Oaikhena Zekeri Esezobor[2]

*Address all correspondence to: amraibureodia@yahoo.com

1 Department of Chemistry, Ambrose Alli University, Ekpoma, Nigeria

2 Faculty of Chemistry and Pharmacy, University of Regensburg, Regensburg, Germany

References

[1] Meletis CD, Barker JE. Therapeutic uses of amino acids. Alternative and Complementary Therapies. 2005;**11**:24-28

[2] Moran-Palacio EF, Tortoledo-Ortiz O, Yañez-Farias GA, Zamora-Álvarez LA, Stephens-Camacho NA, Soñanez-Organis JG, Ochoa-López LM, Rosas-Rodríguez JA. Determination of amino acids in medicinal plants from Southern Sonora, Mexico. Tropical Journal of Pharmaceutical Research. 2014;**13**(4):601-606

[3] Bardaweel SK. D-Amino acids: Prospects for new therapeutic agents. Journal of Medical and Bioengineering. 2014;**3**(3):195-198

[4] Vollmer W, Blanot D, de Pedro MA. Peptidoglycan structure and architecture. FEMS Microbiological Reviews. 2008;**32**:149-167

[5] Wu G. . Amino acids: Metabolism, functions, and nutrition. Amino Acids. 2009;**37**:1-17

[6] Tamanna N, Mahmood N.. Emerging roles of branched-chain amino acid supplementation in human diseases. International Scholarly Research Notices. 2014;**2014**:1-8

[7] Hochbaum AI, Kolodkin-Gal I, Foulston L, Kolter R, Aizenberg J, Losick R. Inhibitory effects of D-amino acids on *Staphylococcus aureus* biofilm development. Journal of Bacteriology. 2011;**193**(20):5616-5622

[8] Kolodkin-Gal I, Romero D, Cao S, Clardy J, Kolter R, Losick R. D-Amino acids trigger biofilm disassembly. Science. 2010;**328**(5978):627-629

[9] Smith SM, Uslaner JM, Hutson PH. . The therapeutic potential of D-amino acid oxidase (DAAO) inhibitors. The Open Medicinal Chemistry Journal. 2010;**4**:3-9

[10] Ferraris DV, Tsukamoto T. Recent advances in the discovery of D-amino acid oxidase inhibitors and their therapeutic utility in schizophrenia. Current Pharmaceutical Design. 2011;**17**(2):103-111

[11] Soomro AA, Devrajani BR, Ghori RA, Lohana H, Qureshi GA. Role of branched chain amino acids in the management of hepatic encephalopathy. World Journal of Medical Sciences. 2008;**3**(2):60-64

[12] Nair KS, Short KR. Hormonal and signalling role of branched-chain amino acids. The Journal of Nutrition. 2005;**135**(6):1547S-1552S

[13] Kusumoto I. Industrial production of L-glutamine. The Journal of Nutrition. 2001;**131**: 2552S-5S

[14] Kulkarni C, Kulkarni KS, Hamsa BR. L-Glutamic acid and glutamine: Exciting molecules of clinical interest. The Indian Journal of Pharmacology. 2005;**37**(3):148-154

[15] da Luz CR, Nicastro H, Zanchi NE, Chaves DFS, Lancha AH Jr. Potential therapeutic effects of branched-chain amino acids supplementation on resistance exercise based muscle damage in humans. Journal of the International Society of Sports Nutrition. 2011;**8**:23

[16] Nicastro H, Artioli GG, Costa AS, Solis MY, da Luz CR, Blachier F, Lancha AH Jr. An overview of the therapeutic effects of leucine supplementation on skeletal muscle under atrophic conditions. Amino Acids. 2011;**40**(2):287-300

[17] Zanchi NE, Nicastro H, Lancha AH Jr. Potential antiproteolytic effects of Leucine: observations of in vitro and in vivo studies. Nutrition & Metabolism. 2008;**5**:20

[18] Hutson SM, Sweatt AJ, Lanoue KF. Branched-chain amino acid metabolism: Implications for establishing safe intakes. The Journal of Nutrition. 2005;**135**(6 Suppl):1557S-1564S

[19] Garcia-de-Lorenzo A, Ortiz-Leyba C, Planas M, Montejo JC, Nunez R, Ordonez FJ, Aragon C, Jimenez FJ. Parenteral administration of different amounts of branch-chain amino acids in septic patients: Clinical and metabolic aspects. Critical Care Medicine. 1997;**25**(3):418-424

[20] Abbas A, Lichtman A, Pillai S. Basic Immunology Functions and Disorders of the Immune System. 4th ed. Philadelphia, PA: Saunders/Elsevier; 2012. p. 40

[21] Kasai K, Ishikawa T, Nakamura, Miura T. Antibacterial properties of L-amino acid oxidase: Mechanisms of action and perspectives for therapeutic applications. Applied Microbiology and Biotechnology. 2015;**99**(19):7847-7857.

[22] Izidoro LFM, Sobrinho JC, Mendes MM, Costa TR, Grabner AN, Rodrigues VM, da Silva SL, Zanchi FB, Zuliani JP, Fernandes CFC, Calderon LA, Stábeli RG, Soares AM. Snake venom L-amino acid oxidases: Trends in pharmacology and biochemistry. BioMed Research International. 2014;**2014**:1-19

[23] Campillo-Brocal JC, Lucas-Elío P, Sanchez-Amat A. Distribution in different organisms of amino acid oxidases with FAD or a quinone as cofactor and their role as antimicrobial proteins in marine bacteria. Marine Drugs. 2015;**13**:7403-7418

[24] Joseph B, Rahan SS, Jeevitha MV, Ajisha SU. Pharmacological effects of snake venom L-amino acid oxidases. International Journal of Research in Ayurveda and Pharmacy. 2011;**2**(1):114-120

[25] Tan N, Fung S. Snake venom L-Amino acid oxidases and their potential biomedical applications. Malaysian Journal of Biochemistry and Molecular Biology. 2008;**16**(1):1-10

[26] Costa TR, Burin SM, Menaldo DL, de Castro FA, Sampaio SV. Snake venom L-amino acid oxidases: An overview on their antitumor effects. Journal of Venomous Animals and Toxins including Tropical Diseases. 2014;**20**:23

[27] Kumar S, Nath A, Murti K, Sethi MK, Nadeem T, Das P. A novel biochemical and pharmacological agent: L-amino acid oxidase with correlation to cancer management: An overview. IOSR Journal of Pharmacy. 2014;**4**(10):33-38

[28] Georgieva D, Murakami M, Perband M, Arnie R, Betzel C. The structure of a native L-amino acid oxidase, the major component of the *Vipera ammodytes ammodytes* venomic, reveals dynamic active site and quaternary structure stabilization by divalent ions. Molecular Biosystems. 2011;**7**:379-384

[29] Ju Y, Tong S, Gao Y, Zhao W, Liu Q, Gu Q, Xu J, Niu L, Teng M, Zhou H. Crystal structure of a membrane-bound L-amino acid deaminase from *Proteus vulgaris*. Journal of Structural Biology. 2016;**195**(3):306-315

[30] Faust A, Niefin K, Hummel W, Schomburg D. The structure of a bacterial L-amino acid oxidase from *Rhodococcus opacus* gives new evidence for the hydride mechanism for dehydrogenation. Journal of Molecular Biology. 2007;**367**:234-248

[31] Arima J, Sasaki C, Sakaguchi C, Mizuno H, Tamura T, Kashima A, Kusakabe H, Sugio S, Inagaki K. Structural characterization of L-glutamate oxidase from *Streptomyces* sp. X-119-6.The FEBS Journal. 2009;**276**(14):4318-4327

[32] Lee ML, Fung S, Chung I, Pailoor J, Cheah SH, Tan N. King cobra *(Ophiophagus hannah)* venom L-amino acid oxidase induces apoptosis in PC-3 cells and suppresses PC-3

solid tumor growth in a tumor xenograft mouse model. International Journal of Medical Sciences. 2014;**11**(6):593-601

[33] Phua CS, Vejayan J, Ambu S, Ponnudurai G, Gorajana A. Purification and antibacterial activities of an L-amino acid oxidase from king cobra *(Ophiophagus hannah)* venom. Journal of Venomous Animals and Toxins Including Tropical Diseases.2012;**18**(2):198-207

[34] Costa TR, Menaldo DL, Zoccal KF, Burin SM, Aissa AF, de Castro FA, Faccioli LH, Antunes LMG, Sampaio SV. CR-LAAO, an L-amino acid oxidase from *Calloselasma rhodostoma* venom, as a potential tool for developing novel immunotherapeutic strategies against cancer. Scientific Reports. 2017;**7**:42673

[35] Pokrovsky VS, Treshalina HM, Lukasheva EV, Sedakova LA, Medentzev AG, Arinbasarova AY, Berezov TT. Enzymatic properties and anticancer activity of L-lysine α-oxidase from *Trichoderma cf. aureoviride* Rifai BKMF-4268D. Anti-Cancer Drugs. 2013; **24**:846-851

[36] Dantas RT, Jorge ARC, Jorge RJB, Pessoa RRP, de Menezes B, Lima DB, Torres AFC. L-Amino acid oxidase from *Bothrops marajoensis* causes nephrotoxicity in isolated perfused kidney and cytotoxicity in MDCK renal cells. Toxicon (Elsevier). 2015;**104**:52-56

Chapter 2

The Anti-Allergic Effects of the His-Ala-Gln Tripeptide and Constituent Amino Acids

Mamoru Tanaka and Takeaki Okamoto

Additional information is available at the end of the chapter

http://dx.doi.org/10.5772/intechopen.68642

Abstract

The type-1 allergy, as typified by allergic rhinitis, pollen, and food allergies, is defined as a hypersensitivity reaction, and its frequency is increasing worldwide. There is a need to develop therapeutic agents that either prevent sensitization to allergens or suppress the allergic response after initiation. It has been reported that various peptides show anti-allergic effects, but there have been few reports concerning peptides derived from food. Previously, we studied the anti-allergic effect of His-Ala-Gln (HAQ), which is present in CE90GMM, a peptide mixture derived from milk casein. In this chapter, special emphasis is placed on the anti-allergic effects of the HAQ peptide *in vitro* and *in vivo*, and the effect of peptide binding, peptide sequence, number of amino acids, and the electron density of the amino acids is investigated.

Keywords: type-1 allergy, milk casein, peptide, amino acid, anti-allergic effect, *in vitro*, *in vivo*

1. Introduction

The prevalence of allergic diseases has been increasing in recent years, especially in Western countries. More than 25% of people living in developed Western countries suffer from allergies [1]. It has been reported that polyphenolic compounds, such as epigallocatechin gallate, show anti-allergic effects, but there have been few reports regarding the anti-allergenic effect of peptides. It has been reported that bioactive peptides from proteins in foods have several functions, such as reduction in blood pressure [2–4], activation of immune cells [5, 6], antiviral actions [7], and improvement of lipid metabolism [8]. Milk casein yields many peptides through hydrolysis with digestive and other enzymes. Hydrolyzed peptides have been indicated to have a variety of bioactive functions [2–8]. However, there have been few reports on the anti-allergic effects of peptides derived from casein. In the present study, we examined whether peptides derived from casein can inhibit type-1 allergic reactions *in vitro* and *in vivo*. Furthermore, characteristics of peptides with anti-allergic actions are investigated.

2. Mechanism of the type-1 allergic response

Type-1 allergies are hypersensitivity disorders mediated by immunological mechanisms, and type-1 allergic responses are induced by certain types of antigens, such as those from plants, mammals, microbes, foods, drugs, and chemicals. Type-1 allergic responses are known to be evoked by the antigen-induced activation of high-affinity IgE (FcεRI) expressed in mast cells and basophils.

When allergens invade the body, mast cells, basophils, eosinophils, and T cells are activated. Then, various tissues in the body are damaged by the physiologically active substances produced or released. Allergic sensitization involves T cell priming after dendritic cell (DC) activation, and the resultant T-helper (Th) 2 response is characterized by the production of interleukin (IL)-4, IL-5, and IL-13 from $CD4^+$ T cells. This Th2 response leads to IgE production from B cells, and this IgE binds to FcεRI on the surface of mast cells and basophils in the skin, gut, and respiratory and cardiovascular systems, arming them for reactivity upon re-exposure to the allergen. The elicitation of classic allergic symptoms occurs within minutes after allergen exposure, when the IgE-bound mast cells and basophils recognize the allergen and become activated [9]. The bond between antigen and IgE is essential, and is the first step in triggering the signaling cascades that lead to degranulation. These signaling pathways are involved in the activation of protein kinase C, induced by 1,2-diacylglycerol, or Ca^{2+} influx into the intracellular matrix induced by deacylglycerol [10, 11]. The activation of mast cells and basophils triggers the production of many chemical mediators, such as histamine, proteolytic enzymes, prostaglandins, leukotrienes, inflammatory cytokines, and arachidonic acid metabolites, including prostaglandins and leukotrienes [12–15]. These mediators cause immediate allergic reactions. Thus, mast cells and basophils are implicated in the development of diseases, such as asthma, allergic rhinitis, and inflammatory arthritis [13].

Th cells also play a central role in type-1 allergic responses. The cells are classified into two types: Th1 and Th2. These cells are responsible for the modulation of cytokine secretion to maintain homeostasis in the host, and disruption of this balance induces various immunological diseases. Allergic diseases are characterized by an excessive Th2-type immune response. It is generally accepted that IL-4 regulates the differentiation of native $CD4^+$ T cells into Th2 cells and immunoglobulin class switching to the IgG1 and IgE isotypes. Excessive IL-4 production by Th2 cells has been associated with an elevation of IgE levels and allergic reactions. Thus, modulation of the Th1/Th2-balanced immune response is important to control allergic symptoms. There are several phases in the pathogenesis of type-1 allergic responses as seen above, and allergic symptoms may be arrested by blocking the response at any of these points.

3. The search for peptides derived from casein with anti-allergic actions

It has been reported that bioactive peptides from proteins in foods have several functions [2–8]. Milk casein yields many peptides through hydrolysis with digestive and other enzymes.

Hydrolyzed peptides have been indicated to have a variety of bioactive functions [2–8]. Therefore, we focused our search on peptides derived from casein, and examined degranulation-inhibitory activity related to anti-allergic effects. Degranulation of rat basophilic leukemia (RBL-2H3) cells was monitored by measuring the activity of released β-hexosaminidase. For antigen stimulation, DNP-specific IgE-primed RBL-2H3 cells were preincubated for 10 min with various concentrations of peptides, then stimulated with antigen (mouse anti-DNP IgE). After 30 min, the medium was collected and 0.2% Triton X-100 was added to the cells. Levels of β-hexosaminidase released into the medium, and within cells, were determined by colorimetric assay using *p*-nitrophenyl-2-acetamide-2-deoxy-β-glucopyranoside and expressed as the percentage of activity released into the medium compared with total activity. In the first stage of this study, we investigated casein and CE90GMM (average molecular weight 640 kDa), a peptide mixture derived from milk casein, on the degranulation of RBL-2H3 cells and found that degranulation of the cells was not suppressed in the presence of casein. On the other hand, significant inhibition of β-hexosaminidase release was seen with CE90GMM stimulation at 250 μg/mL (**Figure 1**) [16]. Thus, this finding suggested that a peptide capable of inhibiting β-hexosaminidase release is present in CE90GMM. Several varieties of peptides have been identified inCE90GMM [17]. We tested the low-molecular-weight peptides: His-Ala-Gln (HAQ), Glu-Gln-Pro-Ile (EQPI), Asp-Met-Glu-Ser (DMES), and Lys-Ile-Lys-Glu (KIKE), which are present in CE90GMM, using RBL-2H3 cells. With the four peptides, at concentrations ranging from 10 to 500 μg/mL, the inhibition of β-hexosaminidase release was observed, which was dose dependent (**Figure 2**) [16].

We next investigated the effect of the four peptides on cytokine production [tumor necrosis factor (TNF)-α and IL-4] from RBL-2H3 cells after antigen stimulation. After stimulation with

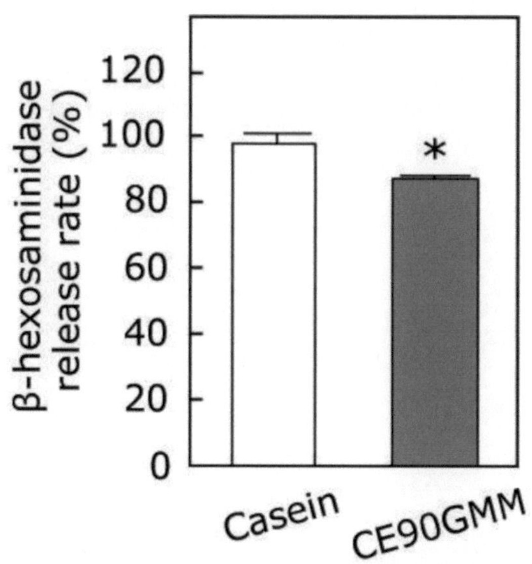

Figure 1. Effects of Casein and CE90GMM on β-hexosaminidase release from RBL-2H3 cells. Data are expressed as the means ± SD values of triplicate determinations. *$p < 0.05$ vs control. DNP-specific IgE sensitized RBL-2H3 cells were challenged with DNP-HSA for 30 min. Casein and CE90GMM were added 10 min before antigen challenge.

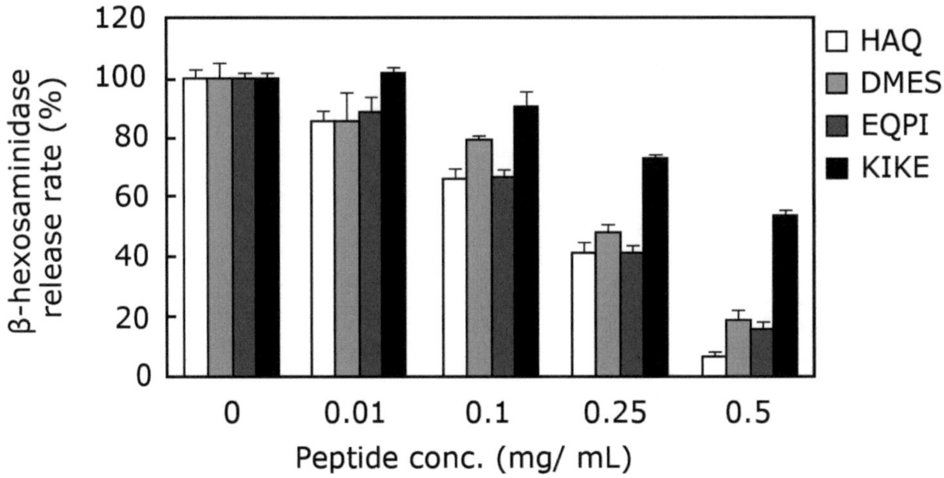

Figure 2. Effects of HAQ, DMES, EQPI and KIKE on β-hexosaminidase release from RBL-2H3 cells. Data are expressed as the means ± SD values of triplicate determinations. DNP-specific IgE sensitized RBL-2H3 cells were challenged with DNP-HSA for 30 min. Peptides were added 10 min before antigen challenge.

anti-DNP-IgE only, control levels of TNF-α and IL-4 were 9.1 and 90.8 pg/mL, respectively. All peptides inhibited the production of the inflammatory cytokines TNF-α and IL-4. In particular, the HAQ and EQPI peptides showed stronger inhibition of TNF-α and IL-4 than was seen with the DMES and KIKE peptides (**Table 1**). In addition, we also investigated the characteristics of peptides with degranulation depression effects [18]. It is known that concentrations of intracellular Ca^{2+}($[Ca^{2+}]i$) in mast cells and basophils rise through signaling after the cross-linkage of antigen to FcεRI through IgE for degranulation. Thus, the degranulation-suppressing effect of the HAQ peptide on $[Ca^{2+}]i$ was examined using fluo-3 AM. The $[Ca^{2+}]i$ from RBL-2H3 cells was significantly increased by treatment with the HAQ peptide without affecting the proliferation and viability of the cells (**Figure 3**). These results imply that the HAQ peptide suppressed the elevation of $[Ca^{2+}]i$ induced by intracellular-signaling pathways caused by the antigen-antibody interaction.

	TNF-α	**IL-4**
Control	100 ± 8 (9.1 pg/mL)	100 ± 5 (90.8 pg/mL)
HAQ	46 ± 6	52 ± 1
DMES	48 ± 3	80 ± 8
EQPI	33 ± 9	56 ± 5
KIKE	61 ± 10	86 ± 3

Table 1. Amount of cytokine production.

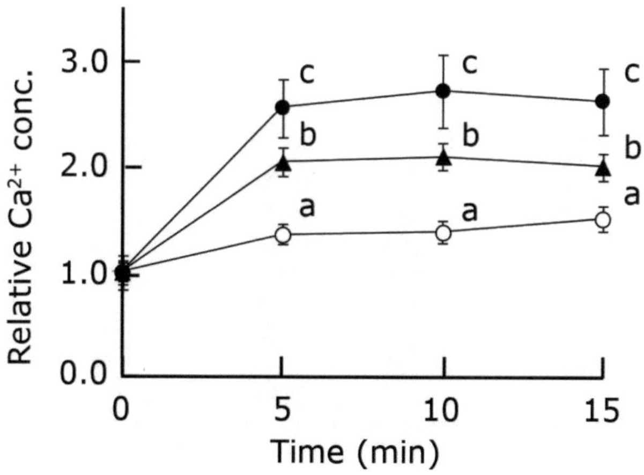

Figure 3. Effects of HAQ peptide on the release of $[Ca^{2+}]i$ of RBL-2H3 cells. Data are presented as the means ± SD (n = 5). $^{*}p < 0.05$ between each group. $[Ca^{2+}]i$ was measured by Calcium Kit-Fluo-3. Anti-DNP IgE-sensitized cells were incubated with Fluo-3 AM for 1 h and then incubated with 500 μM HAQ or PBS for 10 min. Then, the treated cells were stimulated with DNP-HSA, and the fluorescence intensity was measured. (○), non-HAQ-treated cells not sensitized with anti-DNP IgE; (▲), HAQ-treated cells stimulated with antigen; (●), non-HAQ-tpeptide-treated cells stimulated with antigen.

4. Characteristics of the amino acids in peptides with anti-allergic actions

We hypothesized that the amino acids constituting the peptide may be the reason that the degranulation-inhibitory activity differs depending on the type of peptide. It is known that proteins are broken down into peptides in the stomach, which are then broken down into amino acids, dipeptides, and tripeptides that are absorbed in the bloodstream through the small intestine [19]. Thus, the amino acids making up the HAQ peptide were investigated for degranulation-inhibitory activity, cytokine production, and electron density [16].

We measured β-hexosaminidase levels released during degranulation in the antigen-antibody reaction in the presence of amino acids, and evaluated the anti-allergic effects. For the amino acids making up the HAQ peptide, an inhibitory effect of degranulation was found with L-histidine stimulation, but L-alanine and L-glutamine did not inhibit β-hexosaminidase release (**Figure 4**). L-histidine contains an imidazole group, one of the hetero-aromatic rings in the side chain, whereas L-alanine and L-glutamine do not have hetero-aromatic rings in the side chain. Polyphenols have multiple benzene rings as a construction feature. Therefore, we considered that amino acids with aromatic rings may show anti-allergic effects. To confirm this hypothesis, we investigated the effect of amino acids with aromatic rings, L-histidine, L-tryptophan, L-phenylalanine, and L-tyrosine, on the degranulation of RBL-2H3 cells. A significant inhibitory effect on the release of β-hexosaminidase was found with L-tryptophan, L-phenylalanine, and L-histidine, depending on the dose ($p < 0.05$), but L-tyrosine did not inhibit β-hexosaminidase release (**Figure 5**).

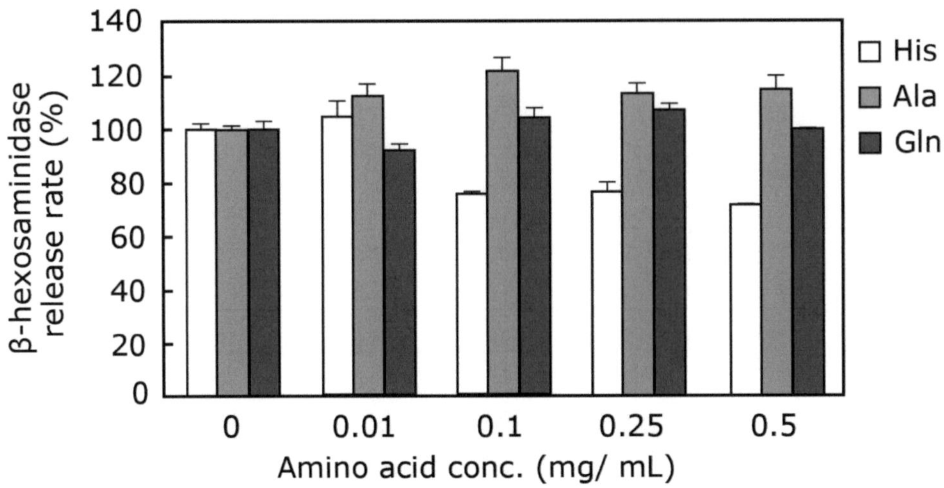

Figure 4. Effects of amino acids (His, Ala, Gln) on β-hexosaminidase release from RBL-2H3. Data are expressed as the means ± SD values of triplicate determinations. DNP-specific IgE sensitized RBL-2H3 cells were challenged with DNP-HSA for 30 min. Amino acids were added 10 min before antigen challenge.

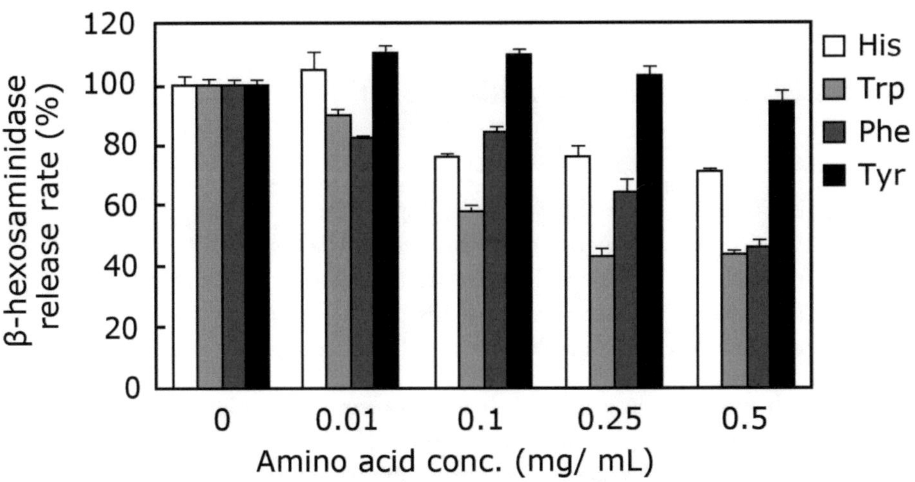

Figure 5. Effects of amino acids (His, Trp, Phe, Tyr) on β-hexosaminidase release from RBL-2H3. Data are expressed as the means ± SD values of triplicate determinations. DNP-specific IgE sensitized RBL-2H3 cells were challenged with DNP-HSA for 30 min. Amino acids were added 10 min before antigen challenge.

We also performed *ab initio* molecular orbital calculations for the aromatic amino acids, to elucidate the mechanism of the anti-allergic action. Geometrical optimization was carried out and the molecular electrostatic potential was calculated for each amino acid. The electrostatic potential maps enabled visualization of the charge distributions of the molecules (**Figure 6**). The aromatic ring of L-tyrosine, which does not show an anti-allergic effect, is positively charged. L-tyrosine has a hydroxyl group on the side chain of the aromatic ring, which is not present in L-histidine, L-tryptophan, or L-phenylalanine. The charge of this hydroxyl group

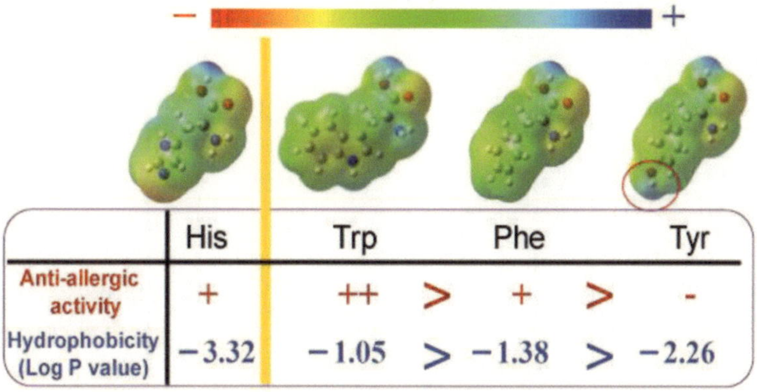

Figure 6. Charge distribution on four kinds of aromatic amino acids (His, Trp, Phe, Tyr).

is positive. Furthermore, according to the octanol/water partition coefficient (LogP), L-tyrosine is less hydrophobic than the other amino acids containing aromatic rings. One reason why degranulation is not observed with L-tyrosine may be that it is difficult for L-tyrosine to pass through the membrane lipid bilayer and incorporate into the cell. On the other hand, L-histidine is a hydrophilic amino acid and shows degranulation-inhibitory activity. In addition to this, the charge distribution on the aromatic ring of L-histidine is largely negative. L-tryptophan and L-phenylalanine are hydrophobic amino acid and the charge distribution on the aromatic ring of L-tryptophan and L-phenylalanine is largely neutral. Consequently, the mechanism of the degranulation-suppressing effect of L-histidine is likely to be different from that of L-tryptophan and L-phenylalanine. These results indicate that amino acids with an anti-allergic action have aromatic rings and neutral or negatively charged side chains.

5. Importance of peptide binding, number of amino acid residues, and peptide sequence on the anti-allergic effect

The degranulation-inhibitory activity differs depending on the type of amino acids. Thus, we examined the effect of peptide binding, number of amino acid residues, and peptide sequence on degranulation inhibition [20, 21].

Initially, we revealed that the level of degranulation-inhibitory activity depends on the peptide binding. In other words, a significant inhibitory effect on the release of β-hexosaminidase was found with the HAQ peptide, depending on the dose, but mixtures of the amino acids in the HAQ peptide did not show an inhibitory effect (**Figure 7**). It has been reported that bioactive peptides from proteins in foods have several functions [2–8]. Our results, indicating that the HAQ peptide can suppress the antigen-induced degranulation of RBL-2H3 cells, are consistent with those observed for other bioactive peptides.

We further found that the level of degranulation-inhibitory activity depended on the number of amino acid residues in the peptide. Our group has previously reported that peptides

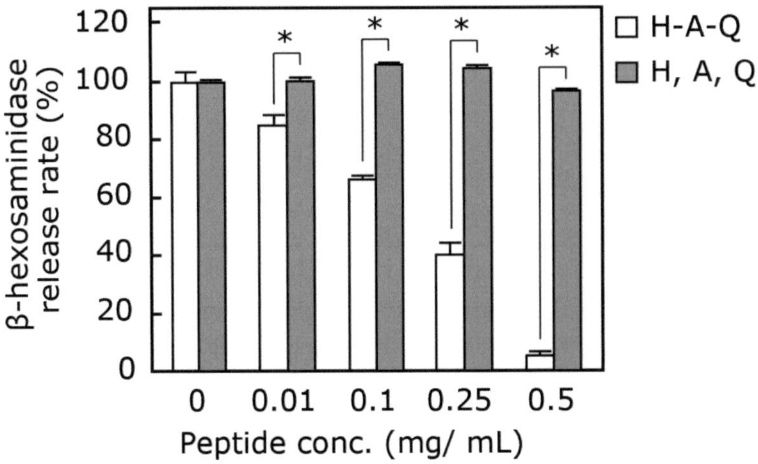

Figure 7. Effects of HAQ and amino acid mixtures made up of HAQ on *β*-hexosaminidase release from RBL-2H3 cells. Data are expressed as the means ± SD values of triplicate determinations. **p* < 0.05. DNP-specific IgE sensitized RBL-2H3 cells were challenged with DNP-HSA for 30 min. HAQ and amino acid mixtures made up of HAQ were added 10 min before antigen challenge.

with an anti-allergic action contain amino acids with aromatic rings and neutral or hydrophobic side chains [16]. Thus, we used peptides containing L-histidine, with an imidazole functional group, to assess the inhibitory effect of amino acid residues on the IgE-induced allergic response in IgE-sensitized RBL-2H3 cells. Imidazole peptides containing only L-histidine decreased the release of β-hexosaminidase. In particular, the tri- and tetra-peptides demonstrated degranulation-inhibitory activity (**Figure 8**).

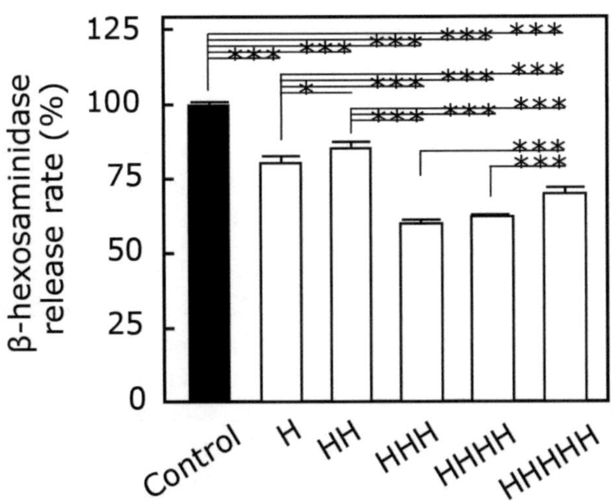

Figure 8. Effects of His and imidazole peptides of HAQ on *β*-hexosaminidase release from RBL-2H3 cells. Data are expressed as the means ± SD values of triplicate determinations. **p* < 0.05. DNP-specific IgE sensitized RBL-2H3 cells were challenged with DNP-HSA for 30 min. His and imidazole peptides were added 10 min before antigen challenge.

Finally, we investigated the amino acid sequence of the HAQ peptide. The activity of released β-hexosaminidase was evaluated for three peptides, with the L-histidine residue at different positions, HAQ, Ala-Gln-His (AQH), and Gln-His-Ala (QHA). The level of degranulation-inhibitory activity was found to depend on the peptide sequence (**Figure 9**). Nishida et al. reported that metallic ion concentration was important for the degranulation-inhibitory reaction and cytokine production from mast cells and basophils [22]. Therefore, we hypothesized that the degranulation from mast cells and basophils is related to an exaggerated effect of chelation. It has been reported that the HAQ peptide has a high affinity for metal ions, and peptides with histidine at the *N*-terminus of the sequence show a strong chelation effect for copper ions compared with peptides with L-histidine at the *C*-terminus [23]. Therefore, the suppression of degranulation may be related to an increase in ability to chelate metal ions, because the HAQ peptide having an L-histidine residue at the *N*-terminus exhibited a stronger inhibitory action against β-hexosaminidase than either AQH or QHA. In conclusion, we propose that the level of degranulation-inhibitory activity depends on the peptide binding, the number of amino acid residues, and the peptide sequence.

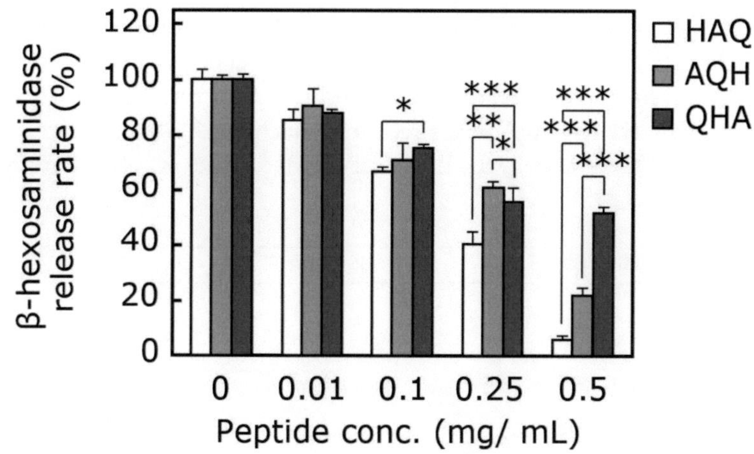

Figure 9. Effects of HAQ, AQH and QHA on *β*-hexosaminidase release from RBL-2H3 cells. Data are expressed as the means ± SD values of triplicate determinations. $^{*}p < 0.05$. DNP-specific IgE sensitized RBL-2H3 cells were challenged with DNP-HSA for 30 min. HAQ, AQH and QHA were added 10 min before antigen challenge.

6. Effect of the HAQ peptide on antibody production in mice

Th1 and Th2 polarization occurs according to cytokine patterns, which begin when antigen-presenting cells interact with naïve T cells and polarize into type 1 and type 2 cells in response to the type of antigen encountered [12–15]. Th1 cells produce IFN-γ and TNF-α which induce T-cell-mediated immunity and IgG2a production and downregulate Th2 cells. On the other hand, IgE production in mice is induced by IL-4 and IL-5 secreted by Th2 cells. IgE response is accompanied by IgG1 production, which is also induced by IL-4, resulting in allergic diseases.

IgA antibodies play an essential role in mucosal protection. Several properties of IgA antibodies, including an ability to be secreted when linked to secretory components, resistance to proteolysis, and an inability to trigger the complement cascade, allow the antibodies to clear antigens from mucosal surfaces in a process called immune exclusion [24]. Antigen delivery on surfaces may induce either immunization or unresponsiveness. Oral tolerance is usually defined as the suppression of humoral and cellular immune responses after mucosal presentation of a putative protein antigen [25]. It has been reported that an increase in specific IgA antibodies occurs concomitantly with the systemic suppression induced by oral tolerance [26]. One proposed explanation is that the induction of regulatory T cells (Treg) causes not only enhancement of the Th1 response and suppression of the Th2 response but also results in an allergy-suppressive mechanism. Thus, we examined immune responses (antibody and cytokine production) to continuous ingestion of the HAQ peptide in a mouse model of type-1 allergy to ovalbumin (OVA) [27]. BALB/c mice were randomly divided into phosphate-buffered saline (PBS)-PBS, PBS-OVA, and HAQ-OVA groups. An HAQ peptide-added diet was orally administered to BALB/c mice for 35 days. The mice were immunized intraperitoneally with OVA on days 8, 18, and 25. Blood, feces, and spleen lymphocytes were obtained from the mice on day 35 or 36. The levels of total IgA, IFN-γ, IL-4, and OVA-specific IgE, IgG1, IgG2a, and IgA were analyzed by enzyme-linked immunosorbent assay (ELISA) (**Table 2** and **Figure 10**). BALB/c mice that were orally administered the HAQ peptide exhibited lower OVA-specific IgE and IgG1 secretion, whereas OVA-specific IgG2a levels remained unchanged in the serum. An increase in OVA-specific IgA was observed in feces, whereas total IgA levels remained unchanged. An increase in IgG1, IgG2a, and IFN-γ levels was observed, while the IL-4 level was decreased, in spleen lymphocytes, in the

	PBS-PBS	PBS-OVA	HAQ-OVA
Serum			
Specific IgE (A490)	0.05 ± 0.02^{a}	1.15 ± 0.57^{b}	0.82 ± 0.18^{c}
Specific IgG1 (A490)	0.05 ± 0.01^{a}	1.33 ± 0.10^{b}	1.25 ± 0.07^{b}
Specific IgG2a (A490)	0.00 ± 0.00^{a}	0.57 ± 0.10^{b}	0.54 ± 0.06^{b}
Faces			
Specific IgA (A490)	0.05 ± 0.03^{a}	0.93 ± 0.13^{b}	1.66 ± 0.18^{c}
Total IgA (8g/g)	3.25 ± 0.74^{a}	3.48 ± 0.74^{b}	3.07 ± 0.74^{c}
Spleen lymphocytes			
Specific IgE (A490)	0.00 ± 0.01	0.02 ± 0.00	0.05 ± 0.1
Specific IgG1 (A490)	0.02 ± 0.00^{a}	0.28 ± 0.05^{b}	0.72 ± 0.07^{c}
Specific IgG2a (A490)	0.01 ± 0.01^{a}	0.06 ± 0.02^{b}	0.13 ± 0.02^{c}

Data are expressed as the mean ± SEM. a-b, a-c, b-c: $p < 0.05$, b-d: $p = 0.09$. Groups of that had received orally administered HAQ or PBS during sensitization to OVA or PBS were challenged intraperitoneally. Unsensitized animals received the vehicle alone. Blood was obtained from the orbital veins of mice at 35 or 36 days after injection and subjected to ELISA specific for IgE or IgG1 against OVA.

Table 2. Effect of continuous ingestion of HAQ on antibodies production from OVA-sensitized mice [27].

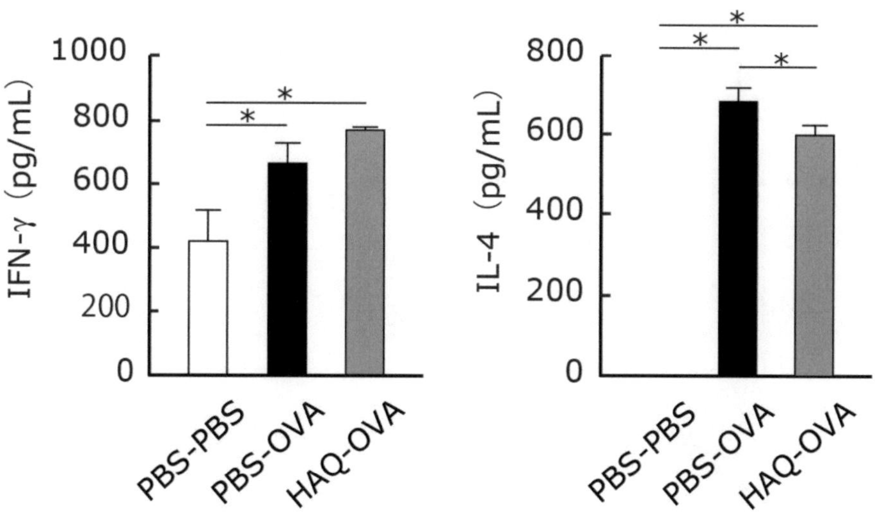

Figure 10. Effect of continuous ingestion of HAQ on IFN-γ and IL-4 production by mouse spleen lymphocytes (72 h). Data are expressed as the mean ± SD values of triplicate determinations. $^{*}p < 0.05$.

presence of the HAQ peptide. These findings suggest that the HAQ peptide may cause a shift from a Th2-type immune response toward a Th1-type response. Thus, the HAQ peptide has a potential regulatory effect on antibody production in a type-1 allergic response.

7. Effect of the HAQ peptide on allergic symptoms in mice

To investigate the effects of the HAQ peptide on the allergic reaction, we performed animal experiments using a murine type-1 allergy model. C3H/HeJ mice were randomly divided into HAQ-LHE, PBS-PBS, and PBS-LHE groups. The PBS-PBS and PBS-LHE groups were orally administered PBS alone. The HAQ-LHE and PBS-LHE groups were initially injected intraperitoneally with 100 μg lysozyme from hen egg white (LHE) and 4 mg aluminum hydroxide in 0.2-mL PBS on day 1. LHE was then reduced to 50 μg and injected intraperitoneally with 4 mg aluminum hydroxide in 0.2 mL PBS on day 8. To assess the sensitization, blood was obtained from the orbital veins of mice under light anesthesia, 11 days after the initial injection, and subjected to ELISA. The HAQ-LHE group was orally administered HAQ peptide (1 mg/day) in 0.2 mL PBS by gavage throughout the experimental period of 14 days.

We evaluated the effect of orally administered LHE on the suppression of allergic reactions in a murine model. The score assessment commonly increases and whole-body temperature is commonly reduced during systemic anaphylaxis [28, 29]. Score assessment and body temperature measurements were therefore conducted to assess anti-allergic effects in mice. The HAQ-LHE and PBS-LHE groups were orally administered 10 mg LHE in 0.5 mL PBS, and then score assessment and rectal temperature measurements were performed to evaluate allergic symptoms (**Figure 11**). The PBS-PBS group was orally administered 0.5 mL PBS alone. Score assessments

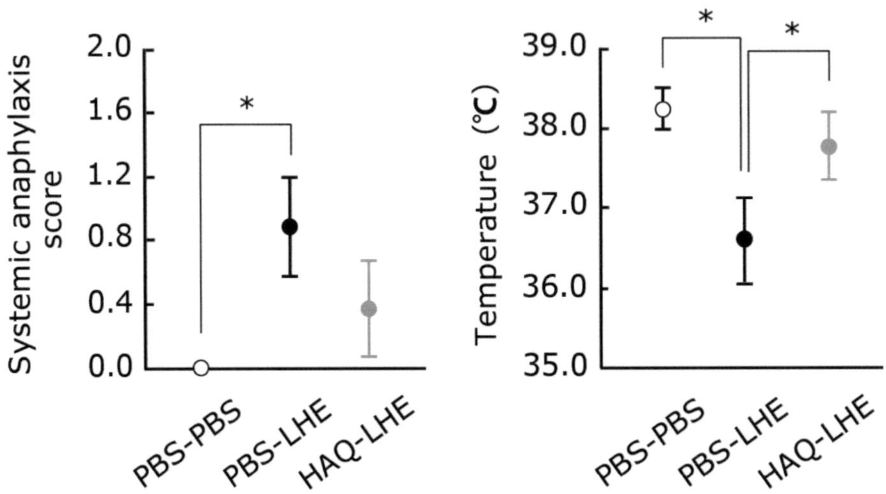

Figure 11. Allergic symptoms after oral challenge in mice presensitized in the presene of HAQ peptide. ○, PBS-PBS PBS orally administered group (n = 5); •, PBS-LHE, PBS orally administered group (n = 9); •, HAQ-LHE, HAQ orally administered group (n = 8). Data are expressed as the mean ± SEM. $^{*}p < 0.05$ between each group. Groups of mice that had received orally administered HAQ peptide or PBS during sensitization to LHE or PBS were challenged intraperitoneally. (A) Symptom scores of systemic anaphylaxis. Anaphylaxis-like symptoms were scored at 20 min after oral challenge on a scale from 0 (no symptoms) to 5 (death) as described in the Materials and Methods. (B) Rectal temperature. The rectal temperature was measured at 30 min after oral challenge.

were performed at 20 min after challenge with LHE using the scoring system as described by Li et al. [28]. To evaluate body temperature, the rectal temperature was measured at 30 min after oral administration of LHE. In the score assessment, the PBS-LHE group score (0.9 ± 0.3) was significantly higher than the PBS-PBS group (0.0 ± 0.0) ($p < 0.05$), but not the HAQ-LHE group (0.4 ± 0.3). In terms of rectal temperature, the PBS-LHE group (36.6 ± 0.5°C) had a significant decrease in body temperature compared with the PBS-PBS (38.2 ± 0.3°C) and HAQ-LHE (37.8 ± 0.4°C) groups ($p < 0.05$). These data demonstrate that continuous administration of the HAQ peptide to sensitized mice suppresses the consistent allergic symptoms induced by antigen stimulation. These results indicate that the HAQ peptide has anti-allergic effects *in vivo* as well as *in vitro* [16].

8. Conclusion

The frequency of allergic disorders is increasing worldwide, and this causes serious issues, including escalation of medical costs, reduction in health levels, and decline in labor productivity. We studied the anti-allergic effects of the HAQ peptide, which is present in CE90GMM, a peptide mixture derived from milk casein. In these studies, we observed five major findings *in vitro* and *in vivo* as follows:

1. CE90GMM and four peptides (HAQ, EQPI, DMES, and KIKE) inhibited the degranulation of RBL-2H3 and the effect varied with the dose.

2. The level of degranulation-inhibitory activity depended on peptide binding, peptide sequence, and the number of amino acids.
3. Peptides with anti-allergic actions possess aromatic rings and neutral or hydrophobic side chains.
4. The HAQ peptide has a potential regulatory effect on antibody and cytokine production in type-1 allergic responses.
5. Continuous administration of HAQ peptide suppressed the mild allergic symptoms in a murine model of type-1 allergy.

Our studies suggest a possible use of the HAQ peptide in preventing type-1 allergic responses. To use HAQ clinically for the prevention of allergic disease, the optimal dosage, efficacy, and adverse effects in humans should be determined.

We expect that the findings obtained in these studies will contribute to the prevention, and improve the treatment, of type-1 allergies.

Acknowledgements

This study was supported by JSPS KAKENHI (grant number 25750051). The authors thank Kenji Yamagishi, Takuya Sugahara, Junko Hirano, and Rina Hirata for assistance with experiments.

Author details

Mamoru Tanaka[1]* and Takeaki Okamoto[2]

*Address all correspondence to: m-tanaka@cc.u-kochi.ac.jp

1 Department of Nutrition, University of Kochi, Kochi, Japan

2 Faculty of Education, Ehime University, Matsuyama, Ehime, Japan

References

[1] Galli SJ, Tsai M, Piliponsky AM. The development of allergic inflammation. Nature. 2008;**454**:445-454

[2] Nakamura Y, Yamamoto N, Sakai K, et al. Purification and characterization of angiotensinI-converting enzyme inhibitors from sour milk. Journal of Dairy Science. 1995;**78**:777-783

[3] Sato M, Hosokawa T, Yamaguchi T, et al. AngiotensinI-converting enzyme inhibitory peptides derived from Wakame (Undaria pinnatifida) and their antihypertensive effect in spontaneously hypertensive rats. Journal of Agriculture and Food Chemistry. 2002;**80**:6245-6252

[4] Tokunaga KH, Yoshida C, Suzuki KM, et al. Antihypertensive effect of peptides from royal jelly in spontaneously hypertensive rats. Biological and Pharmaceutical Bulletin. 2004;**27**:189-192

[5] Hata I, Higashiyama S, Otani H. Identification of a phosphopeptide in bovine αs1-casein digest as a factor influencing proliferation and immunoglobulin production in lymphocyte cultures. Journal of Dairy Research. 1998;**65**:569-578

[6] Otani H, Watanabe T, Tashiro Y. Effects of bovine β-casein (1-28) and its chemically synthesized partial fragments on proliferative responses and immunoglobulin production in mouse spleen cell culture. Bioscience, Biotechnology, and Biochemistry. 2001;**65**:2489-2495

[7] Civra A, Giuffrida MG, Donalisio M, et al. Identification of Equine Lactadherin-derived peptides that inhibit rotavirus infection via integrin receptor competition. Journal of Biological Chemistry. 2015;**290**:12403-12414

[8] Yoshikawa M, Fujita H, Matoba N, et al. Bioactive peptides derived from food proteins preventing lifestyle-related diseases. Biofactors. 2000;**12**:143-146

[9] Vickery BP, Chin S, Burks AW. Pathophysiology of food allergy. Pediatric Clinics of North America. 2011;**58**:363-376

[10] Siraganian RP. Mast cell signal transduction from the high-affinity IgE receptor. Current Opinion in Immunology. 2003;**15**:639-646

[11] Blank U, Rivera J. The ins and outs of IgE-dependent mast-cell exocytosis. Trends in Immunology. 2004;**25**:266-273

[12] Church MK, Levi-Schaffer F. The human mast cell. Journal of Allergy and Clinical Immunology. 1997;**99**:155-160

[13] Wedemeyer J, Tsai M, Galli SJ. Role of mast cells and basophils in innate and acquired immunity. Current Opinion in Immunology. 2000;**12**:624-631

[14] Kidd P. Th1/Th2 balance: The hypothesis, its limitations, and implications for health and disease. Alternative Medicine Review. 2003;**8**:223-246

[15] Mosmann TR, Cherwinski H, Bond MW, et al. Two types of murine helper T cell clone. I. definition according to profiles of lymphokine activities and secreted proteins. Journal of Immunology. 1986;**175**:5-14

[16] Tanaka M, Yamagishi K, Sugahara T, et al. Impact of peptides from casein and peptide-related amino acids on degranulation in rat basophilic leukemia cell line RBL-2H3. Nippon Shokuhin Kagaku Kaishi (in Japanese). 2012;**59**:556-561

[17] Hashimoto K, Sato K, Nakamura Y, et al. Development of a large-scale (50L) apparatus for ampholyte-free isoelectric focusing (autofocusing) of peptides in enzymatic hydrolysates of food proteins. Journal of Agricultural Food Chemistry. 2005;**53**:3801-3806

[18] Tanaka M, Watanabe H, Yoshimoto Y, et al. Anti-allergic effects of His-Ala-Gln tripeptide *in vitro* and *in vivo*. Bioscience, Biotechnology, and Biochemistry. 2016;**81**:380-383

[19] Silk DB. Protein, peptides and amino acid: Which and what? Nestle Nutrition Workshop Series: Clinical and Performance Programme. 2000;**3**:257-271

[20] Okamoto T, Hirano J, Tagashira A, et al. Anti-allergic effect of imidazole peptides in vivo and in vitro. In: The 6th Asian Congress of Dietetics, Taipei, Taiwan; 2016. p. 223

[21] Tanaka M, Watanabe H, Kozai H, et al. Degranulation inhibitory activity may depend on peptide binding and peptide sequence. In: The 12th Asian Congress of Nutrition, Yokohama, Japan; 2017. p. 311

[22] Nishida K, Hirano T. Role of zinc/zinc transporter in allergic response. Seikagaku (in Japanese). 2010;**9:**814-824

[23] Muramoto K, Chen HM, Saito K, et al. Synergistic effect of antioxidative peptides from a soybean protein digest on the antioxidative activity of nonpeptidic antioxidants. Soy Protein Research. 1997;**18**:55-61

[24] Challacombe SJ, Rahman D, Jeffery H, Davis SS, O'Hagan DT. Enhanced secretory IgA and systemic IgG antibody responses after oral immunization with biodegradable microparticles containing antigen. Immunology. 1992;**76**:164-168

[25] Garside P, Mowat AM. Oral tolerance. Seminars in Immunology. 2001;**13**:177-185

[26] Challacombe SJ, Tomasi TB, Jr. Systemic tolerance and secretory immunity after oral immunization. Journal of Experimental Medicine. 1980;**152**:1459-1472

[27] Okamoto T, Hirata R, Tagashira A, et al. Effect of His-Ala-Gln peptide on antibody producibility in mice. Journal of Shikoku Public Health Society (in Japanese). 2017;**62**:63-70

[28] Li XM, Schofield BH, Huang CK, et al. A murine model of IgE-mediated cow's milk hypersensitivity. Journal of Allergy and Clinical Immunology. 1999;**103**:206-214

[29] Tanaka M, Nagano T, Yano H, et al. Impact of ω-5 gliadin on wheat-dependent exercise-induced anaphylaxis in mice. Bioscience, Biotechnology, and Biochemistry. 2011;**75**: 313-317

Chapter 3

Amino Acids Modification to Improve and Fine-Tune Peptide-Based Hydrogels

Stefan Loic

Additional information is available at the end of the chapter

http://dx.doi.org/10.5772/intechopen.68705

Abstract

Among all the materials used in industry, gels play an increasingly important role. These so-called soft-matter materials are defined by their ability to fix a large amount of solvent, either organic (organogels) or aqueous (hydrogels). The large majority of hydrogels are made of natural or synthetic polymers, or natural proteins. However, a new kind of hydrogel has appeared: the peptide-based hydrogels, developed from short amino acids sequences (<20 amino acids). Due to their exceptional qualities in term of biocompatibility, biodegradability, and atom economy, these peptide-based hydrogels open new horizons in term of applications. They are mainly considered in the biomedical domain as injectable hydrogels, or as an extracellular culture matrix to support cell culture. While important, the possibilities of peptide design can exponentially grow using modified and non-natural amino acids instead of the "only" twenty natural ones. Thus, chemical modifications virtually offer infinite opportunities both to improve applications window and to fine-tune properties of the resulting hydrogels. In this context, this chapter proposes to review peptide and amino acid modifications reported to impact the resulting hydrogel.

Keywords: bioinspired materials, modified amino acids, gelator, peptide-based hydrogel, self-assembly, soft matter, supramolecular chemistry

1. Introduction

1.1. Soft matter and hydrogels as powerful materials

At the tenuous frontier between the solid and the liquid states, soft matter is focusing research interest which has exponentially increased since the beginning of the 1990s (<20 peer-reviewed articles published in 1990 versus >700 for 2014, 2015, and 2016). Interestingly, it is in perfect accordance with the Nobel Prize in Physics received by Pr. Pierre-Gilles de Gennes in 1991, who merely entitled his Nobel lecture "Soft Matter" [1]. He has definitely contributed to popularize this terminology. Soft matter includes a wide variety of materials, including colloidal suspensions, surfactants, liquid crystals, polymers, or gels. These latter ones constitute the major thread of this chapter.

Interestingly, the gel state can be described both as a solid state, because of the ability of gels to self-support their own weight (which is a solid state characteristic), and as a liquid, on account of the gel composition made of a large majority of a liquid (generally >98% w/w) and only a small amount of solid (<2% w/v) [2–4]. Thus, gels are defined by their ability to fix a large amount of solvent, either organic (organogels) [5] or aqueous (hydrogels) [6, 7]. Hydrogels are of particular interest and find plenty of applications from medical treatments (*e.g.*, wound healing, dental care, cartilage repair, tissue engineering) [8] to cosmetics, agriculture, and water treatments [9]. The commercially available hydrogels are made of several kinds of starting materials, including natural polymers (*e.g.*, xanthan or Arabic gums, agar-agar), synthetic polymers (*e.g.*, polyurethane (PU), polyethylene glycol (PEG), polyacrylic acid (PAA)), or natural proteins (*e.g.*, collagen, sink fibroin, elastin) [10].

Depending on the nature of the interactions inside the matter, hydrogels can be discriminated between both chemical gels and physical gels [11]. For the first one, covalent bonds are created to form a network from the starting building block (also termed "gelator"), while for the latter, dynamic cross-links based on non-covalent interactions (mainly hydrogen bonds, π-π interactions, and Van der Waals interactions) control the supramolecular self-assembly of molecules of gelator [2–4].

Physical hydrogels, also called supramolecular hydrogels, draw scientific community's attention due to their ability to be, *in theory*, dynamically assembled and disassembled several times, thanks to the formation and break of non-covalent interactions. However, some examples reveal that some hydrogels are too strong to be broken without degradation of the starting gelator. Regardless, the simplicity of the formation of supramolecular hydrogels, consisting of mixing the gelator and water followed by the application of a stimulus to trigger the gelation process, has made these materials of primary interest for a broad range of applications [8, 12, 13]. The stimuli applied are, depending of the system, of different nature including temperature, ultrasound, salt addition, addition of a specific chemical, pH, enzyme, light, electromagnetic field, etc. [2–4].

In the development of high-efficient hydrogels, the place taken by peptide-based hydrogels drastically increased during the last two decades. Based on the self-assembly of peptides, these hydrogels are particularly interesting for applications in biological and medical contexts [14–19]. Great expectations are placed on these innovative materials by the scientific community, and first commercially available systems have emerged, when others are in clinical research trials.

1.2. Peptide-based hydrogelators as innovative and efficient materials

Smaller than proteins, polypeptides and peptides are comprised of amino acids linked by amide bonds, the terminology depending on the length of the chain (respectively ≥100, <100, and <10 amino acids as generally accepted). In a biological context, common natural amino

http://dx.doi.org/10.5772/intechopen.68705

acids (α-amino acids to be more specific) are 20 and are characterized by their molecular structures and their physicochemical properties including their isoelectric point, hydrophobicity, pKa, etc. (**Table 1**) [20]. As a function of their sequence and conditions, proteins and peptides have the ability to self-assemble in a non-covalent way, forming secondary structures termed β-sheets (parallel and antiparallel), α-helix, 3_{10} helix, or π-helix. These structures are fundamental for physiological process and are mainly due to specific intra- or inter-molecular hydrogen bond interactions between the carbonyls and the protons of the amide groups. Other types of interactions are also crucial for peptide self-assembly, including van der Waals forces, electrostatic forces, π-π interactions, and hydrophobic affinity [2–4].

An overwhelming majority of the peptide-based hydrogels rests on peptide β-sheet assemblies. Indeed, starting gelators (*i.e.*, peptides) mainly self-assemble into fibrils, which subsequently combine each other via supramolecular interactions to form fibers. Altogether, these fibers form a network acting as the frame of the hydrogel, essentially self-supported by hydrogen bonds (**Figure 1**) [21]. This analogy to building construction is shared with numer-

Name	Three-letter code	One-letter code	M ($g.mol^{-1}$)	Nature (lateral chain)	Isoelectric point (IP)	pKa (lateral chain)	Hydropathy index	P_β
Alanine	Ala	A	89.09	Aliphatic	6.01	N/A	1.8	0.76
Glycine	Gly	G	75.07	Aliphatic	5.97	N/A	−0.4	0.71
Isoleucine	Ile	I	131.17	Aliphatic	6.02	N/A	4.5	1.73
Leucine	Leu	L	131.17	Aliphatic	5.98	N/A	3.8	1.23
Proline	Pro	P	115.13	Aliphatic	6.48	N/A	1.6	0.43
Valine	Val	V	117.15	Aliphatic	5.97	N/A	4.2	1.82
Phenylalanine	Phe	F	165.19	Aromatic	5.48	N/A	2.8	1.48
Tryptophan	Trp	W	204.23	Aromatic	5.89	N/A	−0.9	1.24
Tyrosine	Tyr	Y	181.19	Aromatic	5.66	10.07	−1.3	1.43
Aspartic acid	Asp	D	133.10	Acid	2.77	3.65	−3.5	0.52
Glutamic acid	Glu	E	147.13	Acid	3.22	4.25	−3.5	0.67
Asparagine	Asn	N	132.12	Amide	5.41	N/A	−3.5	0.53
Glutamine	Gln	Q	146.15	Amide	5.65	N/A	−3.5	0.78
Arginine	Arg	R	174.20	Basic	10.76	12.48	−4.5	0.86
Histidine	His	H	155.16	Basic	7.59	6.00	−3.2	1.01
Lysine	Lys	K	146.19	Basic	9.74	10.53	−3.9	0.82
Serine	Ser	S	105.09	Hydroxyl	5.68	N/A	−0.8	0.92
Threonine	Thr	T	119.12	Hydroxyl	5.87	N/A	−0.7	1.19
Cysteine	Cys	C	121.16	Sulfur-containing	5.07	8.18	2.5	1.26
Methionine	Met	M	149.21	Sulfur-containing	5.74	N/A	1.9	1.26

Table 1. Key properties of the main amino acids (with P_β for amino acids β-sheet propensity).

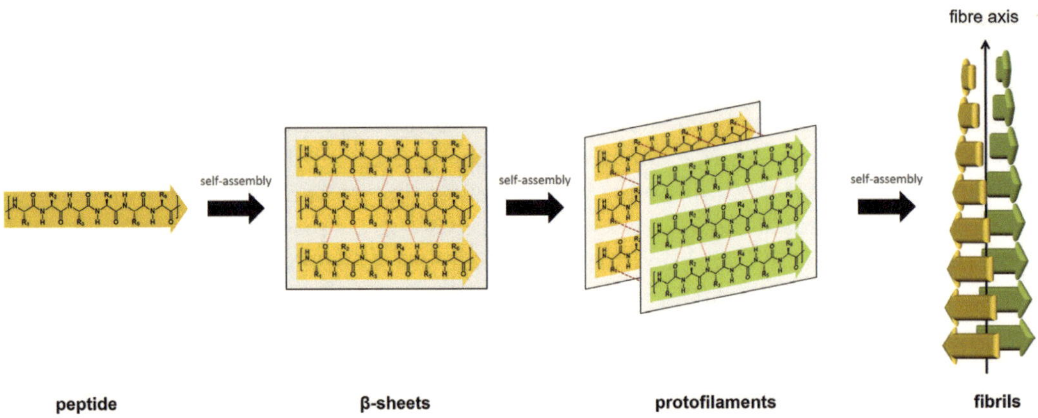

Figure 1. Schematic representation of peptide self-assembling to form fibrils.

ous vital biological self-assemblies (*e.g.*, antibody/antigen, RNA/protein, DNA/DNA, etc.), showing the pivotal role of supramolecular assemblies. Because peptides (*i.e.*, short proteins) natively self-assemble each other, peptide-based hydrogels can be considered as Nature-inspired materials [22].

While β-sheet assemblies are the predominant way peptides interact to form hydrogels, a handful of these soft matter materials are based on α-helical coiled-coil structures [23, 24]. Briefly, these architectures involve repeating heptapeptides (or heptads) mainly based on the sequence (HPPHPPP) in which H is a hydrophobic amino acid and P a polar one. Among them, few examples of coiled-coil-based hydrogels can be cited, including the tri heptad repeats **(Ile-Lys-Gln-Leu-Glu-Ser-Glu)$_3$** and **(Ile-Ala-Gln-Leu-Glu-Tyr-Glu)$_3$** [25], a 34-mer based on the **Gln-Leu-Ala-Arg-Glu-Leu(Gln-Gln-Leu-Ala-Arg-Glu-Leu)$_4$** [26], a 19-mer (**Ac-Leu-Lys-Glu-Leu-Ala-Lys-Val-Leu-His-Glu-Leu-Ala-Lys-Leu-Val-Ser-Glu-Ala-Leu-His-Ala-NH$_2$**) [27], and the hydrogel formed by two 28-residue peptides termed hSAF$_{AAA}$ used as substrate for cell growth [28]. However, these gels based on coiled-coil peptide architecture are limited by both the length of the sequence (from ~20 to 40 amino acid residues) and the concentration of peptide (from 1 to 12% w/v) required. Comparatively, hydrogels based on peptide self-assembling via β-sheet formation can be formed with ultrashort peptides (≤7 amino) and with smaller concentrations (≤0.1% w/v), making those indubitably better candidates in terms of efficiency and atom economy. Thus, the large majority of peptide-based hydrogels developed in the last two decades are based on β-sheet assemblies [2–4].

Contrastingly to the use of polymers in hydrogel formation, peptides have indisputable advantages making them remarkably more attractive. In particular, they are both biocompatible and easily metabolized by proteolysis, making them perfect candidates for biomedical and therapeutic applications [15]. Unlike polymers, peptides are chemically defined and synthesized in a high-purity with high reproducibility. Moreover, due to their own structure, peptides are perfectly biodegradable and seem to be an *ad hoc* alternative to polymers when suitable. Nonetheless, it is fair to remind that polymer science still has an indisputable advantage because of its in-depth knowledge, while peptide-based hydrogels are still at their infancy. Fundamental research strongly supports the interest of these innovative peptide-based materials, trying to decipher the subtle mechanisms of self-assembly and the way(s) to optimize and fine-tune their thermodynamic and kinetic properties. Following this trend, applications have substantially increased and first commercially available peptide-based hydrogels have started to be sold, demonstrating their real interest [29].

1.3. Applications, commercial innovations, and outlook of peptide-based hydrogels

Due to their bioinspired architecture, peptide-based hydrogels are biocompatible, non-immunogenic, and potentially non-inflammatory [30], making them perfect materials for biomedical applications [31]. As a result, several systems were developed:

- as extracellular culture medium (or ECM) for cancer cells [16], stem cells, or neuronal SN4741 cells [32].
- as ECM to control $\alpha_5\beta_1$ integrin expression of endothelial cells [33].
- as injectable hydrogel for controlled-release of opioids [34] or of embryonic stem cell secretome [35].
- as implanted hydrogel-containing stem cells to treat spinal cord injury [36] or for nerve repair after traumatic injury in the nervous system [37]. *In vivo* implantation of RADA-16 peptide derivatives enhances extensive bone regeneration of mice femurs [38].
- as injectable hydrogel incorporating a Gadolinium MRI agent to follow its degradation *in vivo* [35].
- as adjuvant for vaccines, for instance against West Nile virus [39].
- for rapid hemostasis [40].
- for controlled-release of growth factor [41], pindolol, quinine, timolol [42], 5-fluorouracil [43], vancomycin and vitamin B12 [44], tanshinones [45], microRNA [46], or proteins (*e.g.*, BSA, IgG) [47].
- for immobilization of biocatalysts for chemical transformations [48].
- for removal of toxic dyes and heavy metal ions from waste water [49].
- to produce nanostructured silica [50] or to encapsulate carbon nanotubes [51].
- as antibacterial hydrogels [52], like the 20-mer MARG1 against methicillin-resistant *Staphylococcus aureus* [53], or **(Lys-Ile-Gly-Ala-Lys-Ile)$_3$-NH$_2$**, #1 **Figure 2**) against *E. coli* [54].

This laundry list clearly illustrates the potential of peptide-based hydrogels and puts out a new avenue in term of polyvalence, mainly in the biomedical domain. It has not escaped the attention of generalist media, like for instance Times of India [55], Asian Scientist [56], LaboratoryNews [57], Phys Org [58], Medical News Today [59], Medical Daily [60], and so forth.

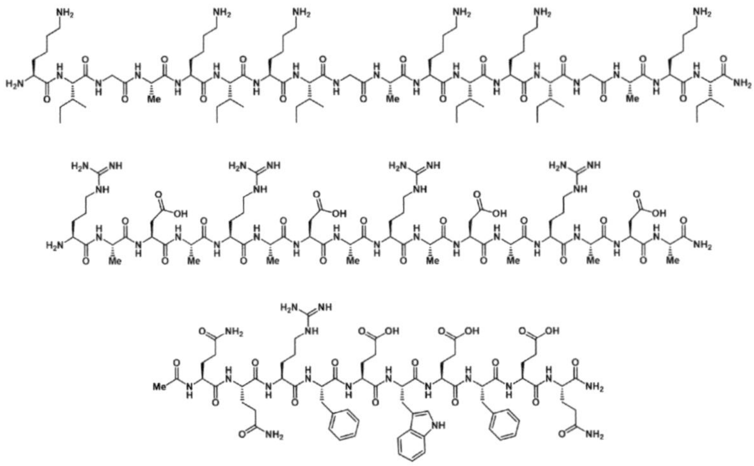

Figure 2. Chemical structures of peptides forming hydrogels.

Undoubtedly, peptide-based hydrogels are far from the laboratory curiosity, as evidenced by the new commercial market concerning these innovative materials. To the best of our knowledge, five products are commercially available:

- Hydromatrix™, developed by Sigma-Aldrich® (Saint Louis, USA) as medium for cell culture. The average price to prepare a 5 mL solution (0.1% w/v) is around 270 USD [61].
- PuraMatrix®, developed by 3-D Matrix Group (Tokyo, Japan) as medium for cell culture is a 16-mer (sequence **Ac-(Arg-Ala-Asp-Ala)$_4$-NH$_2$**, #2 **Figure 2**). The average price to prepare a 5 mL solution (0.1% w/v) is around 230 USD [62].
- Peptigel, developed by PeptiGel Design (Cheshire, UK) as medium for cell culture. The average price to prepare a 1 mL solution is around 130 USD [63].
- PGmatrix™, developed by PepGel LLC (Manhattan, USA) as medium for cell culture. The average price to prepare a 6 mL solution (0.1% w/v) is 375 USD [64].
- Curolox® technology, developed by Credentis (Windisch, Switzerland) as professional dental product for regeneration of enamel. It is not sold as it is, but as both toothpaste and a formulated product. The peptide used is an 11-mer termed P$_{11}$-4 (sequence **Ac-Gln-Gln-Arg-Phe-Glu-Trp-Glu-Phe-Glu-Gln-NH$_2$**, #3 **Figure 2**) [65].

It is a good bet that the number of commercially available peptide-based hydrogelator will increase in a near future.

As described hereinbefore, although the 20 natural amino acids offer a multitude of combinations, *i.e.*, a multitude of peptide-based hydrogel designs, the use of modified and non-canonical amino-acids drastically shoots up the possibilities and offers infinite opportunities both to improve the application window and to fine-tune properties of the resulting hydrogels. The next parts of this chapter are dedicated to the modifications developed to this end and illustrate the tremendous modularity of these materials.

2. Addition or insertion of organic moieties at the extremities or inside the amino acids sequences

2.1. Modifications at both N- and C-terminal ends

As discussed hereinbefore, gelation process is due to peptide self-assembly, driven by hydrogen bonds, electrostatic interactions, etc. In particular, π-π interactions and hydrophobic affinity are key parameters to design efficient hydrogelators, along with the subtle ability of the peptide to be partially soluble in water (mainly thanks to the presence of charged amino acids), and partially insoluble to form fibers. Thus, aromatic amino acids, mainly phenylalanine (**Phe**) and tyrosine (**Y**), are perfect candidates to balance these constraints. This is why a large majority of the peptide-based gelators are comprised of either these aromatic amino acids or other aromatic moieties.

Among them, it is not uncommon to observe the presence of a protecting group at the N-terminus. Indeed, protecting groups are widely used in both liquid and solid phase peptide synthesis (SPPS), and the easiest way to have them is simply by skipping the deprotection step at the end of the synthesis. Undoubtedly, Fmoc (fluorenylmethoxycarbonyl) is the more reported one, functionalizing peptide chains from one amino acid (*e.g.*, Fmoc-**Phe**, #4 **Figure 3**), to two (*e.g.*, **Fmoc-Gly-Ser**, #5 **Figure 3**), four (*e.g.*, **Fmoc-Phe-Arg-Gly-Asp**, #6 **Figure 3**), and more. Contrastingly, the other classic Cbz (carboxybenzyl) protecting group has not been well adopted, due to its smaller aromatic area leading to weaker π-π interactions compared to the Fmoc moiety [66, 67]. Another phenyl group was introduced via a cinnamoyl at the N-terminus of a phenylalanine amino acid, but it required higher concentration to form a gel in water, compared to the **Fmoc-Phe** analog (1.0% versus 0.3% w/v, respectively) [67]. Addition of naphthyl (Nap) derivatives (*e.g.*, 6-bromo-naphthyl (**6Br-Nap**) or 6-cyano-naphthyl (**6CN-Nap**)) has demonstrated high efficiency for several dipeptide gelators, like **Nap-Ala-Ala** (#7 **Figure 3**) or **Nap-Phe-Phe** (#8 **Figure 3**), **(6Br-Nap)-Phe-Phe** (#9 **Figure 3**), or **(6CN-Nap)-Ala-Val** (#10 **Figure 3**) [68]. Interestingly, **Nap-Gly-Ala** (#11 **Figure 3**) is able to form a stable hydrogel at acidic pH (pH = 2) at a concentration as low as 0.07% w/v [69]. Other moieties with larger aromatic areas are reported, even if they are much less common, like phenothiazine (e.g., #12 **Figure 3**) [66], pyrene (e. g., #13 **Figure 3**) [70, 71], spiropyran (e.g., #14 **Figure 3**) [72], or pyridine (#15 **Figure 3**) [73]. In addition, few exotic aromatic groups were used to add specific properties to the peptide, including a fluorescent stilbene chromophore (#16 **Figure 3**) [74] or a photosensitive azobenzene (#17 **Figure 3**) [75]. In this latter, an E conformation leads to hydrogel formation, while the photo-induced isomerization of the double-bond to a Z one triggers a phase change (*i.e.*, a liquid is thus obtained).

Another approach consists to graft a long alkyl chain at the N-terminus to form an amphiphilic hybrid molecule comprised of both a hydrophilic (peptide head) and a hydrophobic parts (alkyl tail). For instance, **$CH_3(CH_2)_{14}$-CO-Gly-Gly-Gly-Ser-Ser-Pro-His-Ser-Arg-Asn-(Ser-Gly)$_5$-Arg-Gly-Asp-Ser-Pro** forms a stable hydrogel from a minimum gelation concentration of ~1.4% w/v [76]. This latter contains a cell attachment site, the so-called **Arg-Gly-Asp** (or RGD) tripeptide, making the resulting hydrogel a good candidate for long-term human umbilical vein endo-

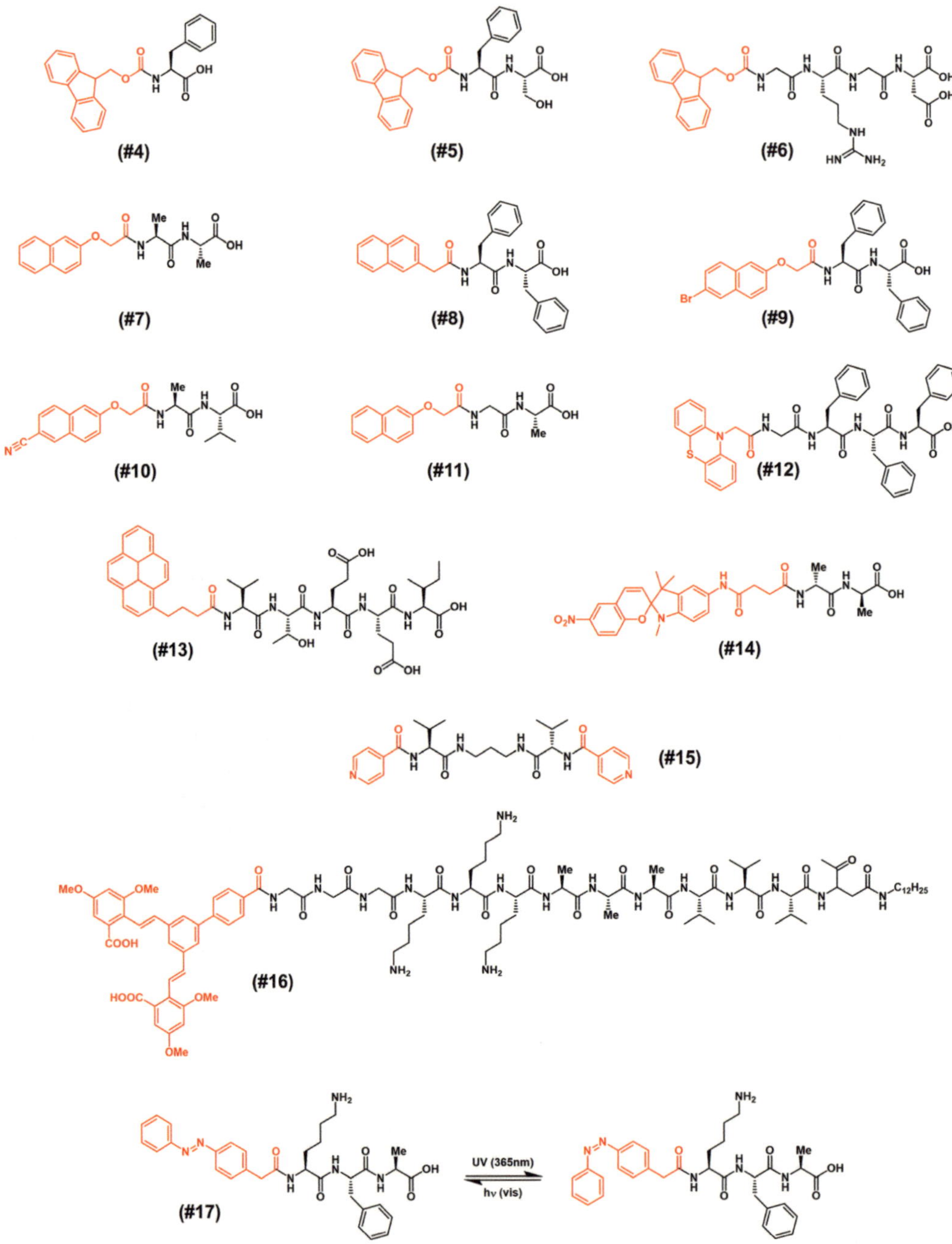

Figure 3. Chemical structures of hydrogel-forming peptides with N-terminal modifications.

thelial cell adhesion and proliferation. According to this model, other amphiphilic peptides containing alkyl chains were designed [77, 78], including **$CH_3(CH_2)_{14}$-CO-Gly-Thr-Ala-Gly-Leu-Ile-Gly-Gln-Glu-Arg-Gly-Asp-Ser** (#18 **Figure 4**) [79] and **$CH_3(CH_2)_{14}$-CO-Val-Val-Val-Ala-Ala-Ala-Glu-Glu-Glu** (#19 **Figure 4**) [80]. In parallel, innovative antibacterial hydrogels

http://dx.doi.org/10.5772/intechopen.68705

were developed from peptides in which the nitrogen of the N-terminus was quaternalized, promoting the *in situ* synthesis of gold nanoparticules (#20, 21 **Figure 4**) [78, 81–83]. In order to favor self-assembly of peptides, addition of nucleobases was investigated and demonstrated promising results to develop biocompatible hydrogels (#22, 23 **Figure 4**) [84, 85]. Acetylation of the terminal primary amine is applied in some cases [86], such as for the design of the efficient KLD12 for cartilage tissue repair (#24 **Figure 4**) [87, 88]. Last of all, several hydrogelators are composed of a non-aromatic *tert*-butyloxycarbonyl (Boc) protecting group at their N-terminus, annihilating the charge of the amine group. In this case, the resulting peptides are weakly water soluble and often find applications as organogels instead of hydrogels [89–92]. However, Baral et al. show its interest in the formation of stable antibacterial hydrogels (#25 **Figure 4**) [93]. Functionalizations with tetraethylene glycol [94] or polyethylene glycol [95] are also reported and increase significantly the global water solubility of the peptide (#26 **Figure 4**).

Regarding the C-terminus, while neutralization of the carboxylic acid thanks to its substitution by an ester (mostly methyl and ethyl esters) [89–92] is widely used to favor organogels, only few modifications are reported for hydrogels. Indeed, the acidic group plays a pivotal role in the peptide self-assembly and solubilization in aqueous solvent. Conversion of the carboxylic acid to the corresponding amide (*i.e.*, amidation) is well documented and tends toward faster hydrogel formation [96–99], but with weaker rigidity [100]. Peptides P_{11}-I (#27 **Figure 5**) [101] and **Ac-Cys-(Phe-Lys-Phe-Glu)$_2$-Cys-Gly-NH$_2$** (#28 **Figure 5**) [102] can be cited as examples.

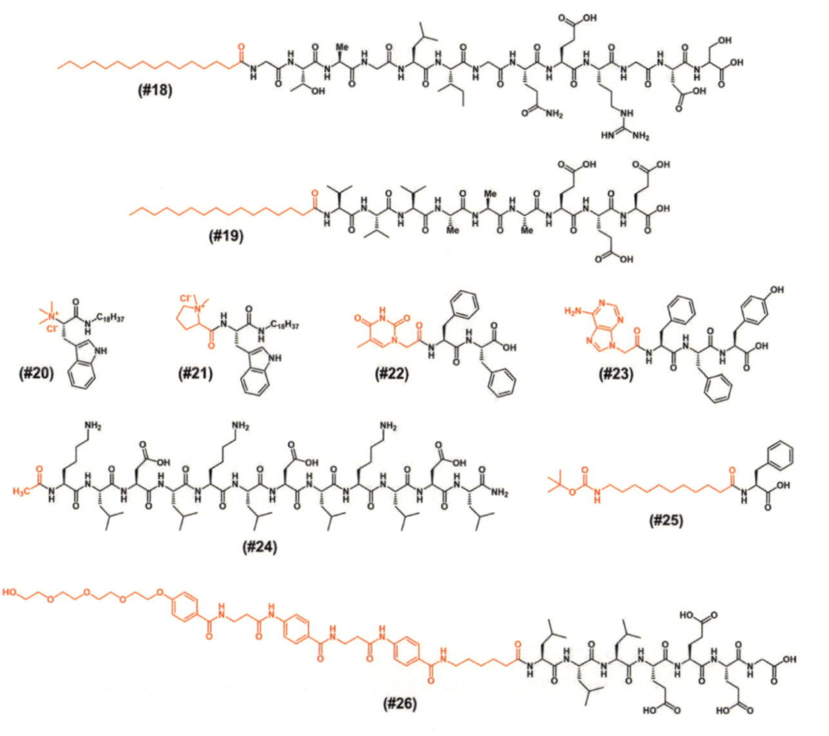

Figure 4. Chemical structures of hydrogel-forming peptides with N-terminal modifications (panel two).

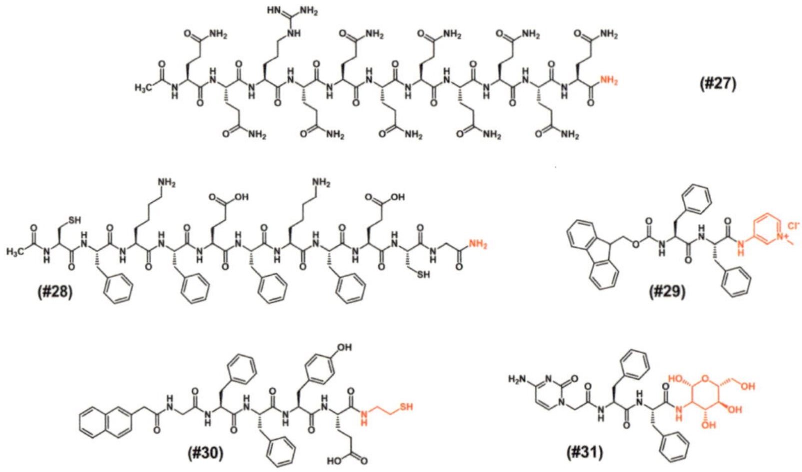

Figure 5. Chemical structures of hydrogel-forming peptides with C-terminal modifications.

Besides, more anecdotal C-terminus modifications were experimented, including pyridinium moieties (*e.g.*, #29 **Figure 5**) [103], thiol ethyl amides (#30 **Figure 5**) [104], tetraethylene glycol [94] or polyethylene glycol [95], kanamycin (an aminoglycoside antibiotic), or glycoside (#31 **Figure 5**) [85].

2.2. Insertion of organic moieties inside the amino acid sequences

Another way to improve hydrogelation and/or to add properties to the resulting gel consists to introduce an organic moiety inside the amino acid sequence. Thus, alkyl chains were inserted between two sequences of one or two amino acids to form a bolaamphiphilic molecule [73]. Indeed, these systems are comprised of a hydrophobic core (*i.e.*, the alkyl chain), twice functionalized at both extremities by hydrophilic amino acids (#32 **Figure 6**) [105]. While several kinds of aggregates are observed for short alkyl chains (*e.g.*, vesicles), longer ones favor hydrogel formation. For instance, with the presence of two histidines on each side (#33 **Figure 6**), optimal length of the alkyl spacer is of 20 methylene groups; in contrast, a length of 10 or 12 methylene leads to too soluble molecules unable to self-assemble to form a gel [106]. Similarly, unsaturated hydrocarbon chains were used, like diacetylene units functionalized by two tetrapeptides on each extremity. This strategy is ingenious because it offers the possibility to add covalent links between the peptide-forming hydrogels (#34, 35 **Figure 6**). Indeed, these materials are simultaneously physical (*i.e.*, by peptide self-assembly) and chemical (*i.e.*, chemical bonds obtained after polymerization) hydrogels, simply obtained by UV radiations (254 nm, 4 W, 30 min) [107].

As described hereinbefore, aromatic moieties play a pivotal role in the peptide self-assembly mechanism and were considered as efficient organic blocks to be inserted inside an amino acid chain. Among them, both naphthalene and perylene (#36 **Figure 6**) diimines demonstrated good results, even if their syntheses were limited by modest yields [108]. However, the same authors described efficient peptide-based hydrogelators comprised of oligophenylenes, oligophenylene vinylenes (#37 **Figure 6**) [109], or oligothiophenes (#38 **Figure 6**).

http://dx.doi.org/10.5772/intechopen.68705

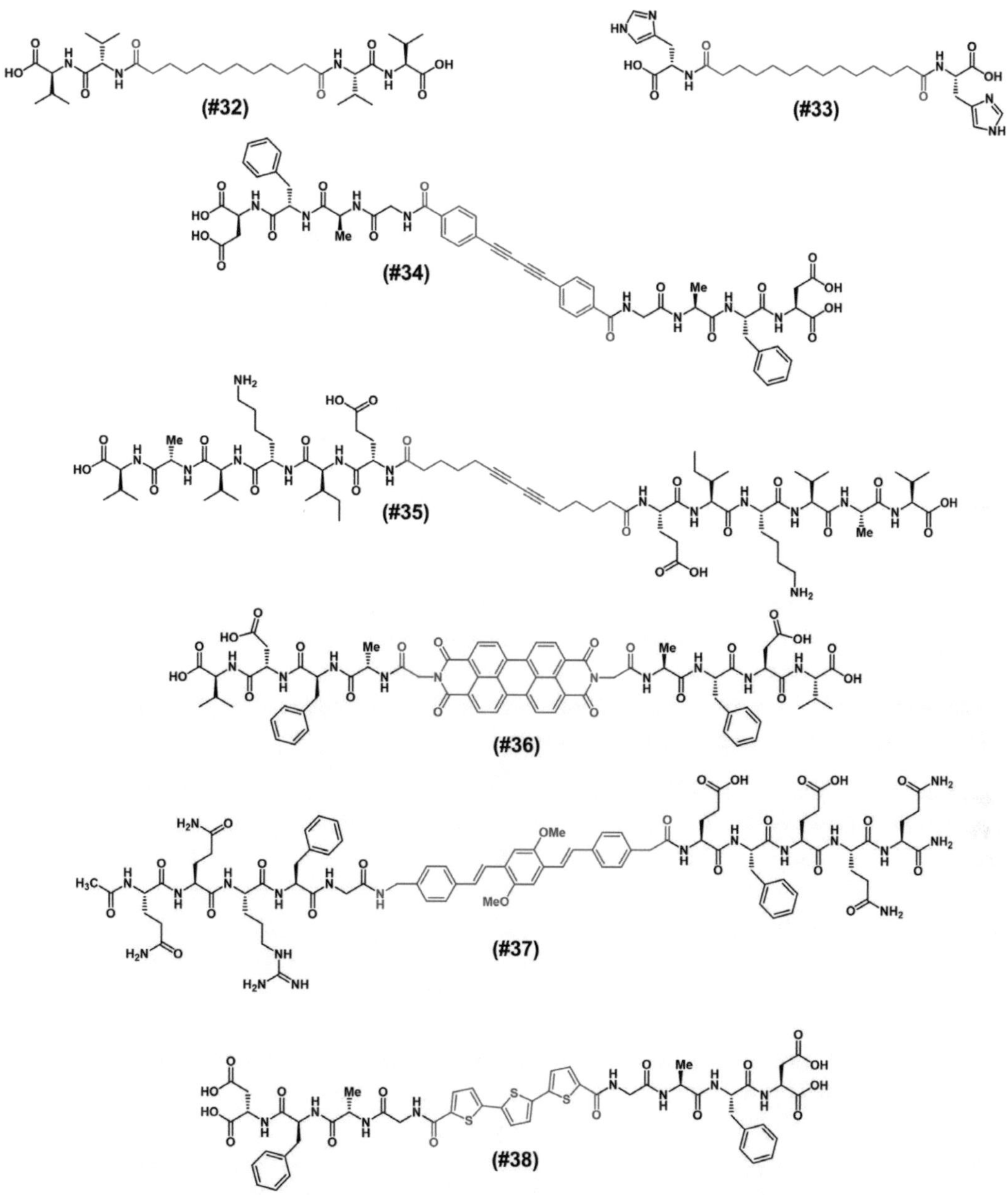

Figure 6. Chemical structures of hydrogel-forming peptides with organic moieties inserted into.

These systems are also very interesting specimens thanks to their photoluminescence properties due to their large π-conjugated surfaces [108, 110, 111]. Last of all, formation of a disulfide bond inside a sequence, mainly between two cysteines, is an easy way to functionalize hydrogelators. Besides, disulfite bond is often associated with protein and peptide self-assembly [112, 113], which plays a fundamental role in the peptide-based gelation process.

3. Using non-canonical and modified amino acids to tune peptide-based hydrogel properties

3.1. Impact of D-amino acids on the peptide-based hydrogel properties

In biological organism, the overwhelming majority of the natural proteins are exclusively composed of amino acids in their L-enantiomer forms. However, the chiral equivalents, *i.e.*, D-enantiomers with a reverse chemical configuration (**Figure 7A**), are present in several cases. For instance, the D-amino acids are found in bacteria in which D-alanine, D-aspartic, and D-glutamic acids are essential constituents for the synthesis of the peptidoglycan, forming the bacteria cell wall. Several antibiotics also contain a D-amino acid, like penicillin G (**δ-($_{L-}$α-aminoadipyl)-$_{L-}$Cys-$_{D-}$Val**), gramicidin, actinomycin, or polymyxins [114]. Not limited to prokaryotes, D-amino acids were also isolated from eukaryotic tissue, including dermophin from frogs, venom toxins from spiders or platypus, or neurohormones from crustaceans. More recently, these D-amino acids have been discovered in various human tissues, especially due to the presence of D-aspartic acid. It is the case for elastin, β-amyloid, α-synuclein, or AB-crystallin. Interestingly, these proteins are involved in several pathologies: arteriosclerosis, Alzheimer's and Parkinson's diseases, and cataract, respectively, demonstrating the undisputable importance of chirality in physiological process [115]. Research on amino acid role *in vivo* (as signaling molecules in the brain or endocrine glands [116], or in age-related diseases) [117] is a current fascinating hot topic, far from the scope of this chapter, and author incites curious readers to have a look to some reviews cited in this paragraph.

Inspired by the existence of D-amino acids, several groups used them to design efficient peptide-based hydrogels. The shorter one is undoubtedly **Fmoc-$_{D-}$Glu**, which forms right-handed helical nanofibers in the presence of **$_{D-}$Lys** (equimolar, #39 **Figure 7B**), while the same mixture with the L-enantiomer equivalents (*i.e.*, **Fmoc-$_{L-}$Glu** + **$_{L-}$Lys**) leads to left-handed helical nanofibers. Interestingly, the hydrogel containing D-enantiomers has slightly better viscoelastic properties than the levorotatory one [118]. This example clearly highlights that the organization at the molecular level (*i.e.*, the chirality) impacts the organization at the micro (*i.e.*, the fibers) and macroscopic scales (*i.e.*, the hydrogel). Series of dipeptides containing D-amino acids were also developed, one with Fmoc N-protecting group, and one with a naphtyl group (Nap). For the first instance, **Fmoc-$_{D-}$Ala-$_{D-}$Ala** (#40 **Figure 7B**) can be cited to be highly efficient to form a hydrogel (>0.13% w/v), slightly better than the corresponding **Fmoc-$_{L-}$Ala-$_{L-}$Ala** (>0.15% w/v), whereas **Fmoc-Gly-$_{D-}$Ala** (#41 **Figure 7B**) required concentration >1.7% w/v [119]. Interestingly, **Nap-Gly-$_{D-}$Ala** (#42 **Figure 7B**) displays exceptional properties with a minimum gelation concentration of only 0.07% w/v in water [69]. In this specific case, hydrogel characteristics are similar to the ones obtained with **Nap-Gly-Ala**, but the circular dichroism signature of each other is the perfect reverse, illustrating the opposite helical fibril structure. Tripeptides are also represented in this list with other Fmoc-derivatized structures containing three phenylalanine: **Fmoc-Phe-Phe-Phe** (#43 **Figure 7B**), **Fmoc-$_{D-}$Phe-$_{D-}$Phe-$_{D-}$Phe** (#44 **Figure 7B**), **Fmoc-Phe-$_{D-}$Phe-$_{D-}$Phe** (#45 **Figure 7B**), and **Fmoc-$_{D-}$Phe-Phe-Phe** (#46 **Figure 7B**) [120]. Interestingly, the fibers obtained in water

http://dx.doi.org/10.5772/intechopen.68705

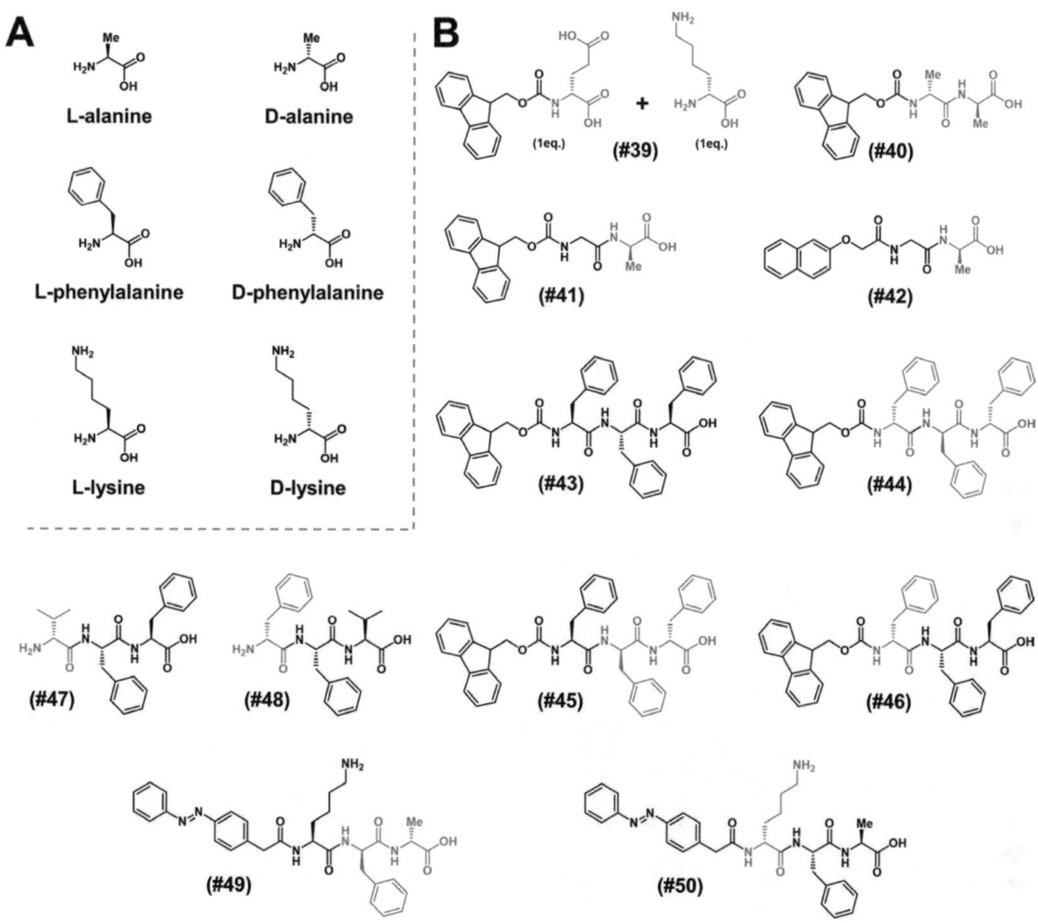

Figure 7. Chemical structures of L- and D-amino acids (A), and chemical structures of hydrogel-forming peptides incorporating D-amino acid(s) (B).

are right-handed for **Fmoc-Phe-Phe-Phe** and **Fmoc-Phe-$_{D-}$Phe-$_{D-}$Phe** and left-handed for the two others. Their viscoelastic properties are also improved by the addition of D-enantiomers, with **Fmoc-$_{D-}$Phe-$_{D-}$Phe-$_{D-}$Phe** about 23% higher than **Fmoc-Phe-Phe-Phe** in terms of storage and loss moduli. Without protecting group at the N-terminus, **Val-Phe-Phe** and **Phe-Phe-Val** peptides were unable to from hydrogels at neutral pH, while the **$_{D-}$Val-Phe-Phe** (#47 **Figure 7B**) and **$_{D-}$Phe-Phe-Val** (*i.e.*, both with the D-amino acid at the N extremity, #48 **Figure 7B**) equivalents did [121]. Another tripeptide, protected with an azobenzene moiety at the N-terminus, was developed from the sequence **Azo-Lys-Phe-Ala**, with D-enantiomers in different positions. Contrary to **Azo-Lys-Phe-Ala** which forms a hydrogel from 3.1% w/v, **Azo-Lys-$_{D-}$Phe-$_{D-}$Ala** (#49 **Figure 7B**) and **Azo-$_{D-}$Lys-Phe-Ala** (#50 **Figure 7B**) require 6.8% w/v, more than twice. Interestingly, the all D-amino acid-containing peptide gel is a little bit more efficient, with a minimum concentration of 3.0% w/v [75].

The last, but clearly not the least, example of the use of D-amino acids to improve hydrogelation is definitively the introduction of the sequence $_{D-}$**Pro-Pro**. Indeed, this latter confers a type II′ turn (a β-hairpin), favoring the contact between the two peptide strands on both sides. This decrease of degrees of freedom drastically favors the hydrogel formation efficiency. Thus, plenty of sequences were developed [122, 123], including the 20-mers MAX1 (sequence **(Val-Lys)$_4$-Val-$_{D-}$Pro-Pro-Thr-(Lys-Val)$_4$-NH$_2$**, #51 **Figure 8**) [124–129] and MAX8 (**(Val-Lys)$_4$-Val-$_{D-}$Pro-Pro-Thr-Lys-Val-Glu-Val-(Lys-Val)$_2$-NH$_2$**) [128, 130], in which the $_{D-}$**Pro-Pro** sequence is just in the middle, contrary to SSP1 (**(Val-Lys)$_2$-Val-$_{D-}$Pro-Pro-Thr-(Lys-Val)$_6$-NH$_2$**, #52 **Figure 8**) or SSP2 (**(Val-Lys)$_3$-Val-$_{D-}$Pro-Pro-Thr-(Lys-Val)$_5$-NH$_2$**) [131, 132]. More complex, a series of three-stranded peptides was developed, including TSS1 (sequence **(Val-Lys)$_4$-Val-$_{D-}$Pro-Pro-Thr-(Lys-Val)$_3$-Lys-$_{D-}$Pro-Pro-(Lys-Val)$_4$-NH$_2$**) [133].

Finally, derivatives from EAK-16 (sequence **Ac-(Ala-Glu-Ala-Glu-Ala-Lys-Ala-Lys)$_2$-NH$_2$**) incorporating D-amino acids were designed. Interestingly, the enantiomers EAK-16 and $_{D}$-EAK-16 (sequence **Ac-($_{D-}$Ala-$_{D-}$Glu-$_{D-}$Ala-$_{D-}$Glu-$_{D-}$Ala-$_{D-}$Lys-$_{D-}$Ala-$_{D-}$Lys)$_2$-NH$_2$**, #53 **Figure 8**) have shown similar hydrogel thermodynamic properties (*e.g.*, storage moduli ≈1 kPa at 1% w/v) [134–136]. In contrast, the corresponding diastereoisomers E$_D$AK-16 (sequence **Ac-($_{D-}$Ala-Glu-$_{D-}$Ala-Glu-$_{D-}$Ala-Lys-$_{D-}$Ala-Lys)$_2$-NH$_2$**, #54 **Figure 8**) and $_D$EA$_D$K (sequence **Ac-(Ala-$_{D-}$Glu-Ala-$_{D-}$Glu-Ala-$_{D-}$Lys-Ala-$_{D-}$Lys)$_2$-NH$_2$**) appear to have extremely modest capability to form stable hydrogels [137].

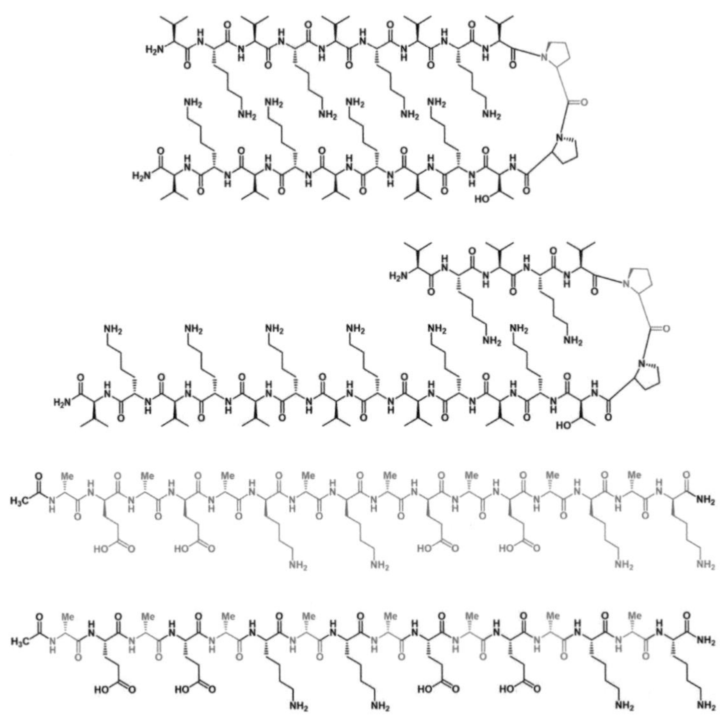

(#51)

(#52)

(#53)

(#54)

Figure 8. Chemical structures of hydrogel-forming peptides incorporating D-amino acid(s).

3.2. Using non-canonical amino acids to design peptide-based hydrogels

As described before, modifications of the peptide may be induced by functionalization of the N- and C-termini, along with the insertion of an organic moiety inside the sequence, or by inversion of the amino acids chirality, using D-amino acids. Furthermore, introduction of non-canonical amino acids (*i.e.*, not from the 20 encoded proteinogenic ones) offers broad new possibilities and has been exploited by researchers to improve and fine-tune hydrogel formation.

Among them, the cyclohexylalanine (**Cha**), a hydrogenated phenylalanine, prevents the formation of π-π interactions and significantly increases the global hydrophobicity of the peptide in which it is inserted. For instance, using the octapeptide **Ac-(Phe-Lys-Phe-Glu)$_2$-NH$_2$**, the substitution of both phenylalanine by two cyclohexylalanine (*i.e.*, **Ac-(Cha-Lys-Cha-Glu)$_2$-NH$_2$**, #55 **Figure 9**) leads to a less soluble peptide requiring a few percentage of HFIP (for hexafluoroisopropanol) to ensure a gelation. However, while a self-supporting hydrogel was obtained at 0.23% w/v for this latter, the Phe-containing peptide necessitates 0.46% w/v [138]. In parallel, experiments on **Ac-(Cha-Lys-Cha-Lys)$_2$-NH$_2$** demonstrate that at the same concentration, **Ac-(Phe-Lys-Cha-Phe)$_2$-NH$_2$** (#56 **Figure 9**) forms a more rigid hydrogel (G′ ≈ 1800 kPa versus 76 Pa) [139]. These experiments have illustrated the pivotal role of hydrophobicity in peptide self-assembly and consequently in the gelation process.

Another non-usual amino acid used is the ornithine (**Orn**), comprised of an aminopropyl lateral chain. It can be considered as a lysine but with only three methylenes in the side chain instead of four. Derived from P_{11} family [101, 140], an 11-mer containing three ornithine, positively charged, was synthesized (sequence **Ac-Gln-Gln-Orn-Phe-Orn-Trp-Orn-Phe-Gln-Gln-Gln-NH$_2$**, #57 **Figure 9**) and mixed with a close sequence, negatively charged, in which Orn were substituted by glutamic acids (sequence **Ac-Gln-Gln-Glu-Phe-Glu-Trp-Glu-Phe-Gln-Gln-Gln-NH$_2$**, #58 **Figure 9**). The equimolar mixture revealed both nematic gels and solutions properties. The assembly is favored by electrostatic interactions between charges carried by the primary amines (**Orn**) and the carboxylic acids (**Glu**) [96]. Other Orn-containing hydrogelators were developed from a decapeptide. More specifically, an oligo(p-phenylvinylene) (termed *OPV*) was functionalized by two pentapeptides on both sides: **Ac-Gln-Gln-Orn-Phe-Orn-*OPV*-Orn-Phe-Gln-Gln-Gln-NH$_2$** and **Ac-Gln-Gln-Arg-Phe-Glu-*OPV*-Glu-Phe-Gln-Gln-Gln-NH$_2$**. They both form a stable hydrogel, the first been more robust than the second one [109]. Working with short sequences of two amino acids without protection either at the N- or at the C-termini, the use of α,β-dehydrophenylalanine (**ΔPhe**) provides distinguished results [141, 142]. Compared to canonical **Phe, ΔPhe** has no chirality on its C^{α} and favors π-π stacking due to its extended electron delocalization. Indeed, while **Phe-Phe** lacks to form a hydrogel, **Phe-ΔPhe** is drastically more efficient forming a stiff one with G′ ≈ 210,000 Pa for only 1% w/v. However, after a full study on **Xxx-ΔPhe** (in which **Xxx** = 1 of the 20 canonical amino acids), only 2 dipeptides were reported as able to gelify: **Leu-ΔPhe** (#59 **Figure 9**) and **Phe-ΔPhe** (#60 **Figure 9**). Recently, the first one was applied *in vivo* in a mouse model as a drug delivery platform for mitoxantrone, an anticancer drug, and seems to be very promising [141].

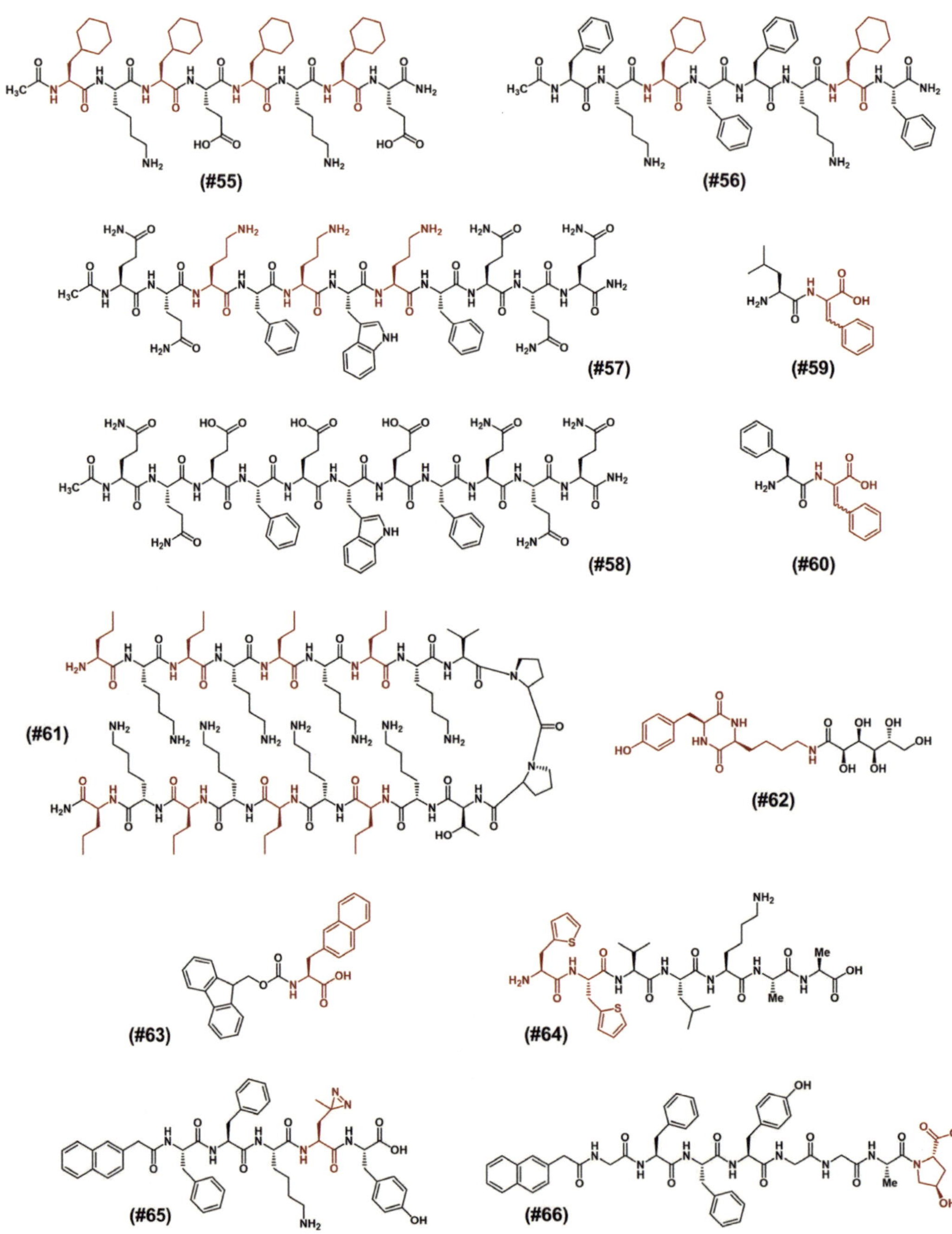

Figure 9. Chemical structures of hydrogel-forming peptides incorporating non-canonical amino acid(s).

Recently, based on the 20-mers MAX1 (sequence **(Val-Lys)$_4$-Val-$_{D-}$Pro-Pro-Thr-(Lys-Val)$_4$-NH$_2$**, see hereinbefore) [124–129], other derivatives were synthesized in which eight valines (excluding the **Val** adjacent to the $_{D-}$**Pro**) were substituted by non-canonical aminobutyric acid (**Abu**), norvaline (**Nva**, #61 **Figure 9**), or norleucine (**Nle**). Results have highlighted that, at 1% w/v, only the peptide-containing norvaline forms a stiffer material than the original MAX1, with G′ ≈ 3300 Pa versus 1800 Pa, respectively [143]. The use of cyclodipeptide was also reported as an efficient method. Indeed, **cyclo($_{L}$-Tyr-$_{L}$-Lys)** (#62 **Figure 9**) and **cyclo($_{L}$-Phe-$_{L}$-Lys)** N$^{\varepsilon}$-acetylated by gluconic acid lead to thixotropic hydrogels, which can be stored for a long period of time as a solution. The gelation is simply triggered by a short and vigorous stirring [144]. In order to increase the aromatic surface available from an ultra-short peptide, Fmoc-β-(2-naphthyl)-L-alanine (#63 **Figure 9**) was evaluated and the resulting hydrogel has exhibited good stability in a large range of pH, from 3 to 12 [145]. Inspired by the pentapeptide fragment **Lys-Leu-Val-Phe-Phe** from the 16 to 20 region of the Aβ protein (involves in Alzheimer's disease) and well-known for its ability to form amyloid fibers, effects of β-2-thienyalanine (**2-Thi**) were evaluated. The obtained **(2-Thi)-(2-Thi)-Val-Leu-Lys-Ala-Ala** (#64 **Figure 9**) has displayed high efficiency in hydrogel formation and liquid crystal properties [146].

Ingeniously, photoleucine, a diazirine-based photo-reactive analog of leucine, was introduced inside a pentapeptide protected with a naphthalene moiety at its N-terminus (#65 **Figure 9**). Thus, the system is applied for protein complex immunoprecipitation (also termed "pull-down" method) and interacts with proteins from 42 to 55 kDa [147].

Derived from the L-proline, the L-4-hydroxyproline (**Hyp**) is one of the amino acids constituting the tropocollagen and, *in fine*, the collagen. Indeed, collagen fibers are composed of the repeating sequence **Gly-Xxx-Hyp** in which **Xxx** = **Lys, Glu, Ser, Ala**, and **Pro** [148]. Based on that, **Nap-Gly-Phe-Phe-Tyr-Gly-Gly-Xxx-Hyp** peptides were studied (#66 **Figure 9**). Depending on the amino acid at the seventh position (*i.e.*, **Xxx**), the minimum gelation concentrations are between 0.04% and 0.10% w/v, demonstrating the assets of this bio-mimicking approach [149].

One of the last categories of modification this part would like to highlight is the functionalization of phenylalanine. Indeed, a convincing body of work has been focused on this pivotal amino acid playing a central role in peptide self-assembly and *in fine*, in hydrogel formation. Thus, working on the short **Fmoc-Phe**, the group of Bradley L. Nilsson studied the impact of the substitution of the hydrogen in the para position by a nitro (-NO_2, #67 **Figure 10**), a cyano (-CN, #68 **Figure 10**), an amino (-NH_2), a hydroxyl (-OH, in this case the amino acid corresponds to a tyrosine), a methoxy (-OMe, #69 **Figure 10**), a trifluoromethyl (-CF_3, #70 **Figure 10**), or a methyl group [100, 150–152]. This modification leads to a redistribution of the electron density of the aromatic ring, influencing the π-π and dipolar interactions between benzyl groups. While **Fmoc-Phe** forms a weak hydrogel at 0.6% w/v (G′ ≈ 40Pa), all these *para* substitutions improve the mechanical behavior of the resulting gels, especially for **Fmoc-(pNO_2-Phe), Fmoc-Tyr**, and **Fmoc-(pNH_2-Phe)**, with storage moduli of ≈410 Pa, 506 Pa, and 527 Pa, respectively. However, the gelation time is not directly linked to the mechanical properties, with t_{gel} ≈ 0.5 min, 5 min, and 10 min for **Fmoc-(pNO_2-Phe), Fmoc-(pCN-Phe)**, and **Fmoc-(pCH_3-Phe)**, respectively.

The effect of halogenation was also investigated on **Fmoc-Phe**, with substitutions on the *ortho*, *meta*, and *para* positions, by fluorine, chlorine, or bromine [153]. In this work, in terms of gelation time, halogenation on the *para* position is drastically shorter (t_{gel} ≈ 0.5 min) compared to both *meta* (t_{gel} ≈ 3–15 min) and *ortho* (t_{gel} ≈ 30–50 min) positions. However, regarding the

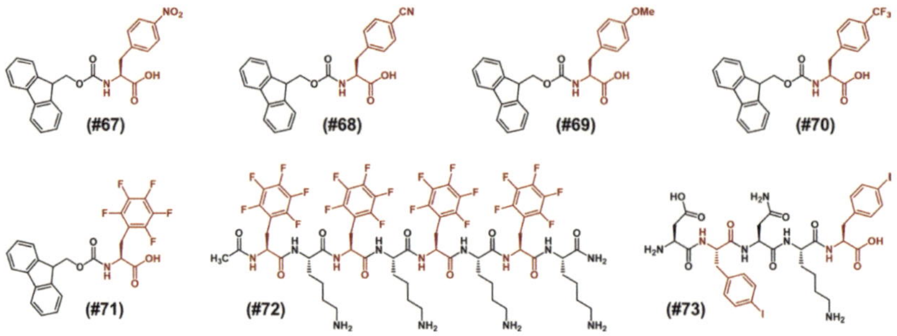

Figure 10. Chemical structures of hydrogel-forming peptides incorporating non-canonical amino acid(s) (panel two).

mechanical properties, the presence of a halogen in the *meta* position is the more efficient, followed by the *ortho* and finally the *para* positions. In parallel, the impact of the atom seems to be determined by its polarizability, with the formation of more rigid hydrogels with fluorine, better than chlorine and then bromine. In this way, the use of a pentafluorinated phenylalanine **Fmoc-(F_5-Phe)** (#71 **Figure 10**) was considered [100, 151]. This perfluorination drastically increases the hydrophobicity of the molecule and drops down the minimum gelation concentration to 0.1% w/v. At 0.2% w/v, the gel has good rheological results, with G′ ≈ 3100 kPa and G″ ≈ 320 Pa, confirming its efficiency compared to the classic **Fmoc-Phe**. Incorporation of **(F_5-Phe)** in a longer peptide chain was a success, as attested by the studies involving **Ac-[(F_5-Phe)-Lys-(F_5-Phe)-Lys]$_2$-NH_2** (#72 **Figure 10**) [139].

In the same way was reported the impact of a single-atom replacement of hydrogen with halogen in the human calcitonin-derived amyloidogenic fragment **Asp-Phe-Asn-Lys-Phe**. Mainly, the substitution of phenylalanine(s) by para-X-Phe (X = Cl, Br, I) led to a drastic improvement of the thermodynamic and kinetic properties. Amyloid structures were confirmed by Atomic Force Microscopy (AFM) and cryo-Transmission Electron Microscopy (cryo-TEM) measurement, and by their green birefringence upon staining with Congo Red, highlighting the **Asp-(pI-Phe)-Asn-Lys-(pI-Phe)** derivative (#73 **Figure 10**) as the most fibrilogenic peptide monomer. As reported in the literature, it was attested that these amyloid fibrils had the ability to form hydrogels with more efficiency compared to the wild-type **Asp-Phe-Asn-Lys-Phe**, illustrated by the 30-fold lower concentration required for the gelification of **Asp-(pI-Phe)-Asn-Lys-Phe**. However, best thermal stability (T = 116°C), lowest gelification time (<10 min), and highest stiffness (storage modulus G′ > 10^4 Pa) were observed for the same bis-iodinated sample (at 15 mM), far better than the bis-bromo and bis-chloro derivatives, respectively. All these results emphasize the positive role of halogenation of peptides, especially iodination, for supramolecular amplification of amyloid self-assembly [154].

3.3. Functionalization of amino acids on their side chain

Among the 20 canonical amino acids, a mere fraction contains a reactive organic group on its lateral chain. These amino acids are mainly the ones with a carboxylic acid: **Asp** and **Glu**, an amine (**Lys**), a guanidine (**Arg**), or a thiol group (**Cys**). Thus, modifications and post-modifications (*i.e.*, after the peptide synthesis) of the hydrogel-forming peptides can be carried out by chemical reactions between the amino acid cited above and an organic compound.

Following this rule, several structures have been developed, offering a new myriad of possibilities to enhance the versatility of the peptide-based hydrogels. Functionalization of lysine is clearly the more often used method. This amino acid has been grafted to a naphthalene diimine (NDI) moiety to increase the aromatic surface available on the peptide in order to favor π-π interactions and subsequently, hydrogelation. The resulting **Fmoc-Lys-Lys(NDI)** (#74 **Figure 11**) self-assembles at low concentration (<1.5% w/v) and can act as a semiconductor [155]. Similarly, lysines were modified by addition of an azobenzene moiety, leading to photoswitchable hydrogels (#75 **Figure 11**) [156, 157]. Presence of a sorbamide group on the Lys side chain offers the possibility to photopolymerize the physical hydrogel obtained by non-covalent interactions in order to create chemical bonds rigidifying the network. This second step improves the mechanical rigidity by 2.5 [158]. Using the same approach, an amphiphilic peptide was developed, composed of a lysine functionalized with an alkyl chain containing a diacetylene segment (#76 **Figure 11**). The subsequent polymerization leads to polyacetylene-containing hydrogel applied for cell culture [159]. Furthermore, substituting the primary amine by an acrylamide group to secondly form polyacrylamide is also an adopted approach (#77 **Figure 11**) [160].

Addition of a hydrazine-containing arm to an amphiphilic peptide was proposed to control the release of ketones. Indeed, ketones can react with hydrazine to form a hydrazone. The high hydrolytic stability of this chemical function leads to a slow release of the ketone-containing compound from the hydrogel (#78 **Figure 11**) [161]. The presence of Boc protecting groups at the N^{ε} of **Lys**, originating from the peptide synthesis, definitely improves the hydrogel properties, as described for **Fmoc-Val-Leu-Lys(Boc) and Fmoc-Lys(Boc)-Leu-Val** (#79 **Figure 11**). The first one is the weakest with $G' \approx 4000$ Pa (at 2% w/v), while the latter is the stiffest with $G' > 100{,}000$ Pa [162]. The same approach was reported with Fmoc protecting-group simply using **Lys(N^{ε}-Fmoc)**. However, the hydrogelation was triggered by addition of one equivalent of either **Fmoc-Phe** or **Fmoc-Leu** and both with two equivalents of Na_2CO_3. Comparing the two mixtures, the hydrogel formulated from **Lys(N^{ε}-Fmoc) + Fmoc-Phe** (#80 **Figure 11**) is drastically more rigid than the one with **Fmoc-Leu**, with storage moduli of 25,000 Pa and 3000 Pa, respectively [163, 164]. Lastly, a lysine-functionalized peptide reported concerns a linear amphiphilic nonapeptide N^{ε}-functionalized in second and last position by a histine and a palmitoyl (including a C_{15} alkyl chain). Gelation properties are interesting, with a minimum concentration required of 0.1% w/v and pH > 6.5 [165].

Working on a 20-mer derived from MAX1, an ingenious zinc-triggered hydrogelation was designed using a 3-amidoethoxyaminodiacetoxy-2-aminopropionic acid instead of a valine at the last position of the peptide sequence. Thus, the two incorporated acid groups perfectly fit to bind metals, in particular Zn^{2+}. While no gelation occurs for peptide alone in solution, addition of $ZnCl_2$ triggers the formation of the hydrogel (#81 **Figure 11**) [166]. In order to produce composite materials comprised of peptide fibers mineralized by calcium, an amphiphilic

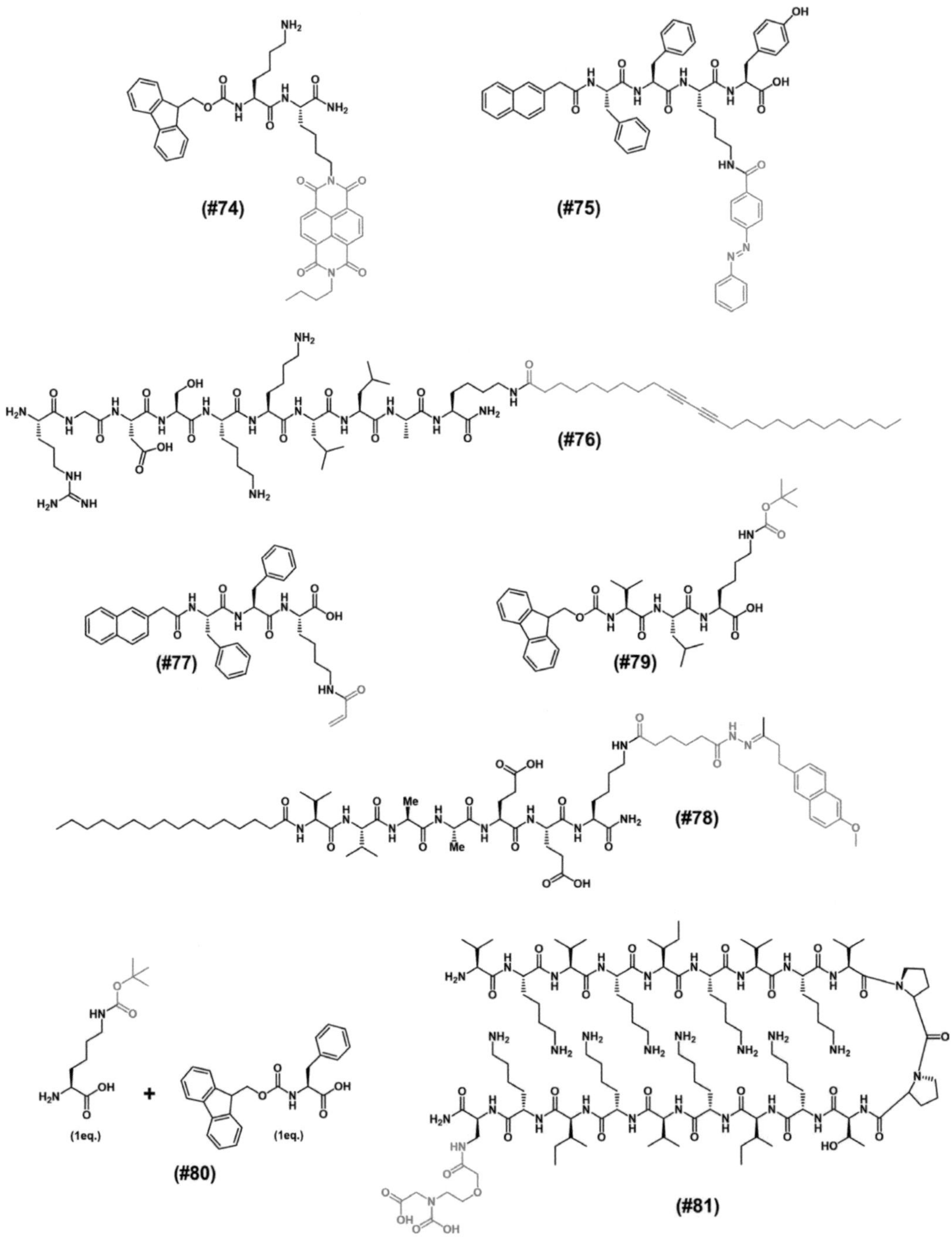

Figure 11. Chemical structures of hydrogel-forming peptides incorporating functionalized side chains.

molecule of 11 amino acids and a long alkyl chain of 15 methylene groups were designed. The key amino acid is a phosphorylated serine, playing a pivotal role in the formation of calcium phosphate minerals. After the self-assembly of the fibers, cysteine thiol groups are oxidized to disulfides before being treated by $CaCl_2$. After 20 min, the fibers start to be coated by crystalline minerals. These composite materials mimic the collagen fibrils and hydroxy-apatite crystals in bone [167]. Concerning the phosphorylated serine, another peptide with the same global structure (alkyl chain and 11-mer) but with a different peptide sequence is able to form hydrogels for which the mechanical properties can be modulated by the cation added (#82 **Figure 12**). Indeed, while a viscous liquid is obtained with Na^+ or K^+, Mg^{2+} and Ca^{2+} lead to hydrogels with moderate storage moduli (around 500 Pa), while with Zn^{2+} or Cu^{2+}, the gels are drastically stiffer ($G' \approx 10{,}000$ Pa) [168].

Working on **Fmoc-Phe-Phe-Gly-Gly-Gly-Tyr**, an innovative approach was experimented, based on $Ru(bpy)_3^{2+}$-catalyzed photo-crosslinking of two tyrosine residues to give dityrosine adduct. The obtained dimer acts as an efficient hydrogelator, while the monomer does not, with an increase of the mechanical stability up to 10,000 times (#83 **Figure 12**) [169]. Interestingly, formation of dityrosine is a strategy used by Nature to improve elastic properties of biomaterials, including the well-known resilin, first identified in *Drosophilia melanogaster* [170].

Last of all, some peptide drugs are able of hydrogelating, offering undoubted perspectives for biomedical applications. Among them, analogs of gonadotropin-releasing hormone (or GnRH) were developed. GnRH is a peptide hormone composed of 10 amino acids (sequence **pyro-Glu-His-Trp-Ser-Tyr-Gly-Leu-Arg-Pro-Gly-NH_2**) and discovered by Guillemin [171] and Schally [172] who received the Nobel Prize in Physiology and Medicine in 1977 with Yalow [173]. Analogs have been continuously and slowly releasing during a long period of time, up to 35 days. In particular, both ganirelix and degarelix show efficient gelation properties, and are composed of non-canonical amino acids (#82 **Figure 12**), including a **pCl-Phe**, a **NaI**, a N^G,$N^{G'}$-diethyl arginine, a (3-pyridyl)-alanine, an N^ε-isopropyl lysine, a para-urea phenylalanine, and a para-6-amido dihydroxyuracil [174]. Finally, the example of vancomycin,

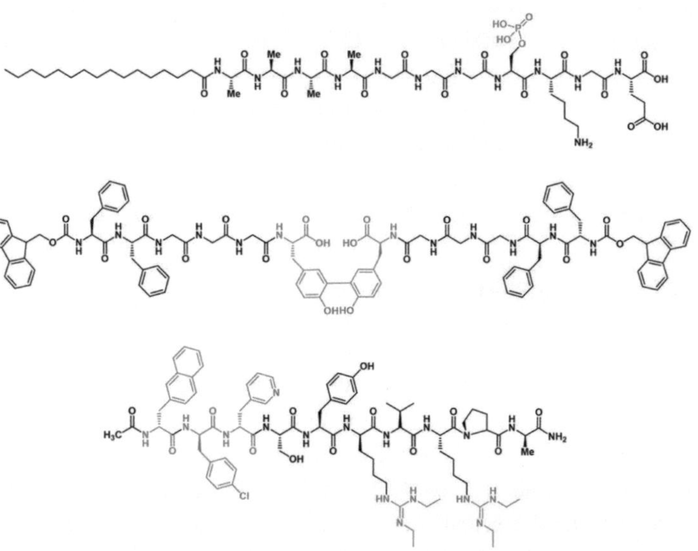

(#82)

(#83)

(#84)

Figure 12. Chemical structures of hydrogel-forming peptides incorporating functionalized side chains (panel two).

a glycopeptide with antibiotic properties, demonstrates the degree of complexity which can be reached. The minimum concentration of this perplexing molecule is around 0.36% w/v and offers new perspective in term of materials for biomedical applications [175]. However, due to the complexity of the system, it has to be considered as a special case, far from the initial idea to develop short peptide-based hydrogels synthesized in a straightforward and fast way.

4. Development of pseudo-peptides and peptidomimetics

4.1. A brief description of peptide analogs

Research on peptide analogs mainly refers to the development of metabolically stable peptide-like structures for biomedical and therapeutic applications. Indeed, the rapid proteolysis of native peptides and their inability to cross cellular membranes have required modifications of their intrinsic structure. Thus, changes of the peptide bond structure have been deeply studied and have led to efficient drugs commercialized by pharmaceutical company. For instance, Ziconotide (brand name Prialt®, sold by Elan Pharms) is a 25-mer indicated for severe chronic pains, Icatibant acetate (brand name Firazyr®, sold by Jerini AG) is a decapeptide indicated for hereditary angioedema, and Afamelanotide (brand name Scenesse®, sold by Clinuvel Pharmaceuticals) is a 13-mer indicated for erythropoietic prophyries [176].

Thus, two terminologies were introduced. The term "pseudopeptide" proposed by Spatola in 1981 strictly refers to peptide analogs in which the peptide backbone has been modified [177, 178]. Briefly, their nomenclature is quite simple, with the use of the Greek letter psi "ψ[…]" in which the term under bracket refers to the "new" peptide link introduced. For example, the dipeptide #85 (**Figure 13A**) is represented as **Phe-Leu** (#85 **Figure 13A**), while **Phe-ψ[CH_2O] Leu** represents the pseudopeptide #86 (**Figure 13A**). In parallel, the term "peptidomimetic" refers to any compound able to mimic the specific action of a peptide (*i.e.*, inhibition, activation, etc.) [179]. Thus a "peptidomimetic" can be an organic molecule without any similarities to a peptide. It is the case of morphine, a "non-peptide" molecule mimicking the opioid peptides [177]. However, the word "pseudopeptide" has become obsolete years after years, and the generic term "peptidomimetic" is used to speak about both [180].

4.2. Peptidomimetics as efficient hydrogelators

First and foremost, the most represented peptidomimetics able to form hydrogels are undoubtedly the ones comprised of one or several β-amino acids. These latter differ from the native α-amino acids due to the presence of an additional carbon atom in the amino acid backbone. Besides, two regioisomers exist, depending on the position of the additional methylene group, termed β^2 or β^3-amino acids (**Figure 13A**, right panel) [181]. With this in mind, a series of β^3-dipeptides was synthesized, protected at the N-terminus by a naphthalene (**Nap**) derivative. Peptidomimetics #87 (**Nap-OCH_2CO-αGly-β^3HAla, Figure 13B**) and #88 (**Nap-CH_2CO-β^3Phg-β^3Phg**, with **Phg** = phenylglycine, **Figure 13B**) have reasonable mechanical properties,

http://dx.doi.org/10.5772/intechopen.68705

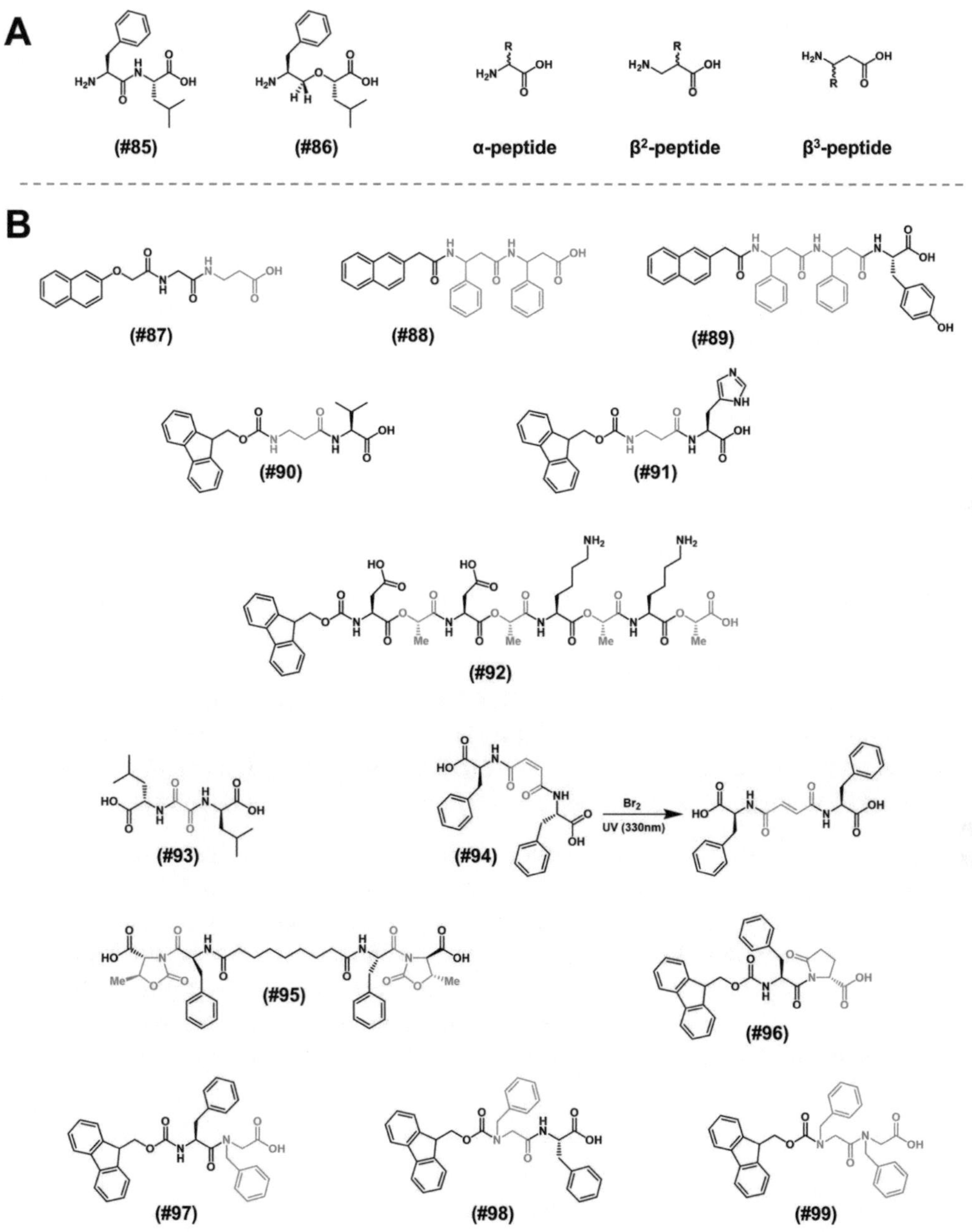

Figure 13. Chemical structures of peptidomimetics (A), and chemical structures of hydrogel-forming peptides incorporating peptidomimetics (B).

the first one being stiffer than the second one [182]. With the same **β³Phg** and naphthalene extremity, a tripeptide (**Nap-CH$_2$CO-β³Phg-β³Phg-αTyr**, #89 **Figure 13B**) was studied as injectable hydrogel *in vitro* and *in vivo*, and offers longer biostability than the corresponding all α-containing-amino acids peptide **Nap-CH$_2$CO-αPhe-αPhe-αTyr** [183]. Other compounds with a Fmoc N-protected group were developed and demonstrated ability to encapsulate and release both vitamins B_2 and B_{12} over a period of more than 48 hours, at physiological pH and temperature. Their structures are **Fmoc-βAla-αVal** (#90 **Figure 13B**) and **Fmoc-βAla-αPhe** and have comparable mechanical properties and release profiles [184]. In parallel, hydrogelation of **Fmoc-βAla-αHis** (#91 **Figure 13B**) is triggered by addition of Zn^{2+} cations [185].

In a peptide, substitution of the native amide function by an ester leads to structures called depsipeptides. The octapeptide Fmoc-DAKA-8 (sequence **Fmoc-Asp-Ala-Asp-Ala-Lys-Ala-Lys-Ala**) was modified, *inter alia,* by the introduction between the first and the second amino acids of an ester bond. Thus, **Fmoc-Asp-ψ[CO-O]Ala-Asp-ψ[CO-O]Ala-Lys-ψ[CO-O]Ala-Lys-Ala]** (#92 **Figure 13B**) self-assembles and forms a hydrogel, with hydrolytic susceptibility (hydrolysis at the ester bond sites), which offer opportunities for applications such as tissue scaffold or drug delivery system [186]. Anecdotally, two peptidomimetics containing an oxalyl function are reported for their ability to form organogel as well as hydrogel, including #93 (**Figure 13B**) [187, 188]. As well, a molecule containing a maleic acid-like function is able to gelify in the presence of light and bromine. Indeed, starting with the Z configuration, UV irradiation (330 nm, 400 W, 30 s) and bromine trigger the photochemical isomerization (formation of a fumaryl-like group). The obtained diastereoisomer is then able to self-assemble (#94 **Figure 13B**) [189].

In parallel, a constrained amino acid derived from oxazolidine was used by the group of C. Tomasini, named *(4R,5S)*-4-carboxy-5-methyl oxazolidin-2-one (3-letter code **Oxd**). The first hydrogelator including an **Oxd** (#95 **Figure 13B**) was composed of a central alkyl chain, two **Phe** (one on each side), and two terminal **Oxd** (one on each side). However, hydrogelation can only occur in a 50/50 water/MeOH mixture [190, 191] or 90/10 water/EtOH [192]. The second system (#96 **Figure 13B**) is definitively more efficient (cpd xx) and exhibits properties of entrapment of aromatic dyes methylene blue (cationic) and eosin Y (anionic) [193].

Last of all, functionalization of the nitrogen of the amide group was proposed as a way to improve hydrogelation. Because of the high efficiency of phenylalanine, N-benzyl glycine (**Nphe**) derivatives synthesized were the following: **Fmoc-Phe-Phe, Fmoc-Phe-Nphe** (#97 **Figure 13B**), **Fmoc-Nphe-Phe** (#98 **Figure 13B**), and **Fmoc-Nphe-Nphe** (#99 **Figure 13B**). While the latter failed to form a hydrogel, the two peptides with one **Nphe** inside turned out to be less stiff than **Fmoc-Phe-Phe** in the same conditions. However, these experiments provide new molecular tools to fine-tune the properties of peptide self-assembly and gelation process [194].

5. Conclusion

Peptide-based hydrogels are innovative materials able to efficiency tackle the challenges of biocompatible and biodegradable soft matter. Mainly, they pave the way to promising medical and biological applications, and it is a safe bet that in the next years, their market share will increase drastically. Furthermore, the first commercially available peptide-based hydrogel has been launched, and new companies have emerged. Nevertheless, all these applications are possible thanks to the plentiful body of work accomplished in fundamental research during the last decades.

In order to improve and fine-tune the hydrogelation process, a critical point to broaden the application window, several strategies have been envisaged, described all along this chapter. Among them, addition of aromatic groups at the N-terminus, mainly via the presence of protecting groups (*e.g.*, Fmoc, Cbz, etc.), clearly favors the hydrogel formation, thanks to the increase of π-π interactions. Introduction of organic moieties inside the peptide sequence is also an efficient strategy, whether they are aromatic or unsaturated. In this latter case, introduction of acetylene groups offers the opportunity to add to the peptide self-assembly polymerization, which reinforces the mechanical properties of the gel. Furthermore, introduction of non-canonical D-amino acids is an efficient approach and leads to more stable hydrogels in term of proteolytic resistance. The development of new amino acids, including α,β-dehydrophenylalanine, β-thienylalanine, or functionalized phenylalanine, especially designed to influence the peptide self-assembly, drastically impacts the final properties of the obtained hydrogel. Functionalization on the lateral chains is another opportunity to support interactions between peptides, and to include new properties, like photoswitchable ones thanks to the introduction of azobenzene moieties. Creation of additional bonds via dityrosine formation or subsequent polymerization is a smart strategy combining both the physical and chemical gels worlds. Last of all, substitution of the inner peptide bond by another one leads to the development of peptidomimetics which are burgeoning and promise a bright future.

This chapter has reviewed the recent advances in peptide-based hydrogel development using modified peptide structures.

Acknowledgements

The author would like to thank Université de Lorraine and the Centre National de la Recherche Scientifique (CNRS) for their support.

Author details

Stefan Loic

Address all correspondence to: dr.stefanloic@gmail.com

Laboratory of Macromolecular Physical Chemistry, University of Lorraine, Nancy, France

References

[1] De Gennes P-G. Soft matter (Nobel Lecture). Angewandte Chemie International Edition [Internet]. 1992;**31**(7):842–845. Available from: http://doi.wiley.com/10.1002/anie.199208421 [Accessed: January 13, 2017]

[2] Tomasini C, Castellucci N. Peptides and peptidomimetics that behave as low molecular weight gelators. Chemical Society Reviews [Internet]. 2013;**42**(1):156–172. Available from: http://xlink.rsc.org/?DOI=C2CS35284B [Accessed: January 23, 2017]

[3] Dasgupta A, Mondal JH, Das D. Peptide hydrogels. RSC Advances [Internet]. 2013;**3**(24): 9117. Available from: http://xlink.rsc.org/?DOI=c3ra40234g [Accessed: January 23, 2017]

[4] De Leon Rodriguez LM, Hemar Y, Cornish J, Brimble MA. Structure-mechanical property correlations of hydrogel forming β-sheet peptides. Chemical Society Reviews [Internet]. 2016;**45**(17):4797–4824. Available from: http://xlink.rsc.org/?DOI=C5CS00941C [Accessed: January 23, 2017]

[5] Vintiloiu A, Leroux J-C. Organogels and their use in drug delivery—A review. Journal of Controlled Release. 2008;**125**(3):179–192

[6] Buwalda SJ, Boere KWM, Dijkstra PJ, Feijen J, Vermonden T, Hennink WE. Hydrogels in a historical perspective: From simple networks to smart materials. Journal of Controlled Release. 2014;**190**:254–273

[7] Estroff LA, Hamilton AD. Water gelation by small organic molecules. Chemical Reviews. 2004;**104**(3):1201–1217

[8] Caló E, Khutoryanskiy VV. Biomedical applications of hydrogels: A review of patents and commercial products. European Polymer Journal. 2015;**65**:252–267

[9] Gulrez SK, Al-Assaf S, Phillips GO. Hydrogels: Methods of preparation, characterisation and applications in molecular and environmental bioengineering. Progress in Molecular and Environmental Bioengineering—From Analysis and Modelling to Technology Applications [Internet]. 2011;646. Available from: http://www.intechopen.com/books/progress-in-molecular-and-environmental-bioengineering-from-analysis-and-modeling-to-technology-applications/hydrogels-methods-of-preparation-characterisation-and-applications [Accessed: January 23, 2017]

[10] Jonker AM, Löwik DWPM, Van Hest JCM. Peptide- and protein-based hydrogels [Internet]. Chemistry of Materials. American Chemical Society. 2012;**24**:759–773. Available from: http://pubs.acs.org/doi/abs/10.1021/cm202640w [Accessed: January 23, 2017]

[11] Sangeetha NM, Maitra U. Supramolecular gels: Functions and uses. Chemical Society Reviews [Internet]. 2005;**34**(10):821–836. Available from: http://xlink.rsc.org/?DOI=b417081b [Accessed: January 23, 2017]

[12] Adler-Abramovich L, Gazit E. The physical properties of supramolecular peptide assemblies: From building block association to technological applications. Chemical Society reviews [Internet]. 2014;**43**(20):6881–6893. Available from: http://xlink.rsc.org/?DOI=C4CS00164H [Accessed: January 23, 2017]

[13] Eskandari S, Guerin T, Toth I, Stephenson RJ. Recent advances in self-assembled peptides: Implications for targeted drug delivery and vaccine engineering. Advanced Drug Delivery Reviews. 2016; in press. DOI: 10.1016/j.addr.2016.06.013

[14] Ruedinger F, Lavrentieva A, Blume C, Pepelanova I, Scheper T. Hydrogels for 3D mammalian cell culture: A starting guide for laboratory practice. Applied Microbiology and Biotechnology [Internet]. 2015;**99**(2):623–636. Available from: http://link.springer.com/10.1007/s00253-014-6253-y [Accessed: January 23, 2017]

[15] Seow WY, Hauser CAE. Short to ultrashort peptide hydrogels for biomedical uses. Materials Today. 2014;**17**:381–388

[16] Worthington P, Pochan DJ, Langhans SA. Peptide hydrogels—Versatile matrices for 3D cell culture in cancer medicine. Frontiers in Oncology [Internet]. 2015 ;**5**(April):92. Available from: http://journal.frontiersin.org/article/10.3389/fonc.2015.00092/abstract [Accessed: January 11, 2017]

[17] Hosseinkhani H, Hong P-D, Yu D-S. Self-assembled proteins and peptides for regenerative medicine. Chemical Reviews [Internet]. 2013;**113**(7):4837–4861. Available from: http://pubs.acs.org/doi/abs/10.1021/cr300131h [Accessed: January 23, 2017]

[18] Li Y, Qin M, Cao Y, Wang W. Designing the mechanical properties of peptide-based supramolecular hydrogels for biomedical applications [Internet]. Science China: Physics, Mechanics and Astronomy. 2014;**57**:849–858. Available from: http://link.springer.com/10.1007/s11433-014-5427-z [Accessed: January 23, 2017]

[19] Jung JP, Gasiorowski JZ, Collier JH. Fibrillar peptide gels in biotechnology and biomedicine. Biopolymers. Wiley Subscription Services, Inc., A Wiley Company [Internet]. 2010;**94**:49–59. Available from: http://doi.wiley.com/10.1002/bip.21326 [Accessed: January 23, 2017]

[20] Bischoff R, Schlüter H. Amino acids: Chemistry, functionality and selected non-enzymatic post-translational modifications. Journal of Proteomics. 2012;**75**:2275–2296

[21] Knowles TPJ, Buehler MJ. Nanomechanics of functional and pathological amyloid materials. Nature Nanotechnology [Internet]. 2011;**6**(8):469–479. Available from: http://www.nature.com/doifinder/10.1038/nnano.2011.102 [Accessed: January 23, 2017]

[22] Petkau-Milroy K, Brunsveld L. Supramolecular chemical biology; bioactive synthetic self-assemblies. Organic & Biomolecular Chemistry [Internet]. 2013;**11**(2):219–232. Available from: http://dx.doi.org/10.1039/C2OB26790J [Accessed: January 23, 2017]

[23] Woolfson DN. Building fibrous biomaterials from alpha-helical and collagen-like coiled-coil peptides. Biopolymers. Wiley Subscription Services, Inc., A Wiley Company [Internet]. 2010;**94:**118–127. Available from: http://doi.wiley.com/10.1002/bip.21345 [Accessed: January 23, 2017]

[24] Kumar VA, Wang BK, Kanahara SM. Rational design of fiber forming supramolecular structures. Experimental Biology and Medicine [Internet]. 2016;**241**(9):1–10. Available from: http://journals.sagepub.com/doi/10.1177/1535370216640941 [Accessed: January 23, 2017]

[25] Dong H, Paramonov SE, Hartgerink JD. Self-assembly of alpha-helical coiled coil nanofibers_Supporting Info. Journal of the American Chemical Society [Internet]. 2008;**130**(41):13691–13695. Available from: http://pubs.acs.org/doi/abs/10.1021/ja8037323 [Accessed: January 12, 2017]

[26] Potekhin SA, Melnik TN, Popov V, Lanina NF, Vazina AA, Rigler P, et al. De novo design of fibrils made of short α-helical coiled coil peptides. Chemistry and Biology. 2001;**8**(11):1025–1032

[27] Fletcher NL, Lockett CV, Dexter AF. A pH-responsive coiled-coil peptide hydrogel. Soft Matter [Internet]. 2011;**7**(21):10210. Available from: http://xlink.rsc.org/?DOI=c1sm06261a [Accessed: January 12, 2017]

[28] Banwell EF, Abelardo ES, Adams DJ, Birchall MA, Corrigan A, Donald AM, et al. Rational design and application of responsive alpha-helical peptide hydrogels. Nature Materials [Internet]. 2009;**8**(7):596–600. Available from: http://www.nature.com/doifinder/10.1038/nmat2479 [Accessed: January 12, 2017]

[29] De Santis E, Ryadnov MG. Peptide self-assembly for nanomaterials: The old new kid on the block. Chemical Society Reviews [Internet]. 2015;**44**(22):8288–8300. Available from: http://xlink.rsc.org/?DOI=C5CS00470E [Accessed: January 23, 2017]

[30] Markey A, Workman VL, Bruce IA, Woolford TJ, Derby B, Miller AF, et al. Peptide hydrogel in vitro non-inflammatory potential. Journal of Peptide Science [Internet]. 2016. Available from: http://doi.wiley.com/10.1002/psc.2940 [Accessed: January 10, 2017]

[31] Seliktar D. Designing cell-compatible hydrogels for biomedical applications. Science. 2012;**336**(6085):1124–1128

[32] Motamed S, Del Borgo MP, Kulkarni K, Habila N, Zhou K, Perlmutter P, et al. A self-assembling [small beta]-peptide hydrogel for neural tissue engineering. Soft Matter [Internet]. 2016;**12**(8):2243–2246. Available from: http://xlink.rsc.org/?DOI=C5SM02902C [Accessed: January 11, 2017]

[33] Pashuck ET, Duchet BJR, Hansel CS, Maynard SA, Chow LW, Stevens MM. Controlled sub-nanometer epitope spacing in a three-dimensional self-assembled peptide hydrogel. ACS Nano [Internet]. 2016;**10**(12):11096–11104. Available from: http://pubs.acs.org/doi/abs/10.1021/acsnano.6b05975 [Accessed: January 10, 2017]

[34] Martin C, Oyen E, Mangelschots J, Bibian M, Ben Haddou T, Andrade J, et al. Injectable peptide hydrogels for controlled-release of opioids. MedChemComm [Internet]. 2016;**7**(3):542–549. Available from: http://xlink.rsc.org/?DOI=C5MD00440C [Accessed: January 11, 2017]

[35] Bakota EL, Wang Y, Danesh FR, Hartgerink JD. Injectable multidomain peptide nanofiber hydrogel as a delivery agent for stem cell secretome. Biomacromolecules [Internet]. 2011;**12**(5):1651–1657. Available from: http://pubs.acs.org/doi/abs/10.1021/bm200035r [Accessed: January 11, 2017]

[36] Li L-M, Han M, Jiang X, Yin X, Chen F, Zhang T-Y, et al. Peptide-tethered hydrogel scaffold promotes recovery from spinal cord transection via synergism with mesenchymal stem cells. ACS Applied Materials & Interfaces [Internet]. 2017. Available from: http://pubs.acs.org/doi/abs/10.1021/acsami.6b12829 [Accessed: January 10, 2017]

[37] Zhang N, He L, Wu W. Self-assembling peptide nanofibrous hydrogel as a promising strategy in nerve repair after traumatic injury in the nervous system. Neural Regeneration Research [Internet]. 2016;**11**(5):717–718. Available from: http://www.ncbi.nlm.nih.gov/pubmed/27335544 [Accessed: January 11, 2017]

[38] He B, Ou Y, Zhou A, Chen S, Zhao W, Zhao J, et al. Functionalized D-form self-assembling peptide hydrogels for bone regeneration. Drug Design, Development and Therapy [Internet]. 2016;**10**:1379–1388. Available from: https://www.dovepress.com/functionalized-d-form-self-assembling-peptide-hydrogels-for-bone-regen-peer-reviewed-article-DDDT [Accessed: January 11, 2017]

[39] Friedrich BM, Beasley DWC, Rudra JS. Supramolecular peptide hydrogel adjuvanted subunit vaccine elicits protective antibody responses against West Nile virus. Vaccine. 2016;**34**(46):5479–5482

[40] Chen C, Zhang Y, Fei R, Cao C, Wang M, Wang J, et al. Hydrogelation of the short self-assembling peptide I3QGK regulated by transglutaminase and use for rapid hemostasis. ACS Applied Materials & Interfaces [Internet]. 2016;**8**(28):17833–17841. Available from: http://pubs.acs.org/doi/abs/10.1021/acsami.6b04939 [Accessed: January 10, 2017]

[41] Lindsey S, Piatt JH, Worthington P, Sönmez C, Satheye S, Schneider JP, et al. Beta hairpin peptide hydrogels as an injectable solid vehicle for neurotrophic growth factor delivery. Biomacromolecules [Internet]. 2015;**16**(9):2672–2683. Available from: http://pubs.acs.org/doi/10.1021/acs.biomac.5b00541 [Accessed: January 12, 2017]

[42] Briuglia ML, Urquhart AJ, Lamprou DA. Sustained and controlled release of lipophilic drugs from a self-assembling amphiphilic peptide hydrogel. International Journal of Pharmaceutics. 2014;**474**(1–2):103–111

[43] Liang L, Yang J, Li Q, Huo M, Jiang F, Xu X, et al. A novel targeting drug delivery system based on self-assembled peptide hydrogel. Journal of Biomaterials and Nanobiotechnology [Internet]. 2011;**2**(5):622–625. Available from: http://www.scirp.org/journal/PaperDownload.aspx?DOI=10.4236/jbnb.2011.225074 [Accessed: January 12, 2017]

[44] Baral A, Roy S, Dehsorkhi A, Hamley IW, Mohapatra S, Ghosh S, et al. Assembly of an injectable noncytotoxic peptide-based hydrogelator for sustained release of drugs. Langmuir [Internet]. 2014;**30**(3):929–936. Available from: http://pubs.acs.org/doi/abs/10.1021/la4043638 [Accessed: January 12, 2017]

[45] Yin Y, Wu C, Wang J, Song F, Yue W, Zhong W. A simply triggered peptide-based hydrogel as an injectable nanocarrier of tanshinone IIA and tanshinones. Chemical Communications [Internet]. 2017;53(3):529–32. Available from: http://xlink.rsc.org/?DOI=C6CC08502D [Accessed: January 10, 2017]

[46] Li J, Kooger R, He M, Xiao X, Zheng L, Zhang Y. A supramolecular hydrogel as a carrier to deliver microRNA into the encapsulated cells. Chemical Communications [Internet]. 2014;**50**(28):3722–3724. Available from: http://xlink.rsc.org/?DOI=c4cc00156g [Accessed: January 12, 2017]

[47] Koutsopoulos S, Unsworth LD, Nagai Y, Zhang S. Controlled release of functional proteins through designer self-assembling peptide nanofiber hydrogel scaffold. Proceedings of the National Academy of Sciences [Internet]. 2009;**106**(12):4623–4628. Available from: http://www.ncbi.nlm.nih.gov/pubmed/19273853 [Accessed: January 12, 2017]

[48] Hickling C, Toogood HS, Saiani A, Scrutton NS, Miller AF. Nanofibrillar peptide hydrogels for the immobilization of biocatalysts for chemical transformations. Macromolecular Rapid Communications [Internet]. 2014;**35**(9):868–874. Available from: http://doi.wiley.com/10.1002/marc.201400027 [Accessed: January 12, 2017]

[49] Nandi N, Baral A, Basu K, Roy S, Banerjee A. A dipeptide-based superhydrogel: Removal of toxic dyes and heavy metal ions from waste-water. Biopolymers [Internet]. 2016. Available from: http://doi.wiley.com/10.1002/bip.22915 [Accessed: January 10, 2017]

[50] Altunbas A, Sharma N, Lamm MS, Yan C, Nagarkar RP, Schneider JP, et al. Peptide-silica hybrid networks: Biomimetic control of network mechanical behavior. ACS Nano [Internet]. 2010;**4**(1):181–188. Available from: http://pubs.acs.org/doi/abs/10.1021/nn901226h [Accessed: January 12, 2017]

[51] Li C, Mezzenga R. The interplay between carbon nanomaterials and amyloid fibrils in bio-nanotechnology. Nanoscale [Internet]. 2013;**5**(14):6207–6208. Available from: http://xlink.rsc.org/?DOI=c3nr01644g [Accessed: January 12, 2017]

[52] Ng VWL, Chan JMW, Sardon H, Ono RJ, García JM, Yang YY, et al. Antimicrobial hydrogels: A new weapon in the arsenal against multidrug-resistant infections. Advanced Drug Delivery Reviews. 2014;**78:**46–62

[53] Salick DA, Pochan DJ, Schneider JP. Design of an injectable ß-hairpin peptide hydrogel that kills methicillin-resistant staphylococcus aureus. Advanced Materials [Internet]. 2009;**21**(41):4120–4123. Available from: http://doi.wiley.com/10.1002/adma.200900189 [Accessed: January 12, 2017]

[54] Liu Y, Yang Y, Wang C, Zhao X. Stimuli-responsive self-assembling peptides made from antibacterial peptides. Nanoscale [Internet]. 2013;**5**(14):6413–6421. Available from: http://xlink.rsc.org/?DOI=c3nr00225j [Accessed: January 12, 2017]

[55] Chatterjee S. IIT-B has Come up with Hydrogels Mimicking Natural Brain Tissue to Cure Parkinson's Disease via Stem Cell Therapy [Internet]. Times of India. 2016. Available from: http://timesofindia.indiatimes.com/city/bengaluru/IIT-B-has-come-up-with-hydrogels-mimicking-natural-brain-tissue-to-cure-Parkinsons-disease-via-stem-cell-therapy/articleshow/55584718.cms [Accessed: January 23, 2017]

[56] Asian Scientist Newsroom. Self-Assembling Peptide Nanogel for Burns [Internet]. Asian Scientist. 2014. Available from: http://www.asianscientist.com/2014/05/in-the-lab/ibn-self-assembling-peptide-nanogel-burns-2014/ [Accessed: January 23, 2017]

[57] The Wonder Stuff—How Peptide Hydrogels Could Change the Face of Biomedicine [Internet]. LaboratoryNews. 2015. Available from: http://www.labnews.co.uk/features/the-wonder-stuff-how-peptide-hydrogels-could-change-the-face-of-biomedicine-20-01-2015/ [Accessed: January 23, 2017]

[58] Tuned Gels Reveal Molecules that Drive Stem Cell Differentiation [Internet]. PhysOrg. 2016. Available from: https://phys.org/news/2016-07-tuned-gels-reveal-molecules-stem.html [Accessed: January 23, 2017]

[59] New Antibacterial Gel Could Revolutionize Treatment of Superbug Infections [Internet]. Medical News Today. 2016. Available from: http://www.medicalnewstoday.com/articles/306299.php [Accessed: January 23, 2017]

[60] To Fight Drug-Resistant Superbugs in Hospitals, Researchers Have Developed New Bio-Film Targeting Antibacterial Gel [Internet]. Medical Daily. 2014. Available from: http://www.medicaldaily.com/fight-drug-resistant-superbugs-hospitals-researchers-have-developed-new-bio-film-targeting-298994 [Accessed: January 23, 2017]

[61] HydroMatrix™ Peptide Cell Culture Scaffold [Internet]. Sigma Aldrich. 2017. Available from: http://www.sigmaaldrich.com/catalog/product/sigma/a6982?lang=fr&region=FR&gclid=CMOP4uWo2NECFQcz0wod1nMD7Q [Accessed: January 23, 2017]

[62] Puramatrix [Internet]. 3D Matrix Group. 2017. Available from: http://puramatrix.com/wp/puramatrix/ [Accessed: January 23, 2017]

[63] PeptiGel [Internet]. PeptiGel Design. 2017. Available from: http://www.peptigeldesign.com/ [Accessed: January 23, 2017]

[64] PGmatrix [Internet]. PepGel LLC. 2017. Available from: http://www.pepgel.com/?page_id=6 [Accessed: January 23, 2017]

[65] Curolox [Internet]. Credentis. 2017. Available from: https://www.curodont.com/en/curolox-technology/ [Accessed: January 23, 2017]

[66] Ou C, Zhang J, Zhang X, Yang Z, Chen M. Phenothiazine as an aromatic capping group to construct a short peptide-based "super gelator". Chemical Communications [Internet]. 2013;**49**(18):1853–1855. Available from: http://xlink.rsc.org/?DOI=c3cc38409h [Accessed: January 11, 2017]

[67] Shi J, Gao Y, Yang Z, Xu B. Exceptionally small supramolecular hydrogelators based on aromatic-aromatic interactions. Beilstein Journal of Organic Chemistry [Internet]. 2011;**7**(1):167–172. Available from: http://www.beilstein-journals.org/bjoc/content/7/1/23 [Accessed: January 11, 2017]

[68] Chen L, Morris K, Laybourn A, Elias D, Hicks MR, Rodger A, et al. Self-assembly mechanism for a naphthalene-dipeptide leading to hydrogelation. Langmuir. 2010;**26**(7):5232–5242

[69] Yang Z, Liang G, Ma M, Gao Y, Xu B. Conjugates of naphthalene and dipeptides produce molecular hydrogelators with high efficiency of hydrogelation and superhelical nanofibers. Journal of Materials Chemistry [Internet]. 2007;**17**(9):850–854. Available from: http://xlink.rsc.org/?DOI=B611255B [Accessed: January 11, 2017]

[70] Ma M, Kuang Y, Gao Y, Zhang Y, Gao P, Xu B. Aromatic–aromatic interactions induce the self-assembly of pentapeptidic derivatives in water to form nanofibers and supramolecular hydrogels. Journal of the American Chemical Society [Internet]. 2010;**132**(8):2719–2728. Available from: http://pubs.acs.org/doi/abs/10.1021/ja9088764 [Accessed: January 16, 2017]

[71] Nanda J, Biswas A, Banerjee A. Single amino acid based thixotropic hydrogel formation and pH-dependent morphological change of gel nanofibers. Soft Matter [Internet]. 2013;**9**(16):4198–4208. Available from: http://xlink.rsc.org/?DOI=c3sm27050e [Accessed: January 17, 2017]

[72] Qiu Z, Yu H, Li J, Wang Y, Zhang Y. Spiropyran-linked dipeptide forms supramolecular hydrogel with dual responses to light and to ligand-receptor interaction. Chemical Communications (Cambridge, England) [Internet]. 2009;**7345**(23):3342–3344. Available from: http://xlink.rsc.org/?DOI=b822840j [Accessed: January 11, 2017]

[73] Nebot VJ, Armengol J, Smets J, Prieto SF, Escuder B, Miravet JF. Molecular hydrogels from bolaform amino acid derivatives: A structure-properties study based on the thermodynamics of gel solubilization. Chemistry—A European Journal [Internet]. 2012;**18**(13):4063–4072. Available from: http://doi.wiley.com/10.1002/chem.201103193 [Accessed: January 11, 2017]

[74] Behanna HA, Rajangam K, Stupp SI. Modulation of fluorescence through coassembly of molecules in organic nanostructures. Journal of the American Chemical Society. 2007;**129**(2):321–327

[75] Huang Y, Qiu Z, Xu Y, Shi J, Lin H, Zhang Y. Supramolecular hydrogels based on short peptides linked with conformational switch. Organic & Biomolecular Chemistry [Internet]. 2011;**9**(7):2149–2155. Available from: www.rsc.org/obc [Accessed: January 11, 2017]

[76] Shroff K, Rexeisen EL, Arunagirinathan MA, Kokkoli E. Fibronectin-mimetic peptide-amphiphile nanofiber gels support increased cell adhesion and promote ECM production. Soft Matter [Internet]. 2010;**6**(20):5064–5072. Available from: http://xlink.rsc.org/?DOI=c0sm00321b [Accessed: January 12, 2017]

[77] Shome A, Dutta S, Maiti S, Das PK. In situ synthesized Ag nanoparticle in self-assemblies of amino acid based amphiphilic hydrogelators: Development of antibacterial soft nanocomposites. Soft Matter [Internet]. 2011;**7**(6):3011–3022. Available from: http://xlink.rsc.org/?DOI=c0sm01087a [Accessed: January 12, 2017]

[78] Dutta S, Shome A, Kar T, Das PK. Counterion-induced modulation in the antimicrobial activity and biocompatibility of amphiphilic hydrogelators: Influence of in-situ-synthesized ag-nanoparticle on the bactericidal property. Langmuir [Internet]. 2011;**27**(8):5000–5008. Available from: http://pubs.acs.org/doi/abs/10.1021/la104903z [Accessed: January 12, 2017]

[79] Jun H-W, Yuwono V, Paramonov SE, Hartgerink JD. Enzyme-mediated degradation of peptide-amphiphile nanofiber networks. Advanced Materials [Internet]. 2005;**17**(21):2612–2617. Available from: http://doi.wiley.com/10.1002/adma.200500855 [Accessed: January 12, 2017]

[80] Greenfield MA, Hoffman JR, De La Cruz MO, Stupp SI. Tunable mechanics of peptide nanofiber gels. Langmuir [Internet]. 2010;**26**(5):3641–3647. Available from: http://pubs.acs.org/doi/abs/10.1021/la9030969 [Accessed: January 12, 2017]

[81] Roy S, Das PK. Antibacterial hydrogels of amino acid-based cationic amphiphiles. Biotechnology and Bioengineering [Internet]. 2008;**100**(4):756–764. Available from: http://doi.wiley.com/10.1002/bit.21803 [Accessed: January 12, 2017]

[82] Mitra RN, Das PK. In situ preparation of gold nanoparticles of varying shape in molecular hydrogel of peptide amphiphiles. Journal of Physical Chemistry C [Internet]. 2008;**112**(22):8159–8166. Available from: http://pubs.acs.org/doi/abs/10.1021/jp712106d [Accessed: January 12, 2017]

[83] Das D, Maiti S, Brahmachari S, Das PK. Refining hydrogelator design: Soft materials with improved gelation ability, biocompatibility and matrix for in situ synthesis of specific shaped GNP. Soft Matter. The Royal Society of Chemistry [Internet]. 2011;**7**:7291–7303. Available from: http://xlink.rsc.org/?DOI=c1sm05608e [Accessed: January 12, 2017]

[84] Li X, Kuang Y, Lin H-C, Gao Y, Shi J, Xu B. Supramolecular nanofibers and hydrogels of nucleopeptides. Angewandte Chemie International Edition [Internet]. 2011;**50**(40):9365–9369. Available from: http://doi.wiley.com/10.1002/anie.201103641 [Accessed: January 13, 2017]

[85] Li X, Kuang Y, Shi J, Gao Y, Lin HC, Xu B. Multifunctional, biocompatible supramolecular hydrogelators consist only of nucleobase, amino acid, and glycoside. Journal of the American Chemical Society [Internet]. 2011;**133**(43):17513–17518. Available from: http://pubs.acs.org/doi/abs/10.1021/ja208456k [Accessed: January 13, 2017]

[86] Ruan L, Zhang H, Luo H, Liu J, Tang F, Shi Y-K, et al. Designed amphiphilic peptide forms stable nanoweb, slowly releases encapsulated hydrophobic drug, and accelerates animal hemostasis. Proceedings of the National Academy of Sciences of the United States of America [Internet]. 2009;**106**(13):5105–5110. Available from: http://www.ncbi.nlm.nih.gov/pubmed/19289834 [Accessed: January 13, 2017]

[87] Caplan MR, Moore PN, Zhang S, Kamm RD, Lauffenburger DA. Self-assembly of a beta-sheet protein governed by relief of electrostatic repulsion relative to van der Waals attraction. Biomacromolecules. 2000;**1**(4):627–631

[88] Kisiday J, Jin M, Kurz B, Hung H, Semino C, Zhang S, et al. Self-assembling peptide hydrogel fosters chondrocyte extracellular matrix production and cell division: Implications for cartilage tissue repair. Proceedings of the National Academy of Sciences [Internet]. 2002;**99**(15):9996–10001. Available from: http://www.ncbi.nlm.nih.gov/pubmed/12119393 [Accessed: January 13, 2017]

[89] Maity S, Jana P, Haldar D, Davis ME, Atwood JL, Barbour LJ, et al. Fabrication of nanoporous material from a hydrophobic peptide. CrystEngComm [Internet]. 2011;**13**(8):3064–3071. Available from: http://xlink.rsc.org/?DOI=c0ce00701c [Accessed: January 13, 2017]

[90] Maity S, Kumar P, Haldar D, Bowerman CJ, Nilson BL, King KN, et al. Sonication-induced instant amyloid-like fibril formation and organogelation by a tripeptide. Soft Matter [Internet]. 2011;**7**(11):5239. Available from: http://xlink.rsc.org/?DOI=c1sm05277b [Accessed: January 13, 2017]

[91] Das AK, Bose PP, Drew MGB, Banerjee A. The role of protecting groups in the formation of organogels through a nano-fibrillar network formed by self-assembling terminally protected tripeptides. Tetrahedron. 2007;**63**(31):7432–7442

[92] Banerjee A, Palui G, Banerjee A. Pentapeptide based organogels: The role of adjacently located phenylalanine residues in gel formation. Soft Matter [Internet]. 2008;**4**(7):1430–1437. Available from: http://xlink.rsc.org/?DOI=b802205b [Accessed: January 13, 2017]

[93] Baral A, Roy S, Ghosh S, Hermida-Merino D, Hamley IW, Banerjee A. A peptide-based mechano-sensitive, proteolytically stable hydrogel with remarkable antibacterial properties. Langmuir [Internet]. 2016;**32**(7):1836–1845. Available from: http://pubs.acs.org/doi/abs/10.1021/acs.langmuir.5b03789 [Accessed: January 13, 2017]

[94] Claussen RC, Rabatic BM, Stupp SI. Aqueous self-assembly of unsymmetric peptide bolaamphiphiles into nanofibers with hydrophilic cores and surfaces. Journal of the American Chemical Society. 2003;**125**(42):12680–12681

[95] Tzokova N, Fernyhough CM, Topham PD, Sandon N, Adams DJ, Butler MF, et al. Soft hydrogels from nanotubes of poly(ethylene oxide)-tetraphenylalanine conjugates prepared by click chemistry. Langmuir [Internet]. 2009;**25**(4):2479–2485. Available from: http://pubs.acs.org/doi/abs/10.1021/la8035659 [Accessed: January 16, 2017]

[96] Aggeli A, Bell M, Boden N, Carrick LM, Strong AE. Self-assembling peptide polyelectrolyteβ-Sheet complexes form nematic hydrogels. Angewandte Chemie International Edition [Internet]. 2003;**42**(45):5603–5606. Available from: http://doi.wiley.com/10.1002/anie.200352207 [Accessed: January 13, 2017]

[97] Zhang S, Lockshin C, Cook R, Rich A. Unusually stable beta-sheet formation in an ionic self-complementary oligopeptide. Biopolymers [Internet]. 1994;**34**(5):663–672. Available from: http://doi.wiley.com/10.1002/bip.360340508 [Accessed: January 13, 2017]

[98] Zhang S, Holmes TC, DiPersio CM, Hynes RO, Su X, Rich A. Self-complementary oligopeptide matrices support mammalian cell attachment. Biomaterials. 1995;**16**(18):1385–1393

[99] Holmes TC, de Lacalle S, Su X, Liu G, Rich A, Zhang S. Extensive neurite outgrowth and active synapse formation on self-assembling peptide scaffolds. Proceedings of the National Academy of Sciences of the United States of America [Internet]. 2000;**97**(12):6728–6733. Available from: http://www.ncbi.nlm.nih.gov/pubmed/10841570 [Accessed: January 13, 2017]

[100] Ryan DM, Doran TM, Anderson SB, Nilsson BL. Effect of C-terminal modification on the Self-Assembly and hydrogelation of fluorinated Fmoc-Phe derivatives. Langmuir [Internet]. 2011;**27**(7):4029–4039. Available from: http://pubs.acs.org/doi/abs/10.1021/la1048375 [Accessed: January 13, 2017]

[101] Aggeli A, Nyrkova IA, Bell M, Harding R, Carrick L, McLeish TC, et al. Hierarchical self-assembly of chiral rod-like molecules as a model for peptide beta-sheet tapes, ribbons, fibrils, and fibers. Proceedings of the National Academy of Sciences of the United States of America [Internet]. 2001;**98**(21):11857–11862. Available from: http://www.ncbi.nlm.nih.gov/pubmed/11592996 [Accessed: January 13, 2017]

[102] Bowerman CJ, Nilsson BL. A reductive trigger for peptide self-assembly and hydrogelation. Journal of the American Chemical Society [Internet]. 2010;**132**(28):9526–9527. Available from: http://pubs.acs.org/doi/abs/10.1021/ja1025535 [Accessed: January 13, 2017]

[103] Debnath S, Shome A, Das D, Das PK. Hydrogelation through self-assembly of fmoc-peptide functionalized cationic amphiphiles: Potent antibacterial agent. Journal of Physical Chemistry B [Internet]. 2010;**114**(13):4407–4415. Available from: http://pubs.acs.org/doi/abs/10.1021/jp909520w [Accessed: January 13, 2017]

[104] Ren C, Song Z, Zheng W, Chen X, Wang L, Kong D, et al. Disulfide bond as a cleavable linker for molecular self-assembly and hydrogelation. Chemical Communications [Internet]. 2011;**47**(5):1619–1621. Available from: http://xlink.rsc.org/?DOI=C0CC04135A [Accessed: January 13, 2017]

[105] Kogiso M, Hanada T, Yase K, Shimizu T. Intralayer hydrogen-bond-directed nano-fiber formation from dicarboxylic valylvaline bolaamphiphiles. Chemical Communications [Internet]. 1998;(2):1791–1792. Available from: http://xlink.rsc.org/?DOI=a803606c [Accessed: January 16, 2017]

[106] Franceschi S, de Viguerie N, Riviere M, Lattes A. Synthesis and aggregation of two-headed surfactants bearing amino acid moieties. New Journal of Chemistry [Internet]. 1999;**23**(4):447–452. Available from: http://xlink.rsc.org/?DOI=a809079c [Accessed: January 16, 2017]

[107] Diegelmann SR, Hartman N, Markovic N, Tovar JD. Synthesis and alignment of discrete polydiacetylene-peptide nanostructures. Journal of the American Chemical Society [Internet]. 2012;**134**(4):2028–2031. Available from: http://pubs.acs.org/doi/abs/10.1021/ja211539j [Accessed: January 16, 2017]

[108] Vadehra GS, Wall BD, Diegelmann SR, Tovar JD. On-resin dimerization incorporates a diverse array of pi-conjugated functionality within aqueous self-assembling peptide backbones. Chemical communications [Internet]. 2010;**46**(22):3947–3949. Available from: http://xlink.rsc.org/?DOI=c0cc00301h [Accessed: January 16, 2017]

[109] Mba M, Moretto A, Armelao L, Crisma M, Toniolo C, Maggini M. Synthesis and self-assembly of oligo(p-phenylenevinylene) peptide conjugates in water. Chemistry - A European Journal [Internet]. 2011;**17**(7):2044–2047. Available from: http://doi.wiley.com/10.1002/chem.201002495 [Accessed: January 16, 2017]

[110] Wall BD, Tovar JD. Synthesis and characterization of π-conjugated peptide-based supramolecular materials. Pure and Applied Chemistry [Internet]. 2012;**84**(4):1039–1045. Available from: http://dx.doi.org/10.1351/PAC-CON-11-10-24 [Accessed: January 16, 2017]

[111] Wall BD, Diegelmann SR, Zhang S, Dawidczyk TJ, Wilson WL, Katz HE, et al. Aligned macroscopic domains of optoelectronic nanostructures prepared via shear-flow assembly of peptide hydrogels. Advanced Materials [Internet]. 2011;**23**(43):5009–5014.

Available from: http://doi.wiley.com/10.1002/adma.201102963 [Accessed: January 16, 2017]

[112] Mossuto MF, Bolognesi B, Guixer B, Dhulesia A, Agostini F, Kumita JR, et al. Disulfide bonds reduce the toxicity of the amyloid fibrils formed by an extracellular protein. Angewandte Chemie International Edition [Internet]. 2011;**50**(31):7048–7051. Available from: http://doi.wiley.com/10.1002/anie.201100986 [Accessed: January 16, 2017]

[113] Li Y, Yan J, Zhang X, Huang K. Disulfide bonds in amyloidogenesis diseases related proteins. Proteins: Structure, Function and Bioinformatics [Internet]. 2013;**81**:1862–1873. Available from: http://doi.wiley.com/10.1002/prot.24338 [Accessed: January 16, 2017]

[114] Heck SD, Faraci WS, Kelbaugh PR, Saccomano NA, Thadeio PF, Volkmann RA. Posttranslational amino acid epimerization: Enzyme-catalyzed isomerization of amino acid residues in peptide chains. Proceedings of the National Academy of Sciences of the United States of America [Internet]. 1996;**93**(9):4036–4039. Available from: http://www.ncbi.nlm.nih.gov/pubmed/8633012 [Accessed: January 16, 2017]

[115] Ollivaux C, Soyez D, Toullec JY. Biogenesis of D-amino acid containing peptides/proteins: Where, when and how? Journal of Peptide Science [Internet]. 2014;**20:**595–612. Available from: http://doi.wiley.com/10.1002/psc.2637 [Accessed: January 16, 2017]

[116] Fuchs SA, Berger R, Klomp LWJ, de Koning TJ. D-amino acids in the central nervous system in health and disease. Molecular Genetics and Metabolism [Internet]. 2005;**85**(3):168–180. Available from: http://www.ncbi.nlm.nih.gov/pubmed/15979028 [Accessed: January 16, 2017]

[117] Kaji Y, Oshika T, Takazawa Y, Fukayama M, Fujii N. Pathological role of D-amino acid-containing proteins and advanced glycation end products in the development of age-related macular degeneration. Anti-Aging Medicine. 2010;**7**(10):107–111

[118] Adhikari B, Nanda J, Banerjee A, Yashima E, Maeda K, Iida H, et al. Multicomponent hydrogels from enantiomeric amino acid derivatives: Helical nanofibers, handedness and self-sorting. Soft Matter [Internet]. 2011;**7**(19):8913. Available from: http://xlink.rsc.org/?DOI=c1sm05907f [Accessed: January 16, 2017]

[119] Zhang Y, Gu H, Yang Z, Xu B. Supramolecular hydrogels respond to ligand-receptor interaction. Journal of the American Chemical Society. 2003;**125**(45):13680–13681

[120] Chronopoulou L, Sennato S, Bordi F, Giannella D, Di Nitto A, Barbetta A, et al. Designing unconventional Fmoc-peptide-based biomaterials: Structure and related properties. Soft Matter [Internet]. 2014;**10**(12):1944–1952. Available from: http://xlink.rsc.org/?DOI=c3sm52457d [Accessed: January 16, 2017]

[121] Marchesan S, Easton CD, Kushkaki F, Waddington L, Hartley PG. Tripeptide self-assembled hydrogels: Unexpected twists of chirality. Chemical Communications [Internet]. 2012;**48**(16):2195–2197. Available from: http://xlink.rsc.org/?DOI=C2CC16609G [Accessed: January 16, 2017]

[122] Haines LA, Rajagopal K, Ozbas B, Salick DA, Pochan DJ, Schneider JP. Light-activated hydrogel formation via the triggered folding and self-assembly of a designed peptide. Journal of the American Chemical Society. 2005;**127**(48):17025–17029

[123] Larsen TH, Branco MC, Rajagopal K, Schneider JP, Furst EM. Sequence-dependent gelation kinetics of β-hairpin peptide hydrogels. Macromolecules [Internet]. 2009;**42**(21):8443–8450. Available from: http://pubs.acs.org/doi/abs/10.1021/ma901423n [Accessed: January 17, 2017]

[124] Schneider JP, Pochan DJ, Ozbas B, Rajagopal K, Pakstis L, Kretsinger J. Responsive hydrogels from the intramolecular folding and self-assembly of a designed peptide. Journal of the American Chemical Society. 2002;**124**(50):15030–15037

[125] Ozbas B, Kretsinger J, Rajagopal K, Schneider JP, Pochan DJ. Salt-triggered peptide folding and consequent self-assembly into hydrogels with tunable modulus. Macromolecules. 2004;**37**(19):7331–7337

[126] Pochan DJ, Schneider JP, Kretsinger J, Ozbas B, Rajagopal K, Haines L. Thermally reversible hydrogels via intramolecular folding and consequent self-assembly of a de novo designed peptide. Journal of the American Chemical Society. 2003;**125**(39): 11802–11803

[127] Kretsinger JK, Haines LA, Ozbas B, Pochan DJ, Schneider JP. Cytocompatibility of self-assembled β-hairpin peptide hydrogel surfaces. Biomaterials. 2005;**26**(25):5177–5186

[128] Branco MC, Nettesheim F, Pochan DJ, Schneider JP, Wagner NJ. Fast dynamics of semiflexible chain networks of self-assembled peptides. Biomacromolecules [Internet]. 2009;**10**(6):1374–1380. Available from: http://pubs.acs.org/doi/abs/10.1021/bm801396e [Accessed: January 17, 2017]

[129] Yucel T, Micklitsch CM, Schneider JP, Pochan DJ. Direct observation of early-time hydrogelation in β-hairpin peptide self-assembly. Macromolecules [Internet]. 2008;**41**(15):5763–5772. Available from: http://pubs.acs.org/doi/abs/10.1021/ma702840q [Accessed: January 17, 2017]

[130] Haines-Butterick L, Rajagopal K, Branco M, Salick D, Rughani R, Pilarz M, et al. Controlling hydrogelation kinetics by peptide design for three-dimensional encapsulation and injectable delivery of cells. Proceedings of the National Academy of Sciences of the United States of America [Internet]. 2007;**104**(19):7791–7796. Available from: http://www.ncbi.nlm.nih.gov/pubmed/17470802 [Accessed: January 17, 2017]

[131] Hule RA, Nagarkar RP, Hammouda B, Schneider JP, Pochan DJ. Dependence of self-assembled peptide hydrogel network structure on local fibril nanostructure. Macromolecules [Internet]. 2009;**42**(18):7137–7145. Available from: http://pubs.acs.org/doi/abs/10.1021/ma9003242 [Accessed: January 17, 2017]

[132] Nagarkar RP, Hule RA, Pochan DJ, Schneider JP. De novo design of strand-swapped β-hairpin hydrogels. Journal of the American Chemical Society. 2008;**130**(13):4466–4474

[133] Rughani RV., Salick DA, Lamm MS, Yucel T, Pochan DJ, Schneider JP. Folding, self-assembly, and bulk material properties of a de novo designed three-stranded β-sheet hydrogel. Biomacromolecules [Internet]. 2009;**10**(5):1295–1304. Available from: http://pubs.acs.org/doi/abs/10.1021/bm900113z [Accessed: January 17, 2017]

[134] Luo Z, Zhao X, Zhang S. Self-organization of a chiral D-EAK16 designer peptide into a 3D nanofiber scaffold. Macromolecular Bioscience [Internet]. 2008;**8**(8):785–791. Available from: http://doi.wiley.com/10.1002/mabi.200800003 [Accessed: January 17, 2017]

[135] Luo Z, Yue Y, Zhang Y, Yuan X, Gong J, Wang L, et al. Designer D-form self-assembling peptide nanofiber scaffolds for 3-dimensional cell cultures. Biomaterials. 2013;**34**(21):4902–4913

[136] Luo Z, Zhao X, Zhang S. Structural dynamic of a self-assembling peptide d-EAK16 made of only D-amino acids. PLoS One [Internet]. 2008;**3**(5):e2364. Available from: http://www.ncbi.nlm.nih.gov/pubmed/18509542 [Accessed: January 17, 2017]

[137] Luo Z, Wang S, Zhang S. Fabrication of self-assembling d-form peptide nanofiber scaffold d-EAK16 for rapid hemostasis. Biomaterials. 2011;**32**(8):2013–2020

[138] Bowerman CJ, Ryan DM, Nissan DA, Nilsson BL. The effect of increasing hydrophobicity on the self-assembly of amphipathic beta-sheet peptides. Molecular BioSystems [Internet]. 2009;**5**(9):1058–1069. Available from: http://xlink.rsc.org/?DOI=b904439f [Accessed: January 17, 2017]

[139] Bowerman CJ, Liyanage W, Federation AJ, Nilsson BL. Tuning beta-sheet peptide self-assembly and hydrogelation behavior by modification of sequence hydrophobicity and aromaticity. Biomacromolecules [Internet]. 2011;**12**(7):2735–2745. Available from: http://pubs.acs.org/doi/abs/10.1021/bm200510k [Accessed: January 17, 2017]

[140] Aggeli A, Bell M, Boden N, Keen JN, McLeish TCB, Nyrkova I, et al. Engineering of peptide β-sheet nanotapes. Journal of Materials Chemistry [Internet]. 1997;**7**(7):1135–1145. Available from: http://xlink.rsc.org/?DOI=a701088e [Accessed: January 17, 2017]

[141] Thota CK, Yadav N, Chauhan VS. A novel highly stable and injectable hydrogel based on a conformationally restricted ultrashort peptide. Scientific Reports [Internet]. 2016;**6**:31167. Available from: http://www.nature.com/articles/srep31167 [Accessed: January 23, 2017]

[142] Panda JJ, Mishra A, Basu A, Chauhan VS. Stimuli responsive self-assembled hydrogel of a low molecular weight free dipeptide with potential for tunable drug delivery. Biomacromolecules [Internet]. 2008;**9**(8):2244–2250. Available from: http://pubs.acs.org/doi/abs/10.1021/bm800404z [Accessed: January 23, 2017]

[143] Micklitsch CM, Medina SH, Yucel T, Nagy-Smith KJ, Pochan DJ, Schneider JP. Influence of hydrophobic face amino acids on the hydrogelation of β-hairpin peptide amphiphiles. Macromolecules. 2015;**48**(5):1281–1288

[144] Xie Z, Zhang A, Ye L, Wang X, Feng Z-G. Shear-assisted hydrogels based on self-assembly of cyclic dipeptide derivatives. Journal of Materials Chemistry [Internet].

2009;**19**(34):6100–6102. Available from: http://xlink.rsc.org/?DOI=b912020c [Accessed: January 17, 2017]

[145] Orbach R, Adler-Abramovich L, Zigerson S, Mironi-Harpaz I, Seliktar D, Gazit E. Self-assembled Fmoc-peptides as a platform for the formation of nanostructures and hydrogels. Biomacromolecules. 2009;**10**(9):2646–2651

[146] Hamley IW, Brown GD, Castelletto V, Cheng G, Venanzi M, Caruso M, et al. Self-assembly of a designed amyloid peptide containing the functional thienylalanine unit. Journal of Physical Chemistry B [Internet]. 2010;**114**(32):10674–10683. Available from: http://pubs.acs.org/doi/abs/10.1021/jp105508g [Accessed: January 17, 2017]

[147] Gao YA, Long MJC, Shi J, Hedstrom L, Xu B. Using supramolecular hydrogels to discover the interactions between proteins and molecular nanofibers of small molecules. Chemical Communications [Internet]. 2012;**48**(67):8404–8406. Available from: http://xlink.rsc.org/?DOI=c2cc33631f [Accessed: January 17, 2017]

[148] Brinckmann J, Notbohm H, Müller PK. Collagen: Primer in structure, processing and assembly [Internet]. In: Brinckmann J, Notbohm H, Müller PK, editors. Berlin: Springer; 2005. p. 252. Available from: https://books.google.com/books?id=4jH_SuLShWcC&pgis=1 [Accessed: January 17, 2017]

[149] Hu Y, Wang H, Wang J, Wang S, Liao W, Yang Y, et al. Supramolecular hydrogels inspired by collagen for tissue engineering. Organic & Biomolecular Chemistry [Internet]. 2010;**8**(14):3267–3271. Available from: http://xlink.rsc.org/?DOI=c002609c [Accessed: January 17, 2017]

[150] Liyanage W, Nilsson BL. Substituent effects on the self-assembly/coassembly and hydrogelation of phenylalanine derivatives. Langmuir [Internet]. 2016;**32**(3):787–799. Available from: http://pubs.acs.org/doi/abs/10.1021/acs.langmuir.5b03227 [Accessed: January 18, 2017]

[151] Ryan DM, Anderson SB, Senguen FT, Youngman RE, Nilsson BL. Self-assembly and hydrogelation promoted by F5-phenylalanine. Soft Matter [Internet]. 2010;**6**(3):475–479. Available from: http://xlink.rsc.org/?DOI=B916738B [Accessed: January 18, 2017]

[152] Ryan DM, Doran TM, Nilsson BL. Complementary π–π interactions induce multicomponent coassembly into functional fibrils. Langmuir [Internet]. 2011;**27**(17):11145–11156. Available from: http://pubs.acs.org/doi/abs/10.1021/la202070d [Accessed: January 18, 2017]

[153] Ryan DM, Anderson SB, Nilsson BL. The influence of side-chain halogenation on the self-assembly and hydrogelation of Fmoc-phenylalanine derivatives. Soft Matter [Internet]. 2010;**6**(14):3220–3231. Available from: http://xlink.rsc.org/?DOI=c0sm00018c [Accessed: January 18, 2017]

[154] Bertolani A, Pirrie L, Stefan L, Houbenov N, Haataja JS, Catalano L, et al. Supramolecular amplification of amyloid self-assembly by iodination. Nature Communications [Internet]. 2015;**6**:7574. Available from: http://gateway.webofknowledge.com/gate-

way/Gateway.cgi?GWVersion=2&SrcAuth=ORCID&SrcApp=OrcidOrg&DestLinkType=FullRecord&DestApp=WOS_CPL&KeyUT=WOS:000357181200001&KeyUID=WOS:000357181200001

[155] Shao H, Parquette JR. A π-conjugated hydrogel based on an Fmoc-dipeptide naphthalene diimide semiconductor. Chemical Communications [Internet]. 2010;**46**(24):4285–4287. Available from: http://xlink.rsc.org/?DOI=c0cc00701c [Accessed: January 18, 2017]

[156] Li X, Gao Y, Kuang Y, Xu B. Enzymatic formation of a photoresponsive supramolecular hydrogel. Chemical Communications [Internet]. 2010;**46**(29):5364–5366. Available from: http://xlink.rsc.org/?DOI=c0cc00163e[Accessed: January 18, 2017]

[157] Li X, Li J, Gao Y, Kuang Y, Shi J, Xu B. Molecular nanofibers of olsalazine form supramolecular hydrogels for reductive release of an anti-inflammatory agent. Journal of the American Chemical Society. 2010;**132**(50):17707–17709

[158] Rughani RV, Branco MC, Pochan DJ, Schneider JP. *De novo* design of a shear-thin recoverable peptide-based hydrogel capable of intrafibrillar photopolymerization. Macromolecules [Internet]. 2010;**43**(19):7924–7930. Available from: http://pubs.acs.org/doi/abs/10.1021/ma1014808 [Accessed: January 18, 2017]

[159] Mata A, Hsu L, Capito R, Aparicio C, Henrikson K, Stupp SI. Micropatterning of bioactive self-assembling gels. Soft Matter [Internet]. 2009;**5**(6):1228–1236. Available from: http://xlink.rsc.org/?DOI=b819002j [Accessed: January 13, 2017]

[160] Zhang Y, Li N, Delgado J, Gao Y, Kuang Y, Fraden S, et al. Post-self-assembly cross-linking of molecular nanofibers for oscillatory hydrogels. Langmuir. 2012;**28**(6):3063–3066

[161] Matson JB, Stupp SI. Drug release from hydrazone-containing peptide amphiphiles. Chemical Communications [Internet]. 2011;**47**(28):7962–7964. Available from: http://xlink.rsc.org/?DOI=c1cc12570b [Accessed: January 18, 2017]

[162] Cheng G, Castelletto V, Moulton CM, Newby GE, Hamley IW. Hydrogelation and self-assembly of Fmoc-tripeptides: Unexpected influence of sequence on self-assembled fibril structure, and hydrogel modulus and anisotropy. Langmuir [Internet]. 2010;**26**(7):4990–4998. Available from: http://pubs.acs.org/doi/abs/10.1021/la903678e [Accessed: January 18, 2017]

[163] Yang Z, Wang L, Wang J, Gao P, Xu B. Phenyl groups in supramolecular nanofibers confer hydrogels with high elasticity and rapid recovery. Journal of Materials Chemistry [Internet]. 2010;**20**(11):2128–2132. Available from: http://xlink.rsc.org/?DOI=b922858f [Accessed: January 18, 2017]

[164] Yang Z, Xu K, Wang L, Gu H, Wei H, Zhang M, et al. Self-assembly of small molecules affords multifunctional supramolecular hydrogels for topically treating simulated uranium wounds. Chemical Communications [Internet]. 2005;**101**(35):4414–4416. Available from: http://xlink.rsc.org/?DOI=b507314f [Accessed: January 18, 2017]

[165] Guler MO, Stupp SI. A self-assembled nanofiber catalyst for ester hydrolysis. Journal of the American Chemical Society. 2007;**129**(40):12082–12083

[166] Micklitsch CM, Knerr PJ, Branco MC, Nagarkar R, Pochan DJ, Schneider JP. Zinc-triggered hydrogelation of a self-assembling β-hairpin peptide. Angewandte Chemie International Edition [Internet]. 2011;**50**(7):1577–1579. Available from: http://doi.wiley.com/10.1002/anie.201006652 [Accessed: January 19, 2017]

[167] Hartgerink JD. Self-assembly and mineralization of peptide-amphiphile nanofibers. Science [Internet]. 2001;**294**(5547):1684–1688. Available from: http://www.sciencemag.org/cgi/doi/10.1126/science.1063187 [Accessed: January 19, 2017]

[168] Stendahl JC, Rao MS, Guler MO, Stupp SI. Intermolecular forces in the self-assembly of peptide amphiphile nanofibers. Advanced Functional Materials [Internet]. 2006;**16**(4):499–508. Available from: http://doi.wiley.com/10.1002/adfm.200500161 [Accessed: January 19, 2017]

[169] Ding Y, Li Y, Qin M, Cao Y, Wang W. Photo-cross-linking approach to engineering small tyrosine-containing peptide hydrogels with enhanced mechanical stability. Langmuir [Internet]. 2013;**29**(43):13299–13306. Available from: http://pubs.acs.org/doi/abs/10.1021/la4029639 [Accessed: January 19, 2017]

[170] Su RSC, Kim Y, Liu JC. Resilin: Protein-based elastomeric biomaterials. Acta Biomaterialia. 2014;**10**:1601–1611

[171] Guillemin R. Peptides in the brain. The new endocrinology of the neuron. Science. 1978; **202**(4366):390–402

[172] Schally AV. Aspects of hypothalamic regulation of the pituitary gland with major emphasis on its implications for the control of reproductive processes. Materia Medica Polona. 1980;**12**(1–2):9–27

[173] Yalow RS. Radioimmunoassays: A probe for fine structure of biologic systems. Medical Physics. 1978;**5**(4):247–257

[174] Maji SK, Schubert D, Rivier C, Lee S, Rivier JE, Riek R. Amyloid as a depot for the formulation of long-acting drugs. In: Weissman JS, editor. PLoS Biology [Internet]. 2008;**6**(2): 240–52. Available from: http://dx.plos.org/10.1371/journal.pbio.0060017 [Accessed: January 19, 2017]

[175] Xing B, Yu CW, Chow KH, Ho PL, Fu D, Xu B. Hydrophobic interaction and hydrogen bonding cooperatively confer a vancomycin hydrogel: A potential candidate for biomaterials. Journal of the American Chemical Society. 2002;**124**(50):14846–14847

[176] Qvit N, Rubin SJS, Urban TJ, Mochly-Rosen D, Gross ER. Peptidomimetic therapeutics: Scientific approaches and opportunities. Drug Discovery Today. 2017;**22**(2):454–462

[177] Venkatesan N, Kim BH. Synthesis and enzyme inhibitory activities of novel peptide isosteres. Current Medicinal Chemistry [Internet]. 2002;**9**(24):2243–2270. Available from: http://www.ncbi.nlm.nih.gov/pubmed/12470245 [Accessed: January 20, 2017]

[178] Avan I, Hall CD, Katritzky AR. Peptidomimetics via modifications of amino acids and peptide bonds. Chemical Society Reviews [Internet]. 2014;**43**(10):3575–3594. Available from: http://xlink.rsc.org/?DOI=c3cs60384a [Accessed: January 20, 2017]

[179] Gante J. Peptidomimetics—Tailored enzyme inhibitors. Angewandte Chemie International Edition in English [Internet]. 1994;**33**(17):1699–1720. Available from: http://doi.wiley.com/10.1002/anie.199416991 [Accessed: January 19, 2017]

[180] Jones JH. Words derived from the noun peptide. Journal of Peptide Science [Internet]. 2006;**12**(2):79–81. Available from: http://doi.wiley.com/10.1002/psc.724 [Accessed: January 20, 2017]

[181] Gopalan RD, Del Borgo MP, Mechler AI, Perlmutter P, Aguilar MI. Geometrically precise building blocks: The self-assembly of beta-peptides. Chemistry and Biology. 2015;**22**:1417–1423

[182] Yang Z, Liang G, Xu B. Supramolecular hydrogels based on beta-amino acid derivatives. Chemical Communications [Internet]. 2006;**97**(7):738–740. Available from: http://xlink.rsc.org/?DOI=b516133a [Accessed: January 20, 2017]

[183] Yang Z, Liang G, Ma M, Gao Y, Xu B. In vitro and *in vivo* enzymatic formation of supramolecular hydrogels based on self-assembled nanofibers of a beta-amino acid derivative. Small [Internet]. 2007;**3**(4):558–562. Available from: http://doi.wiley.com/10.1002/smll.200700015 [Accessed: January 20, 2017]

[184] Nanda J, Banerjee A. β-Amino acid containing proteolitically stable dipeptide based hydrogels: Encapsulation and sustained release of some important biomolecules at physiological pH and temperature. Soft Matter [Internet]. 2012;**8**(12):3380–3386. Available from: http://xlink.rsc.org/?DOI=c2sm07168a [Accessed: January 20, 2017]

[185] Castelletto V, Cheng G, Greenland BW, Hamley IW, Harris PJF. Tuning the self-assembly of the bioactive dipeptide l-carnosine by incorporation of a bulky aromatic substituent. Langmuir [Internet]. 2011;**27**(6):2980–2988. Available from: http://pubs.acs.org/doi/abs/10.1021/la104495g [Accessed: January 20, 2017]

[186] Nguyen MM, Eckes KM, Suggs LJ. Charge and sequence effects on the self-assembly and subsequent hydrogelation of Fmoc-depsipeptides. Soft matter [Internet]. 2014;**10**(15): 2693–2702. Available from: http://xlink.rsc.org/?DOI=c4sm00009a [Accessed: January 20, 2017]

[187] Makarević J, Jokić M, Perić B, Tomišić V, Kojić-Prodić B, Žinić M. Bis(Amino Acid) oxalyl amides as ambidextrous gelators of water and organic solvents: Supramolecular gels with temperature dependent assembly/dissolution equilibrium. Chemistry—A European Journal [Internet]. 2001;**7**(15):3328–3341. Available from: http://doi.wiley.com/10.1002/1521-3765%2820010803%297%3A15%3C3328%3A%3AAID-CHEM3328%3E3.0.CO%3B2-C [Accessed: January 20, 2017]

[188] Jokić M, Makarević J, Žinić M. A novel type of small organic gelators: Bis(amino acid) oxalyl amides. Journal of the Chemical Society Chemical Communications [Internet]. 1995;**17**(17):1723–1724. Available from: http://xlink.rsc.org/?DOI=C39950001723 [Accessed: January 20, 2017]

[189] Frkanec L, Jokić M, Makarević J, Wolsperger K, Žinić M. Bis(PheOH) maleic acid amide-fumaric acid amide photoizomerization induces microsphere-to-gel fiber morphological transition: The photoinduced gelation system. Journal of the American Chemical Society. 2002;**124**(33):9716–9717

[190] Castellucci N, Angelici G, Falini G, Monari M, Tomasini C. L-Phe-D-Oxd: A privileged scaffold for the formation of supramolecular materials. European Journal of Organic Chemistry [Internet]. 2011;**2011**(16):3082–3088. Available from: http://doi.wiley.com/10.1002/ejoc.201001643 [Accessed: January 20, 2017]

[191] Castellucci N, Falini G, Angelici G, Tomasini C. Formation of gels in the presence of metal ions. Amino Acids [Internet]. 2011;**41**(3):609–620. Available from: http://link.springer.com/10.1007/s00726-011-0908-0 [Accessed: January 20, 2017]

[192] Castellucci N, Sartor G, Calonghi N, Parolin C, Falini G, Tomasini C. A peptidic hydrogel that may behave as a "trojan Horse." Beilstein Journal of Organic Chemistry [Internet]. 2013;**9**:417–424. Available from: http://www.beilstein-journals.org/bjoc/content/9/1/44 [Accessed: January 20, 2017]

[193] Milli L, Zanna N, Merlettini A, Di Giosia M, Calvaresi M, Focarete ML, et al. Pseudopeptide-based hydrogels trapping methylene blue and eosin Y. Chemistry—A European Journal [Internet]. 2016;**22**(34):12106–12112. Available from: http://doi.wiley.com/10.1002/chem.201601861 [Accessed: January 20, 2017]

[194] Rajbhandary A, Nilsson BL. Investigating the effects of peptoid substitutions in self-assembly of Fmoc-Diphenylalanine derivatives. Biopolymers [Internet]. 2016;**108**. Available from: http://doi.wiley.com/10.1002/bip.22994 [Accessed: January 20, 2017]

Chapter 4

New Aspects of the Structure of D-Amino Acid Oxidase from Porcine Kidney in Solution: Molecular Dynamics Simulation and Photoinduced Electron Transfer

Arthit Nueangaudom, Kiattisak Lugsanangarm,
Somsak Pianwanit, Sirirat Kokpol,
Nadtanet Nunthaboot, Fumio Tanaka,
Seiji Taniguchi and Haik Chosrowjan

Additional information is available at the end of the chapter

http://dx.doi.org/10.5772/intechopen.68645

Abstract

Mammalian D-amino acid oxidase (DAAO) plays an important role for D-serine metabolism in the brain and regulation of glutamatergic neurotransmission. In the present work, the structures in solution obtained by the methods of molecular dynamic simulation (MDS) and analyses of photoinduced electron transfer (ET) from aromatic amino acids to the excited isoalloxazine (Iso*) are described based upon our recent works, comparing among DAAO dimer, monomer, DAAO-benzoate (DAOB) complex dimer and monomer. The fluorescence lifetimes of DAAO and DAOB in the time domain of picoseconds and femtoseconds are used for the ET analyses as experimental data. The ET parameters (static dielectric constants near isoalloxazine (Iso), standard free energy gap (SFEG) between the photoproducts and reactants), ET rates, and related physical quantities (solvent reorganization energy, net electrostatic energy between the photoproducts and ionic groups in the proteins), in addition to MDS structures, are used to compare the protein structures. The structure of the DAOB dimer in solution obtained by MDS is substantially different from the crystal structure, and the structures of the two subunits are not equivalent in solution. The ET rates and related physical quantities also differ between the two subunits.

Keywords: D-amino acid oxidase from porcine kidney, benzoate complex, molecular dynamics simulation, dimer and monomer structures in solution, analyses of photoinduced electron transfer, rate of photoinduced electron transfer, fluorescence lifetimes

1. Introduction

D-Amino acid oxidase contains flavin adenine dinucleotide (FAD) as a cofactor and exists in a wide range of species from yeasts to humans. The enzyme catalyzes the oxidative degradation of D-amino acids to the corresponding amino acids, ammonium, and hydrogen peroxide. A number of review articles on D-amino acid oxidase (DAAO) from porcine kidney [1–3] and yeast to humans [4–6] have been reported. Mammalian D-amino acid oxidase plays an important role on D-serine metabolism in the brain and regulation of glutamatergic neurotransmission [7, 8]. Various new inhibitors of human D-amino acid oxidase have been found using in silico screening [9]. The crystal structures of DAAO are determined in the DAAO-benzoate (DAOB) complex and DAAO-*o*-aminobenzoate complex [2, 10, 11].

Photochemistry of flavins and flavoproteins [12] and the fluorescence quenching of flavins by various substances [13, 14] have been pioneered by Weber. The quenching mechanism of isoalloxazine (Iso) fluorescence upon complex formation with adenine in FAD is initially resolved by means of fluorescence lifetime measurements [15, 16], and the fluorescence quenching of Iso by indole with Iso-$(CH_2)_n$-indole diads is reported by McCormick [17]. Time-resolved fluorescence spectroscopy of flavins and flavoproteins has been reviewed by van den Berg and Visser [18]. The mechanism of the fluorescence quenching is studied in the systems of riboflavin tetrabutylate and indole, riboflavin tetrabutylate and N,N′-dimethylaniline in organic solvents [19], and flavodoxin from *Desulfovibrio vulgaris* (Miyazaki, F.) [20], by means of a picosecond transient-absorption spectroscopy. The remarkable fluorescence quenching of flavins is ascribed to fast photoinduced electron transfer (ET) from these substances to the excited Iso (Iso*). The ET mechanism in the riboflavin binding protein from egg white is also reported by means of a femtosecond transient-absorption spectroscopy [21]. A number of flavoproteins display very weak fluorescence, which decays with ultrashort lifetimes observed upon excitation with a sub-picosecond pulse laser [22–30]. These experimental results suggest that the valuable and detailed information on the microscopic structures of DAAO can be obtained through analyses of ET rates. We have developed a new method to analyze ET rates from aromatic amino acids to Iso* in flavoproteins using an electron transfer theory and MDS and the fluorescence lifetimes or decays of the flavoproteins as the experimental data [31–36].

In the present chapter, the ET analyses based upon MDS structures have been used to deduce submicroscopic features of various species of DAAO dimer, DAAO monomer, DAOB dimer, and DAOB monomer and compared them among these species.

2. Methods

2.1. Fluorescence spectroscopy of DAAO and DAOB

2.1.1. *Steady-state excitation*

Since fluorescence of free flavins was discovered by Weber [12–14], many workers have been working on its fluorescence characteristics. Kozioł first investigated solvent effects of the fluorescence in organic solvents [37]. However, free flavins are almost insoluble in most organic solvents, so that a number of solvents for the study were limited. Riboflavin tetrabutylate, which is soluble

http://dx.doi.org/10.5772/intechopen.68645

in organic solvents, was synthesized by Yagi's group. Systematic study on the solvent effects of the absorption and fluorescence spectra has been working with riboflavin tetrabutylate [38]. Fluorescence of DAAO was first studied by Massey et al. [39]. McCormic et al. precisely examined on the fluorescence properties of apo- and holo-DAAO [40].

Fluorescence intensity of the bound FAD in DAAO is quite weak compared to that of free FAD, and further fluorescence polarization is also quite different between free and the bound FAD [41]. A relative fluorescence intensity of the bound FAD to free FAD is defined as $R_1 = I/I_0$, where I and I_0 are the fluorescence intensities of the enzyme solution at certain concentration and free FAD at the same concentration with the enzyme sample. A parameter R_2 is defined with experimental polarization anisotropies as $R_2 = (A - A_f) / (A_b - A)$, where A, A_f, and A_b are polarization anisotropies of an enzyme solution, free FAD, and bound FAD, respectively. Dissociation constant of FAD from DAAO (K_d) [42, 43] and relative quantum yield of the bound FAD to the free FAD (r) were obtained with Eqs. (1) and (2) [44–46]:

$$K_d = \frac{R_1}{1 + R_2 - R_1}\left\{[P]_0 - [F]_0 + \frac{R_1}{1 + R_2}[F]_0\right\} \tag{1}$$

$$r = \frac{R_1 R_2}{1 + R_2 - R_1} \tag{2}$$

In Eq. (1), $[P]_0$ and $[F]_0$ are the total concentration of the protein (apoprotein plus holoprotein) and the total concentration of FAD (free and bound FADs) in the enzyme solution.

2.1.2. Fluorescence dynamics

Time-resolved fluorescence of free flavins was first studied by means of a phase-shift method by Weber's group [15, 16]. Transient fluorescence spectroscopy of flavoproteins is most useful experimental tool for the conformational changes of flavoproteins [18]. In 1980, the fluorescence lifetimes of DAAO was first reported by means of a picosecond-resolved fluorescence spectroscopy with a mode-locked Nd:YAG laser (pulse width, 30 ps) and streak camera combination by Nakashima et al. [44, 45]. Later, the fluorescence dynamics was measured with a synchronously pumped, cavity-dumped dye laser and single-photon counting system (pulse width 35 ps) to study a temperature-induced conformational change as described later [46, 47]. The fluorescence lifetimes of DAOB, however, could not be determined in the picosecond time domain [45]. The ultrafast fluorescence dynamics of DAOB was measured in the time domain of femtoseconds by means of a fluorescence up-conversion method (pulse width, 80 fs) [23].

2.2. MDS calculations

The starting structure of the pig kidney DAAO monomer was obtained from using the X-ray structure of the DAAO-benzoate complex dimer (PDB code 1VE9) [10], removing benzoate and/or one of the subunits. All calculations were carried out using the AMBER 10 suite of programs [47]. The parm99 force field [48] was used to describe the protein atoms, whereas the general AMBER force field [49] with the restrained electrostatic potential (RESP) charges [50] was used for the ligand and FAD. The simulated systems were subsequently solvated with a cubic box of ca. 4000 TIP3P water molecules. Electrostatic interactions were corrected by the particle mesh Ewald method [51]. The SHAKE algorithm [52] was employed to constrain all bonds involving hydrogen atoms. Details of the methods are described elsewhere [53–56].

2.3. Method of ET analysis

2.3.1. ET theory

The original Marcus theory [57–59] has been modified in various ways [60–73]. Kakitani and Mataga (KM) theory [66–68] is used for ET phenomena in flavoproteins, because it is applicable both for adiabatic and nonadiabatic ET process and has been found to give satisfactory results for both static [26–30] and dynamic ET analyses [31–36].

Here, the ET rate with the KM model for the DAAO dimer [53] is described as expressed by Eq. (3). The rates are similar for other DAAOs and DAOBs:

$$k_{ET}^{jk}(T) = \frac{\nu_0^q}{1+\exp\{\beta^q(R_{jk}-R_0^q)\}}\sqrt{\frac{k_BT}{4\pi\lambda_{jk}^q}}\exp\left[-\frac{\left\{\Delta G_k^0(T)-e^2/\varepsilon_0^{pk}R_{jk}+\lambda_{jk}^q+E_{Net}^k(j)\right\}^2}{4\lambda_{jk}^q k_BT}\right] \tag{3}$$

where $k_{ET}^{jk}(T)$ is the ET rate from the donor j to the Iso* in subunit k (k = Sub A or Sub B) at temperature T (°C), and q denotes Trp or Tyr. The term ν_0^q is an adiabatic frequency, β^q is the ET process coefficient, and R_{jk} and R_0^q are the donor j-Iso distance in subunit k and its critical distance for the ET process, respectively, and are expressed with Rc (center-to-center distance). The ET process is adiabatic when $R_{jk} < R_0^q$ and nonadiabatic when $R_{jk} > R_0^q$. The temperature (T) is expressed in K unit at the right-hand side. The term $-e^2/\varepsilon_0^{pk}R_{jk}$ is the electrostatic (ES) energy between the donor cation and acceptor anion (ESDA), in which ε_0^{pk} is static dielectric constant. The terms k_B and e are the Boltzmann constant and electron charge, respectively. $E_{Net}^k(j)$ is the net ES (NetES) energy of the donor j in subunit k. The DAAO monomer contains 10 Trp and 14 Tyr residues. In the present work, the ET rates from all of these aromatic amino acids to Iso* are taken into account for the analysis.

Solvent reorganization energy (SROE) [57, 58] of the ET donors q and j (λ_{jk}^q) is expressed in Eq. (4):

$$\lambda_{jk}^q = e^2\left(\frac{1}{2a_{Iso}}+\frac{1}{2a_q}-\frac{1}{R_{jk}}\right)\left(\frac{1}{\varepsilon_\infty}-\frac{1}{\varepsilon_0^{pk}}\right) \tag{4}$$

where a_{Iso} and a_q are the radii of Iso and Trp or Tyr, ε_∞ is the optical dielectric constant, and ε_0^{pk} is the static dielectric constant inside subunit k. The optical dielectric constant used was 2.0. The radii of Iso (a_{Iso}), Trp (a_{Trp}), and Tyr (a_{Tyr}) are 0.224, 0.196, and 0.173 nm, respectively, as previously reported [26–36].

The standard free energy gap (SFEG) between the products and reactants, $\Delta G_k^0(T)$, was expressed with the ionization potential of the ET donor (E_{IP}^q) as in Eq. (5):

$$\Delta G_k^0(T) = E_{IP}^q - G_k^0(T) \tag{5}$$

where $G_k^0(T)$ is the standard free energy gap related to the electron affinity of Iso* in subunit k at temperature T. The values of E_{IP}^q for Trp and Tyr are 7.2 eV and 8.0 eV, respectively [74].

2.3.2. Electrostatic energy between the photoproducts and ionic groups inside the DAAO dimer

The FAD cofactor in DAAO has two negative charges at the pyrophosphate, while DAAO itself contains 22 Glu, 13 Asp, 12 Lys, and 21 Arg residues per subunit as ionic amino acids. The ES energy between the Iso anion or donor cation j and all other ionic groups in subunit k (Sub A or Sub B) is expressed by Eq. (6):

$$\begin{aligned} E_k(j) = & \sum_{i=1}^{44} \frac{C_j C_{Glu}}{\varepsilon_0^{pk} R_j(Glu - i)} + \sum_{i=1}^{26} \frac{C_j C_{Asp}}{\varepsilon_0^{pk} R_j(Asp - i)} + \sum_{i=1}^{24} \frac{C_j C_{Lys}}{\varepsilon_0^{pk} R_j(Lys - i)} \\ & + \sum_{i=1}^{42} \frac{C_j C_{Arg}}{\varepsilon_0^{pk} R_j(Arg - i)} + \sum_{i=1}^{8} \frac{C_j C_P}{\varepsilon_0^{pk} R_j(P - i)} \end{aligned} \tag{6}$$

Here, $j = 0$ is for the Iso anion in subunit k, 1–10 and 11–20 for the Trp cations in Sub A and Sub B, respectively, and 21–34 and 35–48 for the Tyr cations in Sub A and Sub B, respectively. The charge of the aromatic ionic species j (C_j) is $-e$ for $j = 0$ (Iso anion) and $+e$ for $j = 1$–48 (cations of the donors). C_{Glu} (= $-e$), C_{Asp} (= $-e$), C_{Lys} (= $+e$), and C_{Arg} (= $+e$) are the charges of the Glu, Asp, Lys, and Arg residues, respectively. FAD contains two phosphate atoms, each of which binds two oxygen atoms, where the charge of each oxygen atom is $C_P = -0.5e$ and so the total charge of four oxygen atoms is $-2e$. The distances between the aromatic ionic species j and the ith Glu ($i = 1$–44) were denoted as $R_j(Glu - i)$, while the distances between the aromatic ionic species j^{th} and the ith Asp ($i = 1$–26) were denoted as $R_j(Asp - i)$ and so on for the each amino acid residue. The NetES in Eq. (3) is then expressed as in Eq. (7):

$$E_{Net}^k(j) = E_k(0) + E_k(j) \tag{7}$$

where j ranges from 1 to 48 and represents the jth ET donor.

2.3.3. Determination of the ET parameters

The calculated lifetimes of subunit k at temperature (T) are given by Eq. (8):

$$\tau_{Calc}^{Tk} = \frac{1}{\sum_{j=1}^{48} k_{ET}^{jk}(T)} \tag{8}$$

where the fluorescence lifetimes are expressed in ps unit. The physical quantities related to the electronic coupling term (υ_0^q, β^q, and R_0^q) for Trp and Tyr are taken from those reported for flavin mononucleotide binding proteins [32] and are assumed to be independent of temperature within the 10–30°C temperature range. In contrast, the free energy, $G_k^0(T)$, which is related to the electron affinity of Iso*, is assumed to be both temperature and subunit dependent, because $G_k^0(T)$ may be modified with the H-bond structure. The unknown ET parameters are $G_A^0(10)$, $G_B^0(10)$, $G_A^0(30)$, and $G_B^0(30)$ in Eq. (5) and ε_0^A, ε_0^B, and ε_0^{DA}, which are assumed to be independent of temperature. These ET parameters are determined so as to obtain the minimum value of χ^2, as given by Eq. (9):

$$\chi^2 = \frac{\left(\tau_{Calc}^{10A} - \tau_{Obs}^{10}\right)^2}{\tau_{Calc}^{10A}} + \frac{\left(\tau_{Calc}^{10B} - \tau_{Obs}^{10}\right)^2}{\tau_{Calc}^{10B}} + \frac{\left(\tau_{Calc}^{30A} - \tau_{Obs}^{30}\right)^2}{\tau_{Calc}^{30A}} + \frac{\left(\tau_{Calc}^{30B} - \tau_{Obs}^{30}\right)^2}{\tau_{Calc}^{30B}} \tag{9}$$

3. Cooperative binding of FAD associated with the monomer-dimer equilibrium in DAAO

The DAAO exists in a monomer (Mw 39 kDa)-dimer equilibrium state at relatively low concentrations [75–79] and in a dimer-tetramer equilibrium at higher concentrations [80–82]. The protein structures of the DAAO dimer in solution, as obtained by MDS [53, 54], are shown in **Figure 1**. The values of K_d are obtained at various concentrations of holo-DAAO and apo-DAAO [42, 43] according to Eq. (1). **Figure 2** shows K_d vs. DAAO concentration relationship [42]. The values of K_d are remarkably dependent on the protein concentration both in holo-DAAO and apo-DAAO [42], higher at the low concentrations and lower at the high concentrations. **Figure 3**

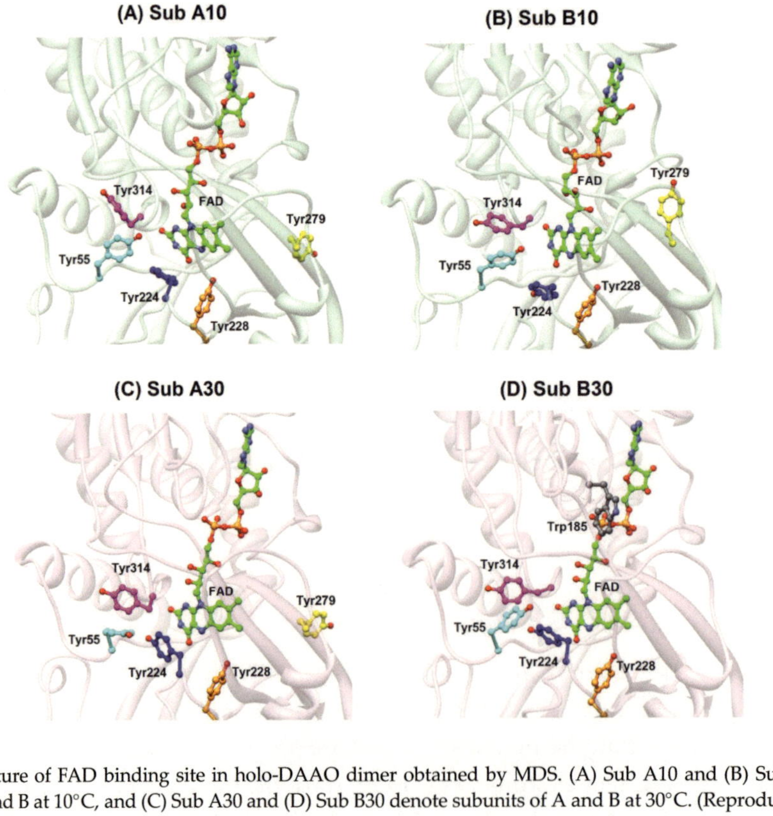

Figure 1. Structure of FAD binding site in holo-DAAO dimer obtained by MDS. (A) Sub A10 and (B) Sub B10 denote subunits of A and B at 10°C, and (C) Sub A30 and (D) Sub B30 denote subunits of A and B at 30°C. (Reproduced from [53] with permission from the PCCP Owner Societies).

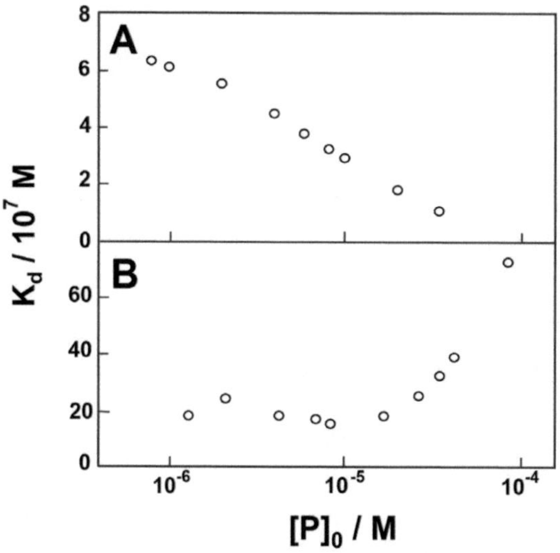

Figure 2. Dependence of K_d on the concentration of DAAO. (A) shows the holo-DAAO and (B) apo-DAAO. DAAO was dissolved into buffer solution at pH 8.3. (Reprinted with permission from [42]. Copyright (1979) American Chemical Society).

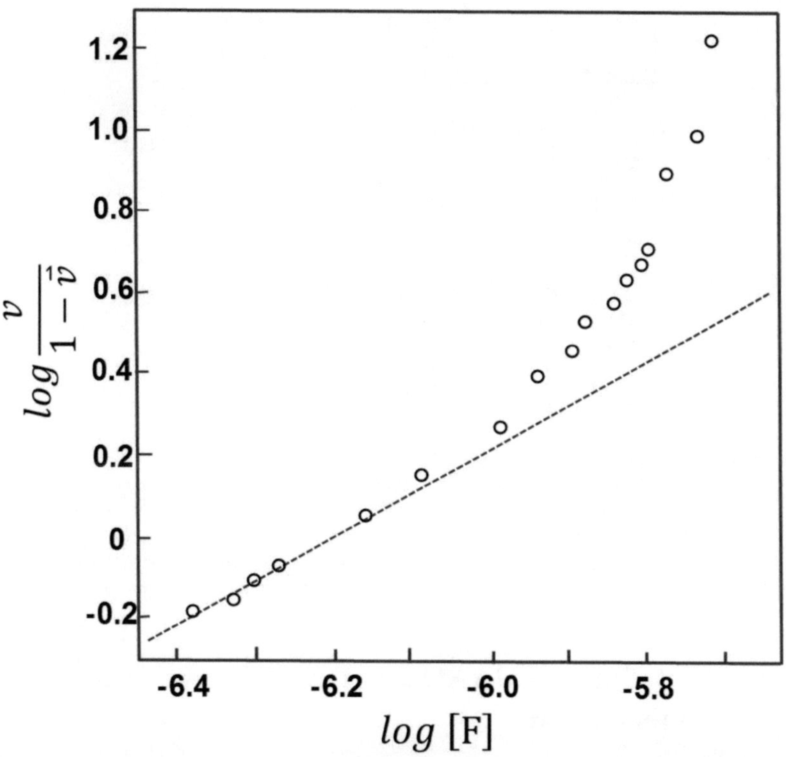

Figure 3. Hill plot of FAD binding in holo-DAAO. Measurements were made at pH 8.3 and 20°C. The binding fraction of FAD is $\bar{v} = [F]_b/[P]_0$ where $[F]_b$ is the concentration of bound FAD. The dashed line indicates a straight line with the Hill coefficient equal to 1. (Reprinted with permission from [42]. Copyright (1979) American Chemical Society).

shows Hill plot for FAD binding, which reveals that the Hill coefficient is nearly 1 at the lower concentrations but appreciably deviates from 1 toward greater than 1 at the higher concentrations of DAAO [42]. The results show that the binding process of FAD is positively cooperative. Approximate relative concentrations of various species of DAAO and the dissociation constants are illustrated in **Figure 4** [42]. The origin of the cooperativity is elucidated to be mainly that Kc (0.01 μM) is much less than Ka (0.74 μM). Namely, the binding of FAD to apo-DAAO monomer induces association of the holo-DAAO monomers into the holo-DAAO dimer, because the protein dissociation constants between holo monomers (K_2 = 3.8 μM) are least comparing to the other protein dissociation constants (K_1 and K_0).

The concept of "allosteric transition" is originally proposed by Monod, Wyman, and Changeux to explain the sigmoidal curve of O_2 binding to hemoglobin [83]. Then, an induced-fit model for the O_2 binding is proposed by Koshland, Némethy, and Filmer [84]. A ligand-induced polymerization of a protein is considered as an alternative model to explain allosteric effect [85–87]. The enzyme activity of the DAAO monomer is 1.5-fold higher than that of the dimer [88]. Under the presence of enough FAD in the brain, DAAO is considered to form the dimer, for which activity is lower than that of the monomer. The enzyme activity may be physiologically regulated through the binding of FAD, which should be significant in schizophrenia, because the activity of DAAO is twofold higher in the patients with schizophrenia [7].

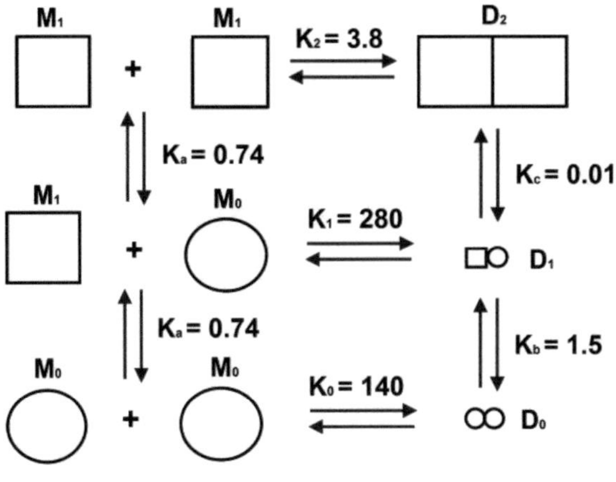

Figure 4. Dissociation equilibrium constants among the various species of DAAO. M_1 and M_0 indicate holomonomer and apomonomer, and D_2, D_1, and D_0 indicate holodimer, heterodimer of holo- and apomonomers, and apodimer, respectively. Ka, Kb, and Kc are dissociation equilibrium constants of FAD from holomonomer, heterodimer, and holodimer, respectively. The dissociation constants are indicated in the unit of μM. The square binds FAD, while the circle has no FAD. The concentration of DAAO is 5 μM. The area of the various species is approximately proportional to their concentrations. (Reprinted with permission from [42]. Copyright (1979) American Chemical Society).

4. Fluorescence lifetimes of DAAO and DAOB in picoseconds-femtoseconds time domain

The dissociation constants of FAD in DAAO are much smaller by 1/74 in the dimer, comparing to the monomer [42, 43] as stated above. This suggests that local structures near Iso binding site are different between the dimer and monomer. The fluorescence lifetimes of

DAAO obtained by Nakashima et al. [44] are 40 ps in the dimer and 130 ps in the monomer. Later, the lifetimes were measured with the new method of single-photon counting instruments and listed in **Table 1** at various concentrations of DAAO and temperatures [46, 47]. The values of lifetimes in DAAO monomer are 228 ps at 10°C and 182 ps at 30°C. The values of the lifetime in the dimer are 44.2 ps at 10°C and 37.7 ps at 30°C [46]. The lifetime of free FAD in water is 2.5 ns [15, 16]. The lifetime in DAAO dimer is shorter by ca. 1/60 times than that in free FAD in water, which is ascribed to fast ET from aromatic amino acids to Iso* [19–21].

The fluorescence lifetime of DAOB is 60 ps in the monomer, and shorter than 5 ps in the dimer, obtained by Nakashima et al. [44]. Time resolution of the lifetime instruments in 1980 was not enough to obtain exact lifetime of DAOB dimer. In 2000 the lifetimes of the DAOB dimer are obtained to be 0.848 and 4.77 ps [23] by means of the up-conversion method, which are much shorter than those in DAAO, and described more in detail later.

T (°C)	Conc. (μM)	τ_0 (ps)	α_0	τ_1 (ps)	α_1	τ_2	α_2	τ_3 (ns)	α_3	χ^2
40	1.6	27.9	−0.877	43.7	0.850	191	0.079	1.91	0.071	1.028
	100	26.5	−0.962	36.2	0.945	162	0.047	1.61	0.008	1.362
	Av.	25.9		40.0		169		1.76		
35	0.78	26.2	−1.030	45.4	0.834	202	0.084	2.10	0.081	0.993
	100	23.5	−0.903	39.9	0.910	161	0.080	1.68	0.010	1.516
	Av.	25.6		41.8		170		1.91		
30	1.6	26.7	−0.989	48.3	0.824	182	0.116	2.23	0.060	1.007
	100	26.7	−1.024	37.7	0.956	169	0.044	1.76	0.005	1.479
	Av.	26.2		43.2		177		2.02		
25	0.78	28.6	−0.842	54.7	0.822	245	0.080	2.47	0.098	1.148
	100	23.5	−0.977	41.0	0.906	165	0.086	1.83	0.008	1.559
	Av.	26.0		43.7		184		2.20		
15	0.78	29.5	−0.841	46.6	0.822	214	0.110	2.78	0.068	1.134
	100	23.5	−1.003	42.9	0.892	179	0.099	1.95	0.009	1.324
	Av.	26.1		45.2		190		2.44		
10	0.78	27.6	−0.980	47.1	0.807	228	0.121	2.93	0.072	1.016
	100	25.3	−0.979	44.2	0.899	194	0.092	2.04	0.009	1.366
	Av.	25.1		48.5		208		2.61		

[a]DAAO was dissolved in 0.017 M pyrophosphate buffer at pH 8.3. The fluorescent species with the lifetimes, τ_1, τ_2, and τ_3, were assigned to be the dimer, monomer, and free FAD, respectively, and α_1, α_2, and α_3 are their fractions. τ_0 is a lifetime for process from an intermediate state to the fluorescent state [46, 47]. χ^2 means a reduced chi-square distribution between the calculate decay function and experimental decay curve. Av. indicates the averaged lifetimes over seven or eight different levels of the enzyme ranging from 100 to 1.6 or 0.78 M. (Reprinted with permission from [47]).

Table 1. Fluorescence decay parameters of FAD in DAAO measured with a synchronously pumped, cavity-dumped dye laser and single-photon counting system.[a].

5. Conformational difference between the DAAO dimer and monomer revealed by MDS and ET analyses

The results of the fluorescence lifetimes of DAAO and DAOB reveal that the local structures differ between the monomers and dimers. However, no structural information can be drawn by the lifetimes alone. Details of the structural difference between DAAO monomer and dimer are obtained through MDS and ET analyses [53, 54].

Table 2 lists the donor-acceptor distances between Iso and the five shortest donors from Iso in the DAAO dimer and monomer. In the dimer, the Rc distances are the shortest in Tyr224 followed by Tyr228, except for Sub A at 30°C where Tyr228 is the shortest followed by Tyr224. In the monomer

Protein	T (°C)	Subunit	Donor [b] (Rc/nm)					
DAAO dimer[c]	10	A	Tyr224	Tyr228	Tyr55	Tyr314	Tyr279	
			(0.74)	(0.82)	(1.07)	(1.11)	(1.32)	
	10	B	Tyr224	Tyr228	Tyr55	Tyr314	Tyr279	
			(0.79)	(0.83)	(0.99)	(1.05)	(1.20)	
	30	A	Tyr 228	Tyr 224	Tyr 314	Tyr 279	Tyr 55	
			(0.85)	(0.90)	(1.06)	(1.30)	(1.47)	
	30	B	Tyr 224	Tyr 228	Tyr 314	Tyr 55	Trp 185	
			(0.72)	(0.81)	(1.06)	(1.06)	(1.14)	
DAAO monomer[d]	10		Tyr224	Tyr228	Tyr314	Trp185	Tyr55	
			(0.82)	(0.88)	(1.06)	(1.27)	(1.64)	
	30		Tyr224	Tyr228	Tyr314	Tyr55	Trp185	
			(0.88)	(0.88)	(1.18)	(1.20)	(1.49)	
DAOB dimer[e]	20	A	Tyr55	Tyr228	Tyr314	Trp185	Tyr224	Benzoate
			(0.95)	(0.96)	(1.06)	(1.10)	(1.32)	(0.66)
	20	B	Tyr228	Tyr314	Tyr224	Tyr55	Trp185	Benzoate
			(0.99)	(1.02)	(1.04)	(1.05)	(1.31)	(0.68)
DAOB monomer[f]	20		Tyr228	Tyr224	Tyr314	Tyr279	Tyr74	Benzoate
			(0.81)	(0.97)	(1.07)	(1.24)	(1.80)	(0.61)

[a]The ET acceptor is Iso*. Mean donor-acceptor distances (Rc) are listed over 5000 snapshots in parentheses. A value of Rc in one snapshot was evaluated as mean distance of all possible pairs between aromatic atoms in Iso and aromatic atoms in a donor.
[b]Five shortest distances between Iso and the aromatic amino acids (plus Bz in DAOB) are listed in order from shorter to longer distances.
[c]Data taken from Ref. [53].
[d]Data taken from Ref. [54].
[e]Data taken from Ref. [56].
[f]Data taken from Ref. [55].

Table 2. ET donor-acceptor distance in DAAO and DAOB.[a].

the distance is also the shortest in Tyr224 at both 10 and 30°C, followed by Tyr228 and Tyr314. The hydrogen bonding (H-bond) structures between Iso and the amino acid residues markedly vary with the protein systems (**Table 3**). At 10°C, in the dimer Iso forms H-bonds with Leu51 (IsoN3H), Thr317 (IsoO2), Gly50 (IsoO4), and Leu51 (IsoO4) in Sub A and with Gly315 (IsoO2), Leu316 (IsoO2), and Thr317 (IsoO2) in Sub B (atom notations are shown in **Chart 1**), while in the monomer, Iso forms H-bond only with Gly50 (IsoO4). At 30°C Iso in the dimer forms H-bonds with Leu51 (IsoN3H) and Thr317 (IsoO2) in Sub A and with Gly315 (IsoO2), Leu316 (IsoO2), and Thr317 (IsoO2) in Sub B, while in the monomer, Iso forms H-bond with Leu316 (IsoO2) and Gly50 (isoO4). The number of H-bonds and kind of H-bond pairs are quite different between the DAAO dimer and monomer, though Iso may also form H-bonds with water molecules as described below.

Protein	Subunit	T (°C)	Iso N3H Leu51 (O)	Iso N5 Ala49 (N)	Iso O2 Gly315 (N)	Iso O2 Leu316 (N)	Iso O2 Thr317 (OG1)	Iso O4 Gly50 (N)	Iso O4 Leu51 (N)
DAAO	A[b]	10	0.29	-	-	-	0.28	0.29	0.29
		30	0.29	-	-	-	0.28	-	-
	B[b]	10	-	-	0.29	0.28	-	-	-
		30	-	-	0.29	0.28	-	-	-
	Monomer[c]	10	0.29	0.29	-	0.29	0.28	0.28	0.29
		30	-	0.29	-	0.28	0.28	0.28	-
DAOB	A[d]	20	0.31	0.31	-	-	0.35	-	-
	B[d]		0.32	0.31	-	-	-	-	-
	Monomer[e]	10	0.29	0.29	-	0.29	0.28	0.28	0.29
		20	0.29	0.29	0.29	-	0.28	-	-
		30	0.29	0.29	-	-	-	0.28	0.29

	Subunit	T (°C)	Bz O1 Arg283 (NH2)	Bz O2 Arg283 (NE)	Bz O2 Arg283 (NH2)
DAOB	A[d]	20	0.29	0.27	0.30
	B[d]		0.28	0.28	0.30
	Monomer[e]	10	0.28	0.28	0.29
		20	0.26	0.28	0.29
		30	0.28	0.28	0.28

[a]The distances in nm units are obtained by averaging over 10,000 MDS snapshots and collected those shorter than 0.3 nm. Atomic notations in Iso are indicated in **Chart 1**. Atom notations of amino acids shown in parentheses are taken from PDB, where N, O, and OG1 denote peptide N and O atoms and O atom of the side chain, respectively. Bz O1 and Bz O2 denote two oxygen atoms of carboxylate in benzoate (Bz).
[b]The data are taken from Ref. [53].
[c]The data are taken from Ref. [54].
[d]The data are taken from Ref. [56].
[e]The data are taken from Ref. [55].

Table 3. Comparison of H-bond distances among the DAAO dimer, monomer, and DAOB dimer and monomer.[a].

As shown in Eq. (3), the ET rate contains several parameters, which are determined by the method described above. **Table 4** lists the ET parameters as ε_0^{DA} and $G_k^0(T)$ in DAAO dimer (Sub A and Sub B), DAAO monomer, DAOB dimer (Sub A and Sub B), and DAOB monomer. Microscopic information can be obtained as the donor-acceptor distances and H-bond distances with MDS and the protein structures. Submicroscopic information can be obtained with the ET parameters, ET rates, and related physical quantities. Among the ET parameters, $G_k^0(T)$ is one of most influential parameters for the ET rate, according to the fluorescence lifetime.

The distribution of the logarithmic ET rates (ln rate) from the five fastest donors to Iso* in the DAAO dimer and monomer is shown in **Figure 5**. At 10°C the three fastest donors are Tyr224, Tyr314, and Tyr228 in Sub A in this order and Tyr314, Tyr224, and Tyr55 in Sub B in the dimer, while they are Tyr224, Tyr314, and Tyr228 in the monomer. At 30°C the three fastest donors are Tyr314, Tyr228, and Tyr224 in Sub A and Tyr224, Tyr314, and Trp185 in Sub B in the dimer, while they are Tyr314, Tyr224, and Tyr55 in the monomer. The values of ET rates are listed in **Table 5**. The ET rates in the dimer are several times faster than those in the monomer.

T (°C)	Protein	Subunit	ε_0^{DA} [b]	$G_k^0(T)$ [c] (eV)	τ (ps)	
					Obs [d]	Calc [e]
10	DAAO dimer[f]	Sub A	5.79	8.61	44.2	44.2
		Sub B	5.82	8.54	-	-
	DAAO monomer[f]	–	5.88	8.69	228	228
30	DAAO dimer[f]	Sub A	5.79	8.73	37.7	37.7
		Sub B	5.82	8.48	-	-
	DAAO monomer[f]		5.89	8.51	182	182
20	DAOB dimer[g]	Sub A	2.53	8.42	4.77	4.77
		Sub B	2.64	8.43	0.848	0.848
	DAOB monomer[h]	-	2.45	8.53	60	60

[a]Static dielectric constants inside the proteins (ε_0^A and ε_0^B) are similar, 5.8–5.9 among all species. The reported values of ET parameters were used for the electronic coupling term ($\nu_0^{Trp} = 1016\ \text{ps}^{-1}$, $\nu_0^{Tyr} = 197\ \text{ps}^{-1}$, $\beta^{Trp} = 21.0\ \text{nm}^{-1}$, $\beta^{Tyr} = 6.25\ \text{nm}^{-1}$, $R_0^{Trp} = 0.663$ nm, and $R_0^{Tyr} = 0.499$ nm) [32].
[b]The static dielectric constant between Iso and the donors within 1 nm from Iso.
[c]Temperature-dependent electron affinity of Iso*.
[d]Experimental fluorescence lifetimes for DAAO dimer and monomer [46] and for DAOB dimer [23] and DAOB monomer [45]. The lifetimes of Sub A and Sub B in DAAO dimer are not experimentally resolved.
[e]Calculated lifetimes.
[f]Data are taken from the work for DAAO dimer [53] and DAAO monomer [54].
[g]Data are taken from the reported work of Ref. [56].
[h]Data are taken from the reported work of Ref. [55].

Table 4. ET parameter in DAAO and DAOB.[a].

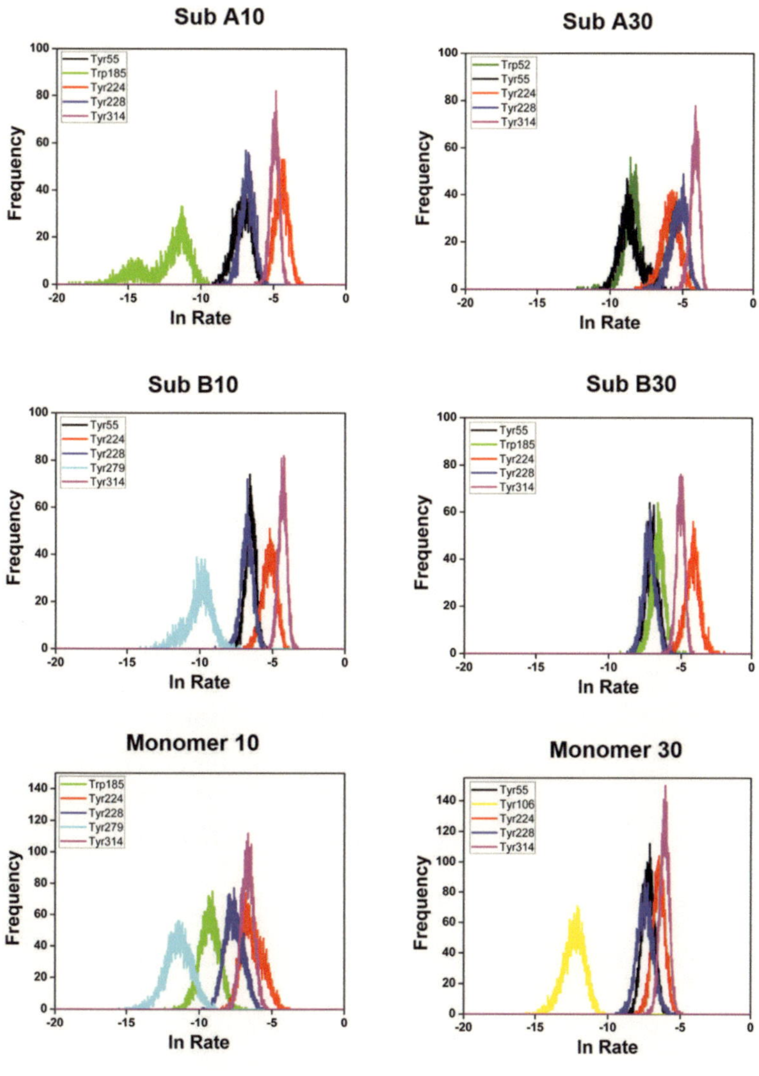

Figure 5. Distribution of logarithmic ET rate from aromatic amino acids to Iso*. Sub A10 and Sub B10 denote Sub A and Sub at 10°C, and Sub A30 and Sub B30 denote Sub A and Sub B at 30°C in DAAO dimer, respectively. Inserts show amino acids with the top fastest ET rates. The distributions for DAAO monomers at 10°C (Monomer 10) and 30°C (Monomer 30) are also shown for comparison. The kinds of the amino acids are different among the six groups including monomer. (Reproduced from [53] with permission from the PCCP Owner Societies).

A NetES sometimes plays an important role on the ET rates and is defined as an ES energy between the photoproducts (Iso anion plus a donor cation), and other ionic groups in the protein [31–36] as described above are also listed in **Table 5**. The NetES has never been numerically evaluated by other research groups. The NetES values in the monomer are greatly modified upon the formation of dimer, which is ascribed to inter-subunit interactions, namely, that the NetES of a donor in Sub A is strongly influenced by that in Sub B and vice versa, because the electrostatic energy is influential over a long range.

Protein	T (°C)	Subunit	Donor	Rate (ps^{-1})	NetES energy (eV)
DAAO dimer[b]	10	A	Tyr224	1.29×10^{-2}	0.044
			Tyr314	7.57×10^{-3}	−0.406
			Tyr228	1.20×10^{-3}	0.146
			Tyr55	9.08×10^{-4}	−0.119
			Trp185	9.71×10^{-6}	−0.104
	10	B	Tyr314	1.38×10^{-2}	−0.479
			Tyr224	5.96×10^{-3}	−0.021
			Tyr55	1.54×10^{-3}	−0.161
			Tyr228	1.25×10^{-3}	0.056
			Tyr279	6.08×10^{-5}	−0.076
	30	A	Tyr314	1.63×10^{-2}	−0.293
			Tyr228	5.86×10^{-3}	0.130
			Tyr224	3.39×10^{-3}	0.108
			Tyr55	2.43×10^{-4}	−0.207
			Trp52	2.15×10^{-4}	−0.593
	30	B	Tyr224	1.68×10^{-2}	−0.038
			Tyr314	6.51×10^{-3}	−0.422
			Trp185	1.43×10^{-3}	−0.465
			Tyr55	9.70×10^{-4}	−0.210
			Tyr228	8.30×10^{-4}	0.097
DAOB dimer[d]	20	A	Tyr228	1.17×10^{-1}	0.075
			Bz	7.50×10^{-2}	−0.085
			Tyr55	1.14×10^{-2}	−0.103
			Trp185	4.65×10^{-3}	−0.434
			Tyr314	1.58×10^{-3}	−0.323
			Trp52	1.68×10^{-5}	−0.113
	20	B	Bz	8.92×10^{-1}	−0.094
			Tyr228	2.80×10^{-1}	0.070
			Tyr314	6.56×10^{-3}	−0.442
			Tyr55	2.64×10^{-4}	−0.159
			Tyr224	5.38×10^{-5}	−0.010
			Trp185	3.72×10^{-5}	−0.183
DAAO monomer[c]	10		Tyr224	2.27×10^{-3}	0.192
			Tyr314	1.38×10^{-3}	−0.073
			Tyr228	5.94×10^{-4}	0.215
			Trp185	1.15×10^{-4}	−0.249

Protein	T (°C)	Subunit	Donor	Rate (ps^{-1})	NetES energy (eV)
			Tyr279	1.57×10^{-5}	0.144
	30		Tyr314	2.35×10^{-3}	−0.434
			Tyr224	1.65×10^{-3}	−0.035
			Tyr55	8.00×10^{-4}	−0.324
			Tyr228	6.85×10^{-4}	0.051
			Tyr106	5.47×10^{-6}	−0.342
DAOB monomer[e]	20		Bz	9.92×10^{-3}	0.898
			Tyr228	4.23×10^{-3}	0.172
			Tyr224	1.93×10^{-3}	0.022
			Tyr314	5.05×10^{-4}	−0.130
			Tyr55	6.59×10^{-5}	−0.171
			Trp185	1.37×10^{-5}	−0.095

[a]Mean ET rates from aromatic amino acids to Iso* and related physical quantities are listed over 10,000 snapshots. The expression of ET rate with KM model is given by Eq. (3). NetES energy denotes electrostatic energy between the photoproducts (Iso anion and a donor cation) and other ionic groups in the proteins given by Eq. (7).
[b]The data are taken from Ref. [53].
[c]The data are taken from Ref. [54].
[d]The data are taken from Ref. [56].
[e]The data are taken from Ref. [55].

Table 5. Comparison of ET rate and NetES energy among DAAO dimer and monomer and DAOB dimer and monomer.[a].

The dependence of the ln Rate on the donor-acceptor distances has been predicted by a Dutton rule to be linear [69]. In DAAO dimer and monomer, the ln Rate linearly decreased with Rc in all cases [53, 54]. This means that the fluorescence lifetimes of FAD in DAAO become longer as the Rc increased. The dimer Rc is mostly shorter than those in the monomer [53]. It is concluded that the shorter lifetimes of the dimer are due to their shorter Rc values compared to the monomer.

6. The two subunits in the DAAO dimer are not equivalent in solution

The conformations of the two subunits in the DAAO dimer are found to be not equivalent in solution [53], as shown in **Figure 1**. The Rc values in Sub A between Iso and the main donors are quite different from those in Sub B (**Table 2**), and the H-bond structure between Iso and the nearby amino acids in Sub A is also quite different from that in Sub B (**Table 3**), though H-bonds between Iso and water molecules are not taken into account. The structural differences led to the nonequivalent ET rate and NetES (**Table 5**), and its related physical quantities as the electrostatic energy between the donor and acceptor (ESDA), and solvent reorganization energy (SROE). The ratio of the ET rate in Sub A/the rate in Sub B is 2.3 in Tyr224, 0.55 in Tyr314, and 0.96 in Tyr228 at 10°C and 0.20 in Tyr224, 2.5 in Tyr314, and 7.1 in Tyr228 at 30°C.

7. Temperature-induced structural transition in DAAO monomer

Massey et al. [39] first reported a temperature-induced conformational change (temperature transition) of DAAO, where the tryptophan fluorescence exhibited a temperature transition at around 15°C. The van't Hoff plot of the enzyme activity is nonlinear and best expressed by two straight lines with different activation energies. The enzyme activities showed a temperature-dependent equilibrium between the high- and low-temperature states [89], while the equilibrium constant of the association of monomers to form dimers exhibited a discontinuous change at 18°C [88]. However, this transition is not found in the specific heat change at the transition temperature by means of a differential scanning microcalorimetry [90]. The temperature transition of DAAO has been studied by monitoring the fluorescence lifetimes [46]. The modified Arrhenius plots of the fluorescence quenching constants of the monomer and dimer based upon the absolute rate theory displayed two linear functions both in the monomer and dimer. The fluorescence quenching in DAAO is ascribed to the ET from aromatic amino acids to Iso* [19–21], as described above. The activation enthalpy gap and the entropy gap for the quenching constants of DAAO displayed different values in the lower and higher temperature ranges than at 16–18°C, but not in the free FAD. The quenching constant of the monomer displayed a more pronounced transition than that of the dimer. No indication of appreciable transition in the specific heat change [90] may be due to the measurements being performed at very high concentrations of DAAO, where the enzyme should be in the dimer or higher association state, and so it might be difficult to detect the transition.

The structural basis for the temperature-induced transition in the DAAO monomer is studied by means of MDS and ET analyses [54]. The Rc values of Tyr224 are 0.82 and 0.88 nm at 10 and 30°C, respectively, and those of Tyr314 are 1.06 and 1.18 nm at 10 and 30°C, respectively, as shown in **Table 2**. H-Bonds are formed between IsoN1 (see **Chart 1** for atom notations of Iso ring) and Gly315N (peptide), between IsoN3H and Leu51O (peptide), and between IsoN5 and

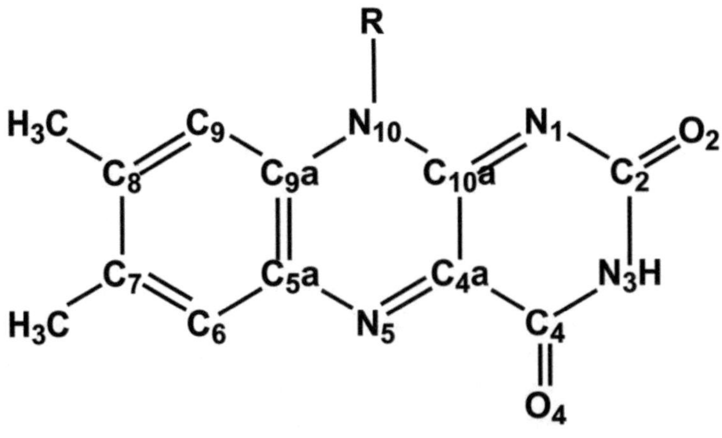

Chart 1. Chemical structure and atom notations of Iso.

Ala49N (peptide) at 10°C, while no H-bond is formed at IsoN1 and IsoN3H at 30°C (**Table 3**). The H-bond of IsoO4 with Leu51N (peptide) at 10°C is switched to Ala49N (peptide) at 30°C. These results may account for the shorter reported fluorescence lifetime of the monomer at 10°C (228 ps) and 30°C (182 ps) [54]. The ET rate from Tyr224 is the fastest among donors at 10°C and the second fastest at 30°C among the donors, while that from Tyr314 is the second fastest at 10°C and the fastest at 30°C (see **Table 5**). The values of NetES in Tyr224 are 0.192 eV at 10°C and −0.035 eV at 30°C, and in Tyr314 are −0.073 eV at 10°C and −0.434 eV at 30°C. The other physical quantities related to the ET rates also displayed appreciable differences at 10 and 30°C. The electron affinities of Iso* are calculated at both temperatures with the semiempirical molecular orbital (MO) method (MOPAC software, PM6 basis set) [54]. The mean calculated electron affinities over 100 snapshots with 0.1 ns intervals are 7.69 eV at 10°C and 7.59 eV at 30°C. Thus, the difference in the observed fluorescence lifetimes between 10 and 30°C is ascribed to the differences in the standard free energy gap and also NetES between the two temperatures.

8. Comparison of the DAOB monomer and dimer structures

Characteristics of monomer and dimer in DAOB and DAAO are compared in **Table 6**.

The Rc values between Iso and Bz are 0.61 nm in the DAOB monomer but 0.66 and 0.68 in Sub A and Sub B, respectively [55, 56], of the dimer as shown in **Table 2**. In the DAOB monomer, the second and third shortest donors are Tyr228 and Tyr224 (0.81 and 0.97 nm,

Physical quantity	DAOB	DAAO
Fluorescence lifetime (ps)		
Monomer	60[b]	130[c], 228 at 10°C[d], 182 ps at 30°C[d]
Dimer	4.8[e]	40[c], 44.2 at 10°C[d], 37.7 at 30°C[d]
Sub A	0.85[e]	
Sub B		
Relative quantum yield of FAD in the enzyme to free FAD [e]	0.0048–0.0077[f]	0.08–0.13[c]
Apparent dissociation constant of FAD (nm)	0.14–0.15[f]	100–300[f]
Dissociation constant of dimer into monomer (μM)	0.4 ± 0.3[f]	3.7[g]

[a]Data are taken with some modifications from [45]. The lifetimes of the DAAO dimer were not separated between the two subunits [44, 46].
[b]Temperature was 20°C. Data are taken from Ref. [45].
[c]Temperature was 20°C. Data are taken from Ref. [44].
[d]Data are taken from Ref. [46].
[e]Data are taken from Ref. [23].
[f]Data are taken from Ref. [45].
[g]Data are taken from Ref. [42, 43].

Table 6. Comparison of characteristics among DAAO monomer and the dimer and DAOB monomer and the dimer.[a].

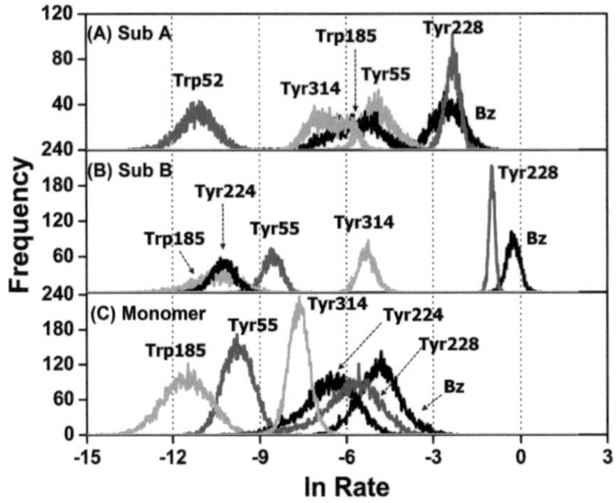

Figure 6. Comparison of the distribution of ln Rate between DAOB dimer and monomer. (A) Sub A and (B) Sub B in the dimer. It was identified that the observed fluorescence lifetime of the dimer, τ_1^{obs} = 0.848 ps, is from Sub B and τ_2^{obs} = 4.77 ps is from Sub A. Inserts denote six fastest ET donors both in Sub A and Sub B. The distribution in the monomer is also shown in (C) Monomer. The data are taken from [56]. (Reproduced by permission of The Royal Society of Chemistry).

respectively), while in the dimer, they are Tyr55 and Tyr228 (0.95 and 0.96 nm) in Sub A and Tyr228 and Tyr314 (0.99 and 1.02 nm) in Sub B. The donor-acceptor Rc distances in the DAOB monomer are, therefore, modified substantially upon formation of the dimer. The H-bond distances between Iso and the nearby amino acids in DAOB are shown in **Table 3**. In the DAOB monomer, IsoN3H forms H-bonds with Leu51, IsoN5 with Ala49, IsoO4 with Leu51, and IsoO2 with Gly315 and Thr317 (see **Chart 1** for the atomic notations). In the dimer, Iso forms H-bonds with Leu51, Asp 49, and Thr317 in Sub A and only with Leu51 and Ala49 in Sub B. The H-bonds of IsoO4 with Leu51 and Gly50 dissociate in the dimer, and in addition the H-bond of IsoO2 with Thr317 dissociates in Sub B as does the H-bond of BzO1 (one of two carboxylate O atoms in Bz) with Tyr228OH. Thus, H-bond structures between Iso or Bz and the nearby amino acids are greatly modified upon dimer formation.

Figure 6 shows comparison of distributions of ln Rate from aromatic amino acids and Bz to Iso* among DAOB monomer and Sub A and Sub B in DAOB dimer [55, 56]. The distribution of Bz in the DAOB monomer shifts to smaller values compared to those of DAOB dimer.

9. Nonequivalent structure between the two subunits in the DAOB dimer in solution

The MDS structures of DAOB dimer and monomer are shown in **Figure 7** [56]. The local structures near Iso display quietly different between the two subunits. The H-bond pairs and

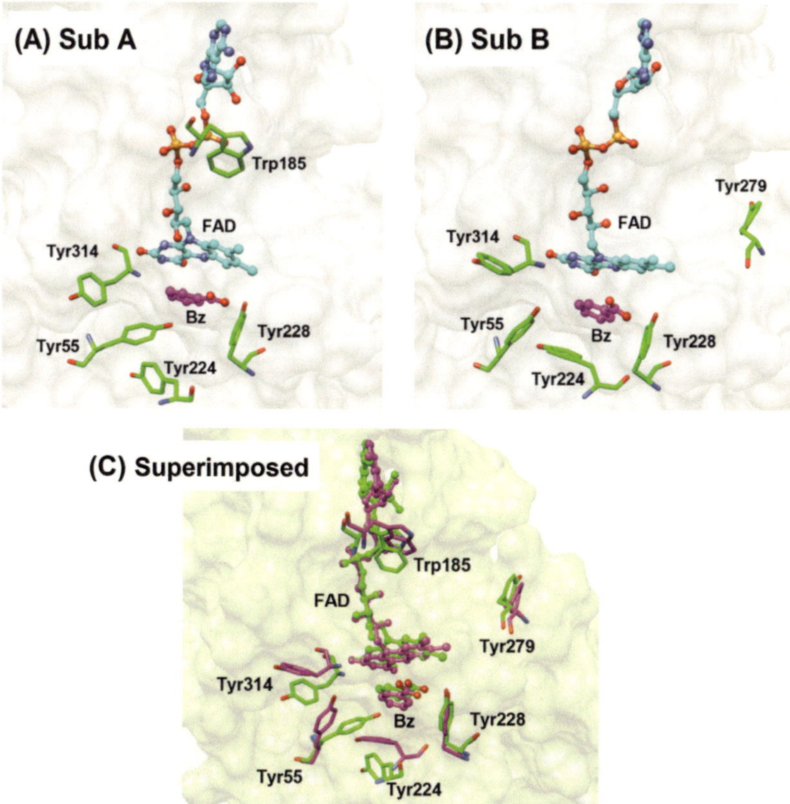

Figure 7. Structure of DAOB dimer at FAD binding site. (A) Sub A, (B) Sub B show subunits A and B in DAOB dimer. (C) Superimposed shows superimposition of Sub A and Sub B. The potential ET donors, Bz, Tyr224, Tyr228, Tyr314, Tyr55, Tyr279, and Trp185, are shown in addition to FAD. In bottom panel FAD and the aromatic amino acids are indicated in green for Sub A, and in magenta for Sub B. MDS calculation was performed at 20 °C. The data are taken from [56]. (Reproduced by permission of the Royal Society of Chemistry).

distances in DAOB also differ between them (see **Table 3**), where the H-bonds between IsoO2 and Thr317 and between BzO1 and Tyr228 in Sub A dissociate in Sub B.

Figure 8 shows ultrafast fluorescence dynamics of DAOB dimer [23]. It is evident that the dimer displayed two lifetime components at any wavelengths monitored. **Table 7** lists the decay parameters at several wavelengths. The mean lifetimes are listed in **Table 6**, 0.848 and 4.77 ps, of which fluorescence is from Sub B and Sub A, respectively [56]. The three main ET donors in the DAOB dimer are Bz, Tyr228, and Tyr55 in Sub A and Bz, Tyr228, and Tyr314 in Sub B, while the three fastest are Bz, Tyr228, and Tyr224 donor in the DAOB monomer. The ET rates and NetES in the DAOB dimer and monomer are listed in **Table 5**. The ET rate from Bz is 7.50×10^{-2} ps^{-1} in Sub A and $8.92 \times 10^{-}1$ ps^{-1} in Sub B of the DAOB dimer. The ET rates from Tyr228 and Tyr55 are also quite different between Sub A and Sub B in the DAOB dimer. Thus, the NetES values are not equivalent in the main donors between Sub A and Sub B.

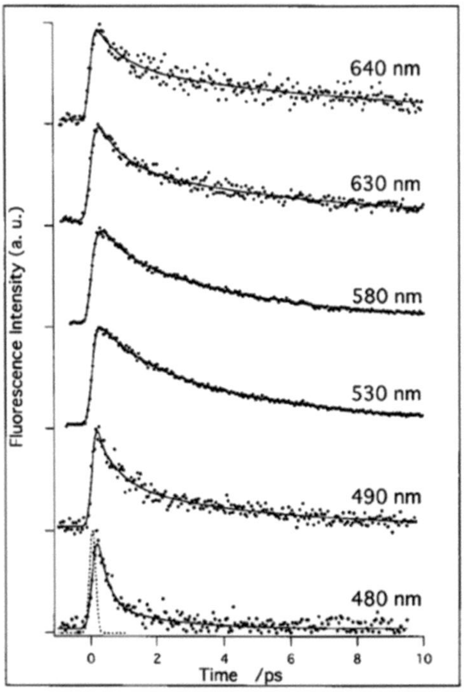

Figure 8. Fluorescence dynamics of DAOB dimer observed at various emission wavelengths. The instrumental response (fwhm ~210 fs) is also indicated with dotted line at the bottom. The decay parameters are listed in **Table 6**. (Reprinted with permission from [23]. Copyright (2000) American Chemical Society).

Wavelength (nm)	a_1	a_2	τ_1 (fs)	τ_2 (ps)	χ^2
480	0.815	0.185	300	1.90	0.74
485	0.710	0.290	420	4.23	0.62
490	0.600	0.400	506	4.46	0.64
510	0.220	0.780	942	4.47	0.10
530	0.337	0.663	1486	4.95	0.04
550	0.360	0.640	1460	5.00	0.05
580	0.260	0.740	940	4.52	0.06
600	0.250	0.750	877	4.40	0.17
630	0.486	0.514	840	6.46	0.35
640	0.470	0.530	713	7.34	1.05

The fluorescence decay functions are expressed by $F(t) = \alpha_1 \exp(-t/\tau_1) + \alpha_2 \exp(-t/\tau_2)$, where τ_1 and τ_2 are lifetimes of the fluorescent components 1 and 2, respectively, and α_1 and α_2 are their respective fractions. The chi-square (χ^2) value between the observed and calculated intensities with the two exponential decay functions is shown. The lifetimes are emission wavelength dependent. (Reprinted with permission from [23]. Copyright (2000) American Chemical Society).

Table 7. Fluorescence decay parameters of the DAOB dimer.

10. Comparison between DAAO and DAOB

The Sub A and Sub B structures of DAOB are almost equivalent in crystal, at least near the FAD binding sites [10]. However, the superimposed MDS-derived Sub A and Sub B structures in solution revealed that the structures near the Iso binding sites are not equivalent [56]. Further, the structures are quite different between the crystal Sub A and MDS-derived Sub A and between the crystal Sub B and MDS-derived Sub B. This may be ascribed that, in the crystal structure, the protein molecules are under the crystal field in the cell units, and so that not many water molecules, while in solution the protein can be relaxed in freely mobile water molecules.

It is evident that the structures near Iso in DAAO are markedly modified upon complex formation with Bz. Absorption spectrum of DAAO is much modified upon binding of Bz. The peak wavelength of the absorption band at around 450 nm of DAAO [39] shifts toward longer wavelength by 13 nm in the complex with vibrational structure [23]. The fluorescence lifetime of the DAOB monomer is 60 ps [45], while ca. 130 [44] or 200 ps [46] in DAAO monomer. The lifetimes of the DAOB dimer stated above [23] are much shorter compared to those of DAAO dimer and DAOB monomer. The remarkably shorter lifetimes in DAOB dimer are mainly ascribed to the ET from Bz to Iso*. To compare the conformation of the DAAO and DAOB using the Rc values of the aromatic amino acids other than Bz, the Rc values in the DAAO dimer at 20°C are taken as the average of those at 10 and 30°C. The Rc values of Tyr224 in the DAAO dimer, 0.82 nm in Sub A, and 0.76 nm in Sub B (**Table 2**) are much smaller than in Sub A (1.32 nm) and Sub B (1.04 nm) in the DAOB dimer. The values of Rc of Tyr228 in the DAAO dimer (0.84 nm in Sub A and 0.82 nm in Sub B) are smaller than in the DAOB dimer (0.96 nm in Sub A and 0.99 nm in Sub B), while those for Tyr55 in the DAAO dimer (1.27 nm in Sub A and 1.03 nm in Sub B) are larger than in Sub A (0.95 nm) but broadly similar to that in Sub B (1.05 nm) in the DAOB dimer. Thus, the Rc values are greatly modified upon the binding of Bz.

Root of mean square fluctuation (RMSF) is considered to be a useful index for protein fluctuation. **Figure 9** shows RMSF values against residue numbers in all four species. The mean RMSF values over all amino acids and FAD are the smallest in the DAOB dimer (0.191 and 0.171 in Sub A and Sub B, respectively) among the four proteins, the DAOB monomer (0.522) and DAAO (0.347, 0.344, and 0.701 in the dimer Sub A, Sub B, and the monomer, respectively). It is well known that the binding of Bz to DAAO greatly stabilizes the protein, and indeed this trait is used in the purification procedure of DAAO [78]. It is also recognized that the DAAO monomer is the most unstable among the DAAO and DAOB species, and so the mean RMSF may be related to protein stability in general. In fact the dissociation constant of FAD is the least in DAOB dimer and the greatest in DAAO monomer [42, 43, 45]. Denaturation of DAAO easily takes place after FAD dissociation.

The static dielectric constants (ε_0^{DA}) between Iso and ET donors within 1 nm from Iso are compared in both DAAO and DAOB [53–56], where the dielectric constants are larger (5.7–5.9) in the DAAO isomers than in the DAOB isomers (2.45–2.64), as shown in **Table 4**.

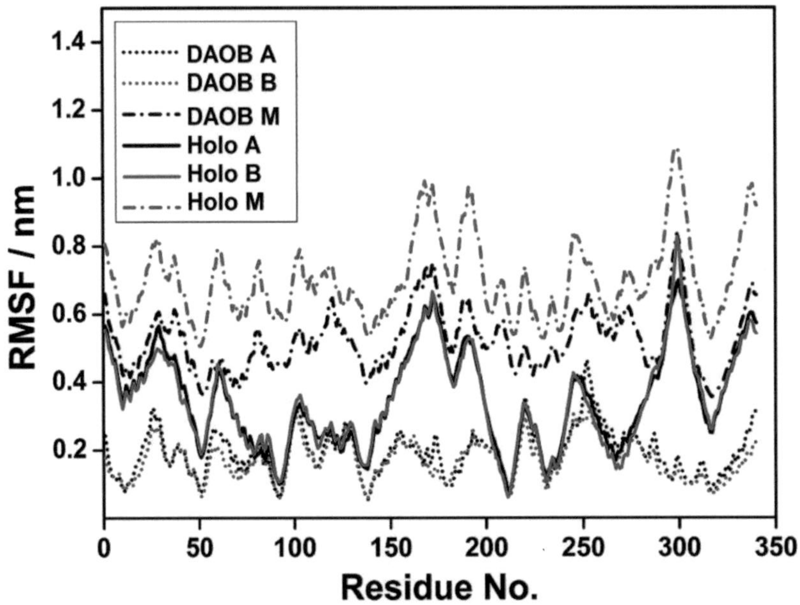

Figure 9. Comparison of root of mean square fluctuations among DAAO dimer, DAAO monomer, DAOB dimer, and DAOB monomer. Root of mean square fluctuations (RMSFs) were obtained by AMBER 10. Holo M, Holo A, and Holo B in the insert denote the DAAO monomer, Sub A, and Sub B of the DAAO dimer, respectively. DAOB M, DAOB A, and DAOB B denote the DAOB monomer, Sub A, and Sub B of the DAOB dimer, respectively. RMSFs of DAAO monomer were taken from [54], those for DAAO dimer from [53] and those of DAOB monomer from [55], and DAOB dimer from [56]. (Reproduced by permission of the Royal Society of Chemistry).

The polarity near Iso is considered to be higher when the value of ε_0^{DA} is higher. The radial distribution functions (RDFs) of water molecules and numbers of water molecules near Iso are reported in DAAO dimer [53] and in DAOB [56]. The RDFs of DAAOs are shown in **Figure 10**. At 10°C approximately 5.5 and 16 molecules are predicted to exist near Iso in Sub A and Sub B, respectively, while at 30°C this switched to 12 and 6 water molecules in Sub A and Sub B, respectively. The number of water molecules could also relate to polarity around Iso. The RDF in DAOB is shown in **Figure 11**, where in the DAOB dimers are few if any, and five water molecules existed near Iso in Sub A and Sub B, respectively. No water molecules are predicted to exist near Iso in the DAOB monomer. Thus, the number of water molecules is much greater in the DAAO dimer than that in DAOB dimer and the monomer, which is in accordance with the ε_0^{DA} results. Stokes shift of the fluorescence spectra in flavoproteins is related to the polarity around Iso. The fluorescence spectra of Iso display at 523 nm of peak wavelength in the DAOB dimer [23] and at 530 nm in DAAO [39]. The ε_0^{DA} values obtained by ET analyses and the RDF of water molecules obtained by MDS are both in accordance with the behavior of the fluorescence spectra.

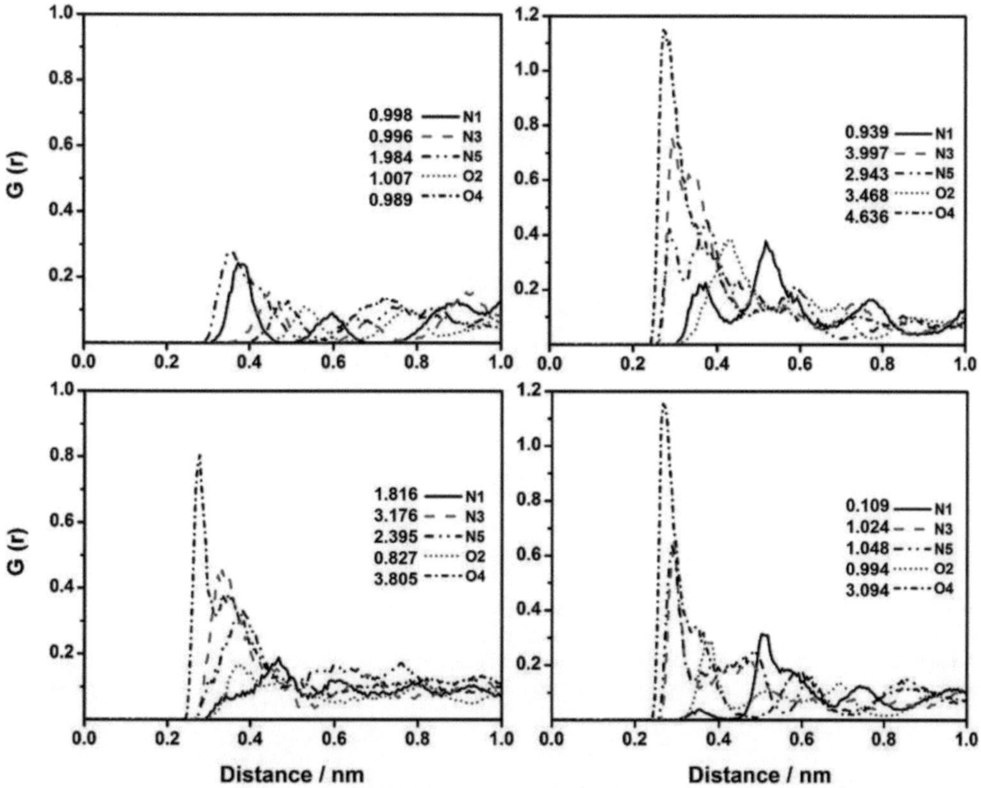

Figure 10. (A) Sub A 10 °C, (B) Sub B 10 °C, (C) Sub A 30 °C and (D) Sub B 30 °C show the radial distribution function derived number of water molecules near hetero atoms in Iso ring in the DAAO dimer. The vertical axes, G(r), denote the radial distribution function. Inserts indicate mean numbers of water molecules at the distances of first layer from the hetero atoms in Iso (see **Chart 1** for atom notations). The data are taken from [53]. Reproduced by permission of the PCCP Owner Societies.

11. Conclusions

MDS is a useful tool to study the structures of DAAO and DAOB in solution, while their experimental fluorescence lifetimes are also a useful index to monitor their structural changes, because the fluorescence lifetimes in flavoproteins are determined by the rates of ET from the aromatic amino acids to Iso*. Thus, combining the MDS structures and the experimental fluorescence lifetimes by ET analysis provides more precise information on the submicroscopic features of the structures of DAAO and DAOB. It is concluded as follows:

1. The origin of the cooperativity in the FAD binding processes is due to much lower (1/74 fold) dissociation constant of FAD in the DAAO dimer than in the monomer. The structural

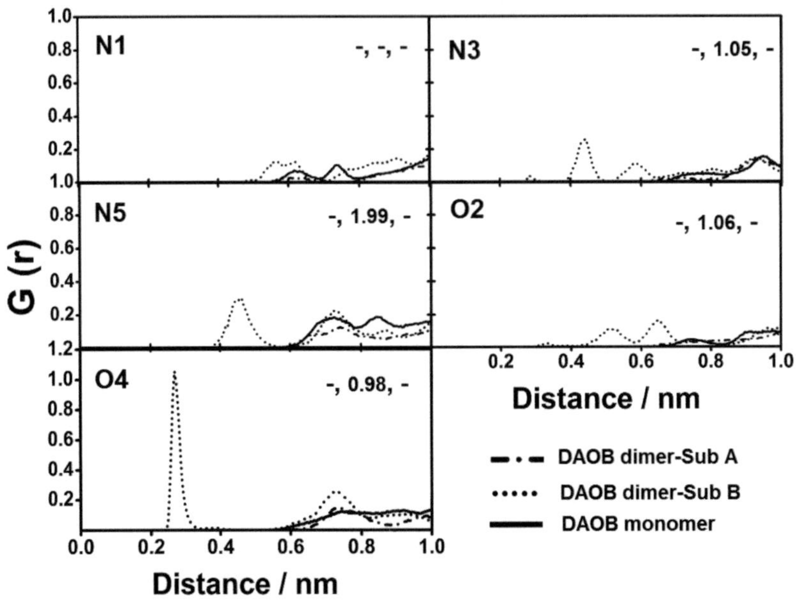

Figure 11. Radial distribution function of water molecules near the heteroatoms of Iso in DAOB. Vertical axes, G(r), denote the radial distribution functions. Red, blue, and black numbers showed the mean number of water molecules in Sub A and Sub B of DAOB dimer, and DAOB monomer is indicated in red, blue, and black, respectively. The data for the DAOB dimer are taken from [56] and for DAOB monomer from [55]. (Reproduced by permission of The Royal Society of Chemistry).

basis for the cooperative binding in DAAO is elucidated by differences in the H-bond structures, the Rc, and the NetES values between the DAAO dimer and the monomer.

2. The temperature-induced transition in the DAAO monomer is ascribed to the differences in the SFEG and NetES between the two temperatures. The change in the SFEG with temperature may be brought about by the change in H-bond structures.

3. The two subunits of the DAAO dimer are not equivalent in solution, as revealed by MDS and ET analyses.

4. The structures of DAOB dimer are almost equivalent for the two subunits in the crystal but are nonequivalent in solution as revealed by the experimental fluorescence lifetimes, MDS structures, and ET analyses.

5. The mean RMSF values over all residues are the smallest in the DAOB dimer and the largest in the DAAO monomer. It is well recognized that the binding of Bz to DAAO greatly stabilizes the protein and the DAAO monomer is the most unstable among the DAAO and DAOB isomers. The mean RMSF may be related to protein stability in general.

6. The ε_0^{DA} values in the DAAO isomers (5.7–5.9) are much larger than those in the DAOB isomers (2.45–2.64), which are elucidated by the number of water molecules near Iso, as derived from the RDF analysis. Water molecules in DAAO are excluded upon the binding of competitive inhibitor of Bz.

7. The Stokes shift of the fluorescence spectra is related to the polarity around Iso, with a change in the emission peak from 524 nm in the DAOB dimer to 530 nm in the DAAO dimer. The ε_0^{DA} values obtained by ET analysis and number of water molecules near Iso obtained by RDF analyses are both in accordance with the observed Stokes shift.

Acknowledgements

A. N. would like to acknowledge the postdoctoral fellowship of Chulalongkorn University. F.T. is thankful for financial support from the Ratchadaphiseksomphot Endowment Fund and a short-term visit grant from Chulalongkorn University. We would like to thank the Computational Chemistry Unit Cell, Chulalongkorn University and The National e-Science Infrastructure Consortium for providing computing resources. S. T. thankful for the Japan Society for the Promotion of Science (Grants-in-Aid for Scientific Research No. 26410029).

Author details

Arthit Nueangaudom[1], Kiattisak Lugsanangarm[2], Somsak Pianwanit[1], Sirirat Kokpol[1], Nadtanet Nunthaboot[3], Fumio Tanaka[1,4*], Seiji Taniguchi[4*] and Haik Chosrowjan[4]

*Address all correspondence to: fumio.tanaka@yahoo.com and taniguchi@ilt.or.jp

1 Department of Chemistry, Faculty of Science, Chulalongkorn University, Bangkok, Thailand

2 Program of Chemistry, Faculty of Science and Technology, Bansomdejchaopraya Rajabhat University, Bangkok, Thailand

3 Department of Chemistry and Center of Excellence for Innovation in Chemistry, Faculty of Science, Mahasarakham University, Mahasarakham, Thailand

4 Division of Laser Biochemistry, Institute for Laser Technology, Osaka, Japan

References

[1] Horiike K, Ishida T, Tanaka H, Arai R. Distribution of D-amino acid oxidase and D-serine in vertebrate brains. Journal of Molecular Catalysis B: Enzymatic 2001;**12**:37–41

[2] Miura R, Setoyama C, Nishina Y, Shiga K, Miyahara I, Mizutani H, Hirotsu K. Porcine kidney d-amino acid oxidase: The three-dimensional structure and its catalytic mechanism based on the enzyme–substrate complex model. Journal of Molecular Catalysis B: Enzymatic 2001;**12**:43-52

[3] Tishkov VI, Khoronenkova SV. d-Amino acid oxidase: Structure, catalytic mechanism, and practical application. Biochemistry (Moscow). 2005;**70**:40-54

[4] Pollegioni L, Piubelli L, Sacchi S, Pilone MS, Molla G. Physiological functions of d-amino acid oxidases: From yeast to humans. Cellular and Molecular Life Sciences 2007;**64**:1373-1394

[5] Kawazoe T, Park HK, Iwana S, Tsuge H, Fukui K. Human d-amino acid oxidase: An update and review. Chemical Record 2007;**7**:305-315

[6] Sacchi S, Caldinelli L, Cappelletti P, Pollegioni L, Molla G. Structure–function relationships in human d-amino acid oxidase. Amino Acids 2012;**43**:1833-1850

[7] Madeira C, Freitas ME, Vargas-Lopes C, Wolosker H, Panizzutti R. Increased brain d-amino acid oxidase (DAAO) activity in schizophrenia. Schizophrenia Research. 2008;**101**:76-83

[8] Boks MPM, Rietkerk T, van de Beek MH, Sommer IE, de Koning TJ, Kahn RS. Reviewing the role of the genes G72 and DAAO in glutamate neurotransmission in schizophrenia. European Neuropsychopharmacology 2007;**17**:567-572

[9] Katane M, Osaka N, Matsuda S, Maeda K, Kawata T, Saitoh Y, Sekine M, Furuchi T, Doi I, Hirono S, Homma H: Identification of novel d-amino acid oxidase inhibitors by in silico screening and their functional characterization in vitro. Journal of Medicinal Chemistry 2013;**56**:1894-1907

[10] Mizutani H, Miyahara I, Hirotsu K, Nishina Y, Shiga K, Setoyama C, Miura R. Three-dimensional structure of porcine kidney d-amino acid oxidase at 3.0 Å resolution. Journal of Biochemistry. 1996;**120**:14-17

[11] Mattevi A, Vanoni MA, Todone F, Rizzi M, Teplyakov A, Coda A, Bolognesi M, Curti B. Crystal structure of d-amino acid oxidase: a case of active site mirror-image convergent evolution with flavocytochrome b2. Proceedings of the National Academy of Sciences of the United States of America. 1996;**93**:7496-7501

[12] Weber G. Fluorescence of riboflavin, diaphorase and related substances [thesis]. Cambridge: University of Cambridge; 1947

[13] Weber G. The quenching of fluorescence in liquids by complex formation. Determination of the mean life of the complex. Transactions of the Faraday Society. 1948;**44**:185-189

[14] Weber G. Fluorescence of riboflavin and flavin-adenine dinucleotide. Biochemical Journal 1950;**47**:114-121

[15] Spencer RD, Weber G. Structure and Function of Oxidation-Reduction Enzymes. In: Åkeson Å, Ehrenberg A, editors. Oxford: Pergamin Press; 1972. p. 393-399.

[16] Weber G, Tanaka F, Okamoto BY, Drickamer HG. The effect of pressure on the molecular complex of isoalloxazine and adenine. Proceedings of the National Academy of Sciences of the United States of America. 1974;**71**:1264-1266

[17] McCormick DB. Interactions of flavins with amino acid residues: Assessments from spectral and photochemical studies. Photochemistry and Photobiology 1977;**26**:169-182

[18] van den Berg PA, Visser AJWG. In: Valeur B, Brochon JC, editors. Tracking molecular dynamics of flavoproteins with time-resolved fluorescence spectroscopy. Berlin: Springer; 2001. p. 457-485

[19] Karen A, Ikeda N, Mataga N, Tanaka F. Picosecond laser photolysis studies of fluorescence quenching mechanisms of flavin: a direct observation of indole-flavin singlet charge transfer state formation in solutions and flavoenzymes. Photochemistry and Photobiology 1983;**37**:495-502

[20] Karen A, Sawada MT, Tanaka F, Mataga N. Dynamics of excited flavoproteins-picosecond laser photolysis studies. Photochemistry and Photobiology 1987;**45**:49-53

[21] Zhong D, Zewail AH. Femtosecond dynamics of flavoproteins: Charge separation and recombination in riboflavine (vitamin B2)-binding protein and in glucose oxidase enzyme. Proceedings of the National Academy of Sciences of the United States of America. 2001;**98**:11867-11872

[22] Mataga N, Chosrowjan H, Shibata Y, Tanaka F. Ultrafast fluorescence quenching dynamics of flavin chromophores in protein nanospace. Journal of Physical Chemistry B 1998;**102**:7081-7084

[23] Mataga N, Chosrowjan H, Shibata Y, Tanaka F, Nishina Y, Shiga K. Dynamics and mechanisms of ultrafast fluorescence quenching reactions of flavin chromophores in protein nanospace. Journal of Physical Chemistry B 2000;**104**:10667-10677

[24] Mataga N, Chosrowjan H, Taniguchi S, Tanaka F, Kido N, Kitamura M. Femtosecond fluorescence dynamics of flavoproteins:comparative studies on flavodoxin, its site-directed mutants, and riboflavin binding protein regarding ultrafast electron transfer in protein nanospaces. Journal of Physical Chemistry B 2002;**106**:8917-8920

[25] Tanaka F, Chosrowjan H, Taniguchi S, Mataga N, Sato K, Nishina Y, Shiga K. Donor-acceptor distance-dependence of photoinduced electron-transfer rate in flavoproteins. Journal of Physical Chemistry B 2007;**111**:5694-5699

[26] Chosrowjan H, Taniguchi S, Mataga N, Tanaka F, Todoroki D, Kitamura M. Comparison between ultrafast fluorescence dynamics of FMN binding protein from desulfovibrio vulgaris, strain miyazaki, in solution vs crystal phases. Journal of Physical Chemistry B 2007;**111**:8695-8697

[27] Chosrowjan H, Taniguchi S, Mataga N, Tanaka F, Todoroki D, Kitamura M. Ultrafast fluorescence dynamics of FMN-binding protein from Desulfovibrio vulgaris (Miyazaki F) and its site-directed mutated proteins. Chemical Physics Letters 2008;**462**:121-124

[28] Chosrowjan H, Taniguchi S, Mataga N, Nakanishi T, Haruyama Y, Sato S, Kitamura M, Tanaka F. Effects of the disappearance of one charge on ultrafast fluorescence dynamics of the FMN binding protein. Journal of Physical Chemistry B 2010;**114**:6175-6182

[29] Chosrowjan H, Taniguchi S, Wongnate T, Sucharitakul J, Chaiyen P, Tanaka F. Conformational heterogeneity in pyranose 2-oxidase from Trametes multicolor revealed by ultrafast fluorescence dynamics. Journal of Photochemistry and Photobiology A 2012;**234**:44-48

[30] Taniguchi S, Chosrowjan H, Wongnate T, Sucharitakul J, Chaiyen P, Tanaka F. Ultrafast fluorescence dynamics of flavin adenine dinucleotide in pyranose 2-oxidases variants and their complexes with acetate: Conformational heterogeneity with different dielectric constants. Journal of Photochemistry and Photobiology A 2012;**245**:33-42

[31] Nunthaboot N, Tanaka F, Kokpol S, Chosrowjan H, Taniguchi S, Mataga N. Simultaneous analysis of ultrafast fluorescence decays of FMN binding protein and its mutated proteins by molecular dynamic simulation and electron transfer theory. Journal of Physical Chemistry B 2008;**112**:13121-13127

[32] Nunthaboot N, Pianwanit S, Kokpol S, Tanaka F. Simultaneous analyses of photoinduced electron transfer in the wild type and four single substitution isomers of the FMN binding protein from Desulfovibrio vulgaris, Miyazaki F. Physical Chemistry Chemical Physics 2011;**13**:6085-6097

[33] Nunthaboot N, Kido N, Tanaka F, Lugsanangarm K, Nueangaudom A, Pianwanit S, Kokpol S. Relationship between rate of photoinduced electron transfer and hydrogen bonding chain of tyrosine-glutamine-flavin in flavin photoreceptors: Global analyses among four TePixDs and three AppAs. Journal of Photochemistry and Photobiology A 2013;**252**:14-24

[34] Lugsanangarm K, Pianwanit S, Kokpol S, Tanaka F, Chosrowjan H, Taniguchi S, Mataga N. Analysis of photoinduced electron transfer in flavodoxin. Journal of Photochemistry and Photobiology A 2011;**217**:333-340

[35] Lugsanangarm K, Pianwanit S, Kokpol S, Tanaka F, Chosrowjan H, Taniguchi S, Mataga N. Photoinduced electron transfer in wild type and mutated flavodoxin from Desulfovibrio vulgaris, strain Miyazaki F.: Energy gap law. Journal of Photochemistry and Photobiology A. 2011;**219**:32-41.

[36] Lugsanangarm K, Pianwanit S, Nueangaudom A, Kokpol S, Tanaka F, Nunthaboot N, Ogino K, Takagi R, Nakanishi T, Kitamura M, Taniguchi S, Chosrowjan H. Mechanism of photoinduced electron transfer from tyrosine to the excited flavin in the flavodoxin from Helicobacter pylori. A comparative study with the flavodoxin and flavin mononucleotide binding protein from Desulfovibrio vulgaris (Miyazaki F). Journal of Photochemistry and Photobiology A 2013;**268**:58-66

[37] Kozioł J. Studies on flavins in organic solvents - III. Spectral behaviour of lumiflavin. Photochemistry and Photobiology. 1969;**9**:45-53

[38] Kotaki A, Yagi K. Fluorescence properties of flavins in various solvents. The Journal of Biochemistry (Tokyo). 1970;**68**:509-516

[39] Massey V, Curti B, Ganther H. A temperature-dependent conformational change in D-amino acid oxidase and its effect on catalysis. Journal of Biological Chemistry 1966;**241**:2347-2357

[40] Wu FY-H, Tu S-C, Wu C-W, McCormick DB. Characteristics of the fluorescence spectra of apoenzyme and flavin portions of D-amino acid oxidase. Biochemical and Biophysical Research Communications 1970;**41**:381-385

[41] Yagi K, Tanaka F, Ohishi N. Structure and function of D-amino acid oxidase IX. Changes in the fluorescence polarization of FAD upon complex formation. Journal of Biochemistry (Tokyo). 1975;**77**:463-468

[42] Tanaka F, Yagi K. Cooperative binding of coenzyme in D-amino acid oxidase. Biochemistry 1979;**18**:1531-1536

[43] Tanaka F, Yagi K. Cooperative binding of flavin adenine dinucleotide accompanied by the change in the quaternary structure of D-amino acid oxidase. In: Yagi K, Yamano T, editors. Flavins and Flavoproteins. Tokyo, Japan: Japan Scientific Press; 1980. p. 1091-1094

[44] Nakashima N, Yoshihara K, Tanaka F, Yagi K. Picosecond fluorescence lifetime of the coenzyme of D-amino acid oxidase. Journal of Biological Chemistry 1980;**255**:5261-5263

[45] Yagi K, Tanaka F, Nakashima N, Yoshihara K. Picosecond laser fluorometry of FAD of D-amino acid oxidase-benzoate complex. Journal of Biological Chemistry 1983;**258**:3799-3802

[46] Tanaka F, Tamai N, Yamazaki I, Nakashima N, Yoshihara K. Temperature-induced changes in the coenzyme environment of D-amino acid oxidase revealed by the multiple decays of FAD fluorescence. Biophysical Journal 1989;**56**:901-909

[47] Tanaka F, Tamai N, Yamazaki I. Picosecond-resolved fluorescence spectra of D-amino acid oxidase. A new fluorescent species of the coenzyme. Biochemistry 1989;**28**:4259-4262

[48] Case DA, Darden TA, Cheatham ITE, Simmerling CL, Wang J, Duke RE, Luo R, Crowley M, Walker RC, Zhang W, Merz KM, Wang B, Hayik S, Roitberg A, Seabra G, Kolossváry I, Wong KF, Paesani F, Vanicek J, Wu X, Brozell SR, Steinbrecher T, Gohlke H, Yang L, Tan C, Mongan J, Hornak V, Cui G, Mathews DH, Seetin MG, Sagui C, Babin V, Kollman PA. AMBER 10. San Francisco: University of California; 2008.

[49] Wang J, Cieplak P, Kollman PA. How well does a restrained electrostatic potential (RESP) model perform in calculating conformational energies of organic and biological molecules? Journal of Computational Chemistry 2000;**21**:1049-1074

[50] Bayly CI, Cieplak P, Cornell W, Kollman PA. A well-behaved electrostatic potential based method using charge restraints for deriving atomic charges: The RESP model. Journal of Physical Chemistry 1993;**97**:10269-10280

[51] Essmann U, Perera L, Berkowitz ML, Darden T, Lee H, Pedersen LG. A smooth particle mesh Ewald method. Journal of Chemical Physics 1995;**103**:8577-8593

[52] Ryckaert J-P, Ciccotti G, Berendsen HJC. Numerical integration of the cartesian equations of motion of a system with constraints: Molecular dynamics of n-alkanes. Journal of Computational Physics 1977;**23**:327-341

[53] Nueangaudom A, Lugsanangarm K, Pianwanit S, Kokpol S, Nunthaboot N, Tanaka F. Non-equivalent conformations of D-amino acid oxidase dimer from porcine kidney between the two subunits. Molecular dynamics simulation and photoinduced electron transfer. Physical Chemistry Chemical Physics 2014;**16**:1930-1944

[54] Nueangaudom A, Lugsanangarm K, Pianwanit S, Kokpol S, Nunthaboot N, Tanaka F. Structural basis for the temperature-induced transition of D-amino acid oxidase from pig kidney revealed by molecular dynamic simulation and photo-induced electron transfer. Physical Chemistry Chemical Physics 2012;**14**:2567-2578

[55] Nueangaudom A, Lugsanangarm K, Pianwanit S, Kokpol S, Nunthaboot N, Tanaka F. The mechanism of photoinduced electron transfer in the D-amino acid oxidase–benzoate complex from pig kidney: Electron transfer in the inverted region. Journal of Photochemistry and Photobiology A 2012;**250**:6-17

[56] Nueangaudom A, Lugsanangarm K, Pianwanit S, Kokpol S, Nunthaboot N, Tanaka F, Taniguchi S, Chosrowjan H. Theoretical analyses of the fluorescence lifetimes of the D-amino acid oxidase-benzoate complex dimer from porcine kidney: Molecular dynamics simulation and photoinduced electron transfer. RSC Advances 2014;**4**:54096-54108

[57] Marcus RA. On the theory of oxidation-reduction reactions involving electron transfer I. Journal of Chemical Physics 1956;**24**:966-978

[58] Marcus RA. Electrostatic free energy and other properties of states having nonequilibrium polarization I. Journal of Chemical Physics 1956;**24**:979-989

[59] Marcus RA. Chemical and electrochemical electron-transfer theory. Annual Review of Physical Chemistry 1964;**15**:155-196

[60] Hush NS. Adiabatic theory of outer sphere electron-transfer reactions in solution. Transactions of the Faraday Society 1961;**57**:557-580

[61] Warshel A, Chu Z, Parson W. Dispersed polar on simulations of electron transfer in photosynthetic reaction centers. Science 1989;**246**:112-116

[62] Warshel A, Parson W. Computer simulations of electron-transfer reactions in solution and in photosynthetic reaction centers. Annual Review of Physical Chemistry 1991;**42**:279-309

[63] Bixon M, Jortner J. Non-Arrhenius temperature dependence of electron-transfer rates. Journal of Physical Chemistry 1991;**95**:1941-1944

[64] Bixon M, Jortner J. Charge separation and recombination in isolated supermolecules. Journal of Physical Chemistry 1993;**97**:13061-13066

[65] Bixon M, Jortner J, Cortes J, Heitele H, Michel-Beyerle ME. Energy gap law for nonradiative and radiative charge transfer in isolated and in solvated supermolecules. Journal of Physical Chemistry 1994;**98**:7289-7299

[66] Kakitani T, Mataga N. New energy gap laws for the charge separation process in the fluorescence quenching reaction and the charge recombination process of ion pairs produced in polar solvents. Journal of Physical Chemistry 1985;**89**:8-10

[67] Kakitani T, Yoshimori A, Mataga N. Effects of the donor-acceptor distance distribution on the energy gap laws of charge separation and charge recombination reactions in polar solutions. Journal of Physical Chemistry 1992;**96**:5385-5392

[68] Matsuda N, Kakitani T, Denda T, Mataga N. Examination of the viability of the Collins-Kimball model and numerical calculation of the time-dependent energy gap law of photoinduced charge separation in polar solution. Chemical Physics 1995;**190**:83-95

[69] Moser CC, Keske JM, Warncke K, Farid RS, Dutton PL. Nature of biological electron transfer. Nature 1992;**355**:796-802

[70] Beratan D, Betts J, Onuchic J. Protein electron transfer rates set by the bridging secondary and tertiary structure. Science 1991;**252**:1285-1288

[71] Bendall DS. Protein Electron Transfer. Oxford: BIOS Scientific; 1996

[72] Gray HB, Winkler JR. Electron transfer in proteins. Annual Review of Biochemistry 1996;**65**:537-561

[73] Warshel A, Parson W. Dynamics of biochemical and biophysical reactions: Insight from computer simulations. Quarterly Reviews of Biophysics. 2001;**34**:563-679

[74] Vorsa V, Kono T, Willey KF, Winograd N. Femtosecond photoionization of ion beam desorbed aliphatic and aromatic amino acids:Fragmentation via α-cleavage reactions. Journal of Physical Chemistry B 1999;**103**:7889-7895

[75] Fonda ML, Anderson BM. D-amino acid oxidase : III. Studies of flavin adenine dinucleotide binding. Journal of Biological Chemistry. 1968;**243**:5635-5643

[76] Henn SW, Ackers GK. Molecular sieve studies of interacting protein systems. V. Association of subunits of D-amino acid oxidase apoenzyme. Biochemistry. 1969;**8**:3829-3838

[77] Henn SW, Ackers GK. Molecular sieve studies of interacting protein systems: IV. Molecular size of the D-amino acid oxidase apoenzyme subunit. Journal of Biological Chemistry. 1969;**244**:465-470

[78] Yagi K, Naoi M, Harada M, Okamura K, Hidaka H, Ozawa T, Kotaki A. Structure and function of d-amino acid oxidase I. Further purification of hog kidney D-amino acid oxidase and its hydrodynamic and optical rotatory properties. Journal of Biochemistry (Tokyo). 1967;**61**:580-597

[79] Yagi K, Ohishi N. Structure and function of D-amino acid oxidase IV. Electrophoretic and ultracentrifugal approach to the monomer-dimer equilibrium. Journal of Biochemistry (Tokyo). 1972;**71**:993-998

[80] Antonini E, Brunori M, Bruzzesi MR, Chiancone E, Massey V. Association-dissociation phenomena of D-amino acid oxidase. Journal of Biological Chemistry 1966;**241**:2358-2366

[81] Tojo H, Horiike K, Shiga K, Nishina Y, Watari H, Yamano T. Self-association mode of a flavoenzyme D-amino acid oxidase from hog kidney. I. Analysis of apparent weight-average molecular weight data for the apoenzyme in terms of models. Journal of Biological Chemistry. 1985;**260**:12607-12614

[82] Tojo H, Horiike K, Shiga K, Nishina Y, Watari H, Yamano T. Self-association mode of a flavoenzyme D-amino acid oxidase from hog kidney. II. Stoichiometry of holoenzyme

association and energetics of subunit association. Journal of Biological Chemistry. 1985;**260**:12615-12621

[83] Monod J, Wyman J, Changeux J-P. On the nature of allosteric transitions: A plausible model. Journal of Molecular Biology 1965;**12**:88-118

[84] Koshland DE, Némethy G, Filmer D. Comparison of experimental binding data and theoretical models in proteins containing subunits. Biochemistry 1966;**5**:365-385

[85] Nichol LW, Jackson WJH, Winzor DJ. A theoretical study of the binding of small molecules to a polymerizing protein system. A model for allosteric effects. Biochemistry. 1967;**6**:2449-2456

[86] Nichol LW, Winzor DJ. Ligand-induced polymerization. Biochemistry 1976;**15**:3015-3019

[87] Selwood T, Jaffe EK. Dynamic dissociating homo-oligomers and the control of protein function. Archives of Biochemistry and Biophysics 2012;**519**:131-143

[88] Shiga K, Shiga T. The kinetic features of monomers and dimers in high- and low-temperature conformational states of D-amino acid oxidase. Biochimica et Biophysica Acta 1972;**263**:294-303

[89] Koster JF, Veeger C. The relation between temperature-inducible allosteric effects and the activation energies of amino-acid oxidases. Biochimica et Biophysica Acta 1968;**167**:48-63

[90] Sturtevant JM, Mateo PL. Proposed temperature-dependent conformational transition in D-amino acid oxidase: A differential scanning microcalorimetric study. Proceedings of the National Academy of Sciences of the United States of America. 1978;**75**:2584-2587

Chapter 5

The "History" of Desmosines: Forty Years of Debate on the Hypothesis That These Two Unnatural Amino Acids May Be Potential Biomarkers of Chronic Obstructive Pulmonary Disease

Simona Viglio, Monica Di Venere,
Maddalena Cagnone, Marco Fumagalli and
Paolo Iadarola

Additional information is available at the end of the chapter

http://dx.doi.org/10.5772/intechopen.68570

Abstract

Desmosine and isodesmosine (collectively known as desmosines), two unnatural amino acids unique to mature elastin in humans, have been widely discussed as being potential biomarkers of disorders, which involve connective tissue and whose clinical manifestations result in elastin degradation. In particular, experimental data accumulated over the last 40 years have demonstrated that patients with chronic obstructive pulmonary disease (COPD) excrete higher amounts of urinary desmosines than healthy controls. Based on this evidence, it has been speculated by several authors that these cross-links may be potential biomarkers of COPD with clinical significance. Nevertheless, a strict correlation between the amount of these amino acids and the severity of the disease still has to be demonstrated. For this reason, the debate on the opportunity to consider desmosines as biomarkers of COPD is still open, and the development of sophisticated methods aimed at obtaining very precise measurement of their concentration is still considered technically challenging. The aim of this chapter is to trace the history of this debate through the presentation and discussion of a large number of articles dealing with the detection and quantification of desmosines in different biological fluids, from early years until the present.

Keywords: unnatural amino acids, desmosines, COPD, biological fluids, biomarkers

1. Introduction

Elastin is a highly elastic protein in connective tissue that allows many tissues in the body to resume their shape after stretching or contracting. Being the main component of the elastic fiber, it provides resiliency to numerous pulmonary structures including the alveolar wall, thus influencing the characteristic shape of alveoli. The strong hydrophobicity of elastin, together with a good number of intermolecular cross-links, produces a highly insoluble network of polypeptide chains working as a perfect elastomer in an aqueous environment [1, 2].

Elastin is synthesized and secreted from fibroblasts and vascular smooth cells as a soluble 75-kDa precursor called tropoelastin, containing 12–13 repeats of alternating hydrophobic and cross-linking domains. While glycine, alanine, proline and valine are the predominant residues of the hydrophobic domains in a random-coil organization, polyalanine tracts containing lysine residues embedded in a rigid α-helical structure are characteristic of the cross-linking domains.

The formation of cross-links occurs extracellularly generating two amino acid isoforms, known as desmosine and isodesmosine (collectively, desmosines or DESs), whose content may be used as a quantitative measurement of insoluble elastin formation [1–4].

Given their uniqueness to mature elastin in mammals, DESs have been discussed as potential biomarkers of disorders, which involve connective tissue and whose clinical manifestations result in elastin degradation. In fact, being the desmosine-containing peptides derived from the destruction of the elastic fibers excreted in the urine, their determination in the body fluids may represent an indirect measurement of extracellular matrix degradation or of elastase activity [5, 6].

Connective tissue destruction is a major problem in chronic obstructive pulmonary disease (COPD), a class of disorders characterized by massive destruction of the elastic fibers of the alveoli with disabling airflow limitation, productive cough and dyspnea [7, 8]. According to the protease/antiprotease hypothesis, elastolytic proteases, in particular human neutrophil elastase (HNE), are responsible for the digestion of alveolar elastin. If, for any reason, the level of this proteolytic enzyme overcomes that of α1-antitrypsin (AAT), the most important defense barrier in the lower respiratory tract, elastin present in the alveolar walls will be degraded with consequent loss of lung functions [9].

One of the current emerging challenges of COPD is the research of parameters that may elucidate the events associated with a given disease condition. The detection of reliable biomarkers that can be correlated with the clinical outcome of this disorder obviously plays a special role in this setting. In spite of the increasing number of biomarkers proposed, currently a useful biomarker for COPD is still lacking. However, being this disorder characterized by the uncontrolled degradation of the extracellular matrix with abnormal excretion of elastin-derived fragments containing DESs, substantial interest has been placed in the development of reliable assays to measure their concentration in body fluids. The fact that these peptides are quantitatively excreted in urine provides the rationale for understanding why DESs may represent an indirect measurement of extracellular matrix degradation.

The question of whether these amino acids may actually be considered surrogate markers of elastin degradation has been extensively debated [4, 6, 10–12]. Experimental evidences accumulated over the last 40 years have demonstrated that smokers as well as patients with COPD or other destructive lung diseases excrete in their urine amounts of DESs higher than non-smokers and healthy controls [13–17]. On the basis of these data, it has been speculated by several authors that these cross-links may be considered clinically significant biomarkers of COPD [11, 12]. As a consequence, efforts have been devoted by different research groups in the world to the determination of DESs excreted in a variety of biological fluids which include urine, plasma, and induced sputum. Despite a large number of articles describing the application of different techniques for the screening and quantification of DESs has been published so far, only in few cases a strict correlation between the amount of these amino acids and the severity of the disease has been demonstrated [13–17]. In particular, what has emerged in the course of the last years is that only very accurate desmosine determinations can help researchers in understanding which is the degree of elastin degradation in COPD at different stages of severity and may allow these amino acids to become a reliable tool either in the differential diagnosis or in the clinical management of the disease [10, 12, 18].

For this reason, the debate on the opportunity to consider DESs as surrogate markers of COPD is still open, and the development of increasingly sophisticated techniques aimed at obtaining very precise measurement of their concentration is considered technically challenging.

Indeed, the significant step forward of both sensitivity/specificity and degree of reproducibility of results provided by these technological advancements allowed researchers to fully utilize the power of such data sets thus improving the understanding of mechanisms involved in elastin degradation. Among the variety of research groups in the world that have focused their attention on DESs determination, a great contribution was provided by our own team. Novel strategies have been developed for the accurate determination of these amino acids in urine, plasma and sputum of a large number of smokers and patients affected by pulmonary diseases [3, 4, 6, 11, 13, 14, 16–19].

Aim of this chapter is to trace the history of this debate through the presentation and discussion of a large number of articles dealing with the detection and quantification of DESs in different biological matrices from early years until the present.

2. Biosynthetic pathway of desmosines

The biosynthesis of these cross-links involves the oxidative deamination of the ε-amino group of four lysine residues, by means of lysyl oxidase, a copper-dependent enzyme. The aldehydic residues which yield from this oxidation can then participate in Schiff-base reaction with the ε-amino group of other lysine residues or in aldol condensations with other similar aldehydic residues. These reactions result in the cyclization of the side-chain groups of the four lysines with the formation of DESs, two isomeric pyridinium cross-links which are characteristic of insoluble elastin [20]. The schematic view of these reactions, together with the final structure of these two amino acids, is shown in **Figure 1**.

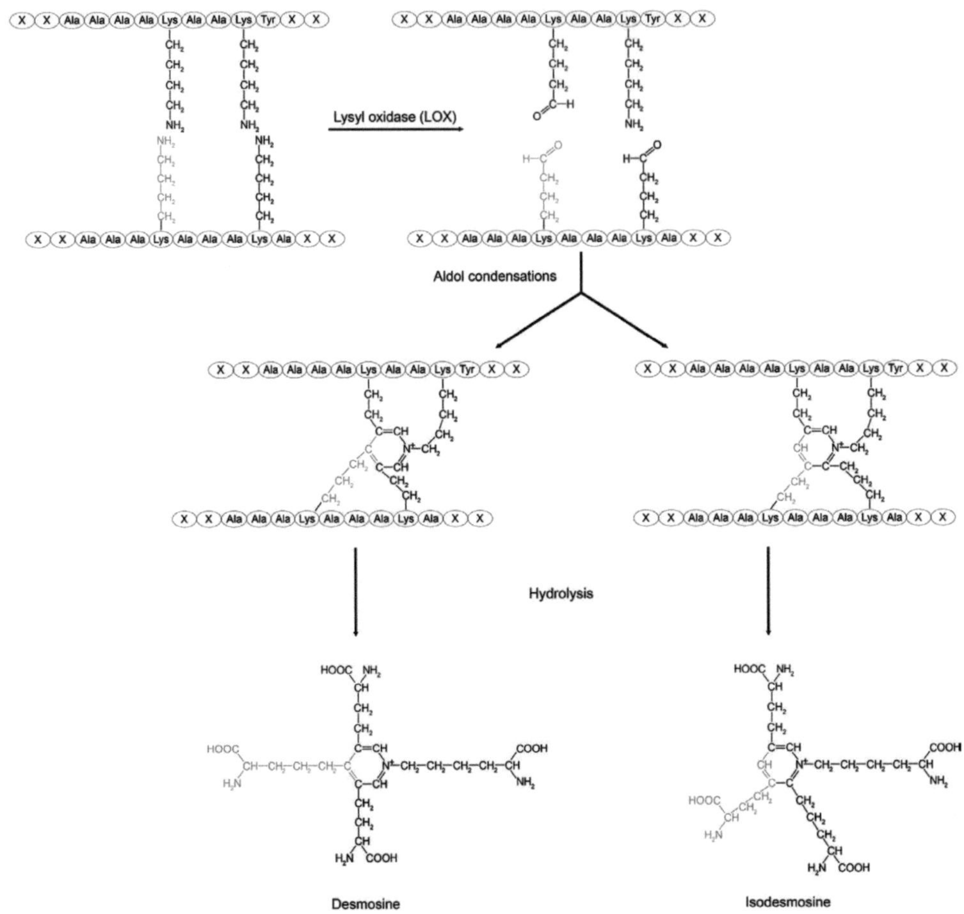

Figure 1. A schematic view of biosynthetic reaction of DESs and IDES.

3. Methodological aspects in the determination of desmosines

The first reports dealing with the determination of DESs in elastin appeared in the 1960s when such amino acids were isolated from purified elastin of bovine *Ligamentum nuchae* [21]. Given their peculiarity in humans, the question of whether these amino acids could be used as clinically significant biomarkers of extracellular matrix disorders was of primary importance. In light of this, it became immediately evident that the only way to answer this question would have been the availability of strategies aimed at calculating the concentration of DESs in different matrices. The research of such strategies resulted in the development of a wide variety of techniques and, as a consequence, in the publication of a large number of articles describing their application to different tissues and fluids. This paragraph is planned to show that while the low technological content of methods available in the 1960s could provide poor results in terms of DESs' clinical significance, the impressive advances achieved with the progression of technology have completely changed the scenario. It will appear that amino acid analysis and the immunochemical and early chromatographic methods, with their intrinsic limitations in terms of specificity and/or sensitivity, have been mainly used to the analysis of tissue hydrolyzates and/or urine. By contrast, thanks to the tremendous increase in sensitivity/specificity of modern methodologies, free and total DESs can currently be detected in all available tissues/fluids. This overall improvement has strengthened the firm belief that DESs might indeed be possible biomarkers of elastin-degrading disorders, in particular lung disorders.

3.1. Amino acid analysis

From a chronological point of view, amino acid analysis was the first method developed for desmosine detection [22–24]. The rationale was that, being amino acids, DESs necessarily were ninhydrin-positive. In brief, desmosine-containing elastin fragments were submitted to a conventional acid hydrolysis and the amino acids separated by ion exchange chromatography prior to be detected through the colorimetric reaction with ninhydrin. Despite appearing as two well-resolved peaks that were integrated with precision, DESs in real samples could be separated only by applying elution conditions that were not coincident with those used for standard compounds. In addition, the quantification of DESs required a chromatographic platform that was not identical to that applied for conventional amino acid analysis. This changeover of platforms was obviously quite laborious and time-consuming for those laboratories in which analysis of elastin hydrolyzates occurred infrequently. To overcome this drawback, an improved procedure was developed which provided excellent resolution and quantification of DESs without the necessity of systems changeover [25]. Despite this improvement, the methodological approach based on amino acid analysis was characterized by a number of limitations, the most important being the poor sensitivity and the need to perform on samples a series of preliminary steps which made the procedure very complex. Nevertheless, amino acid analysis was successfully used for a variety of applications including: (i) determination of DESs concentration in elastin isolated from uterus and skin of young animals and humans [26]; (ii) measurement of elastin turnover in hamsters [27]; (iii) determination of the primary sequence around elastin cross-linking sites and correlation of this information to possible structural "signals" which might modulate or otherwise direct cross-link formation [28]. Changes in elution buffers [29, 30] or in elution mode [31, 32] further improved resolution and sensitivity in the analysis of tissue hydrolyzates.

The application of amino acid analysis to the detection of DESs in human urine was a step forward in the generation of results that could be considered relevant from a diagnostic point of view. For example, the analysis of urine from patients affected by Marfan syndrome during the early development of the disorder revealed that these subjects excreted a consistently lower amount of DESs than that of controls [33]. This was the first evidence of altered elastin cross-linking in a heritable connective-tissue disease. These data were correlated with a low lysyl oxidase activity and/or with an attenuation of the conversion of precursor aldehydes and lysyl cross-links into desmosines. Another study carried out on urines of cystic fibrosis (CF) patients chronically infected with *Pseudomonas aeruginosa* aimed at understanding whether destruction of the lung elastic fibers was an ongoing process in this disorder and whether proteolysis could contribute to the pathological changes in both airways and pulmonary parenchyma [34]. The amino acid analysis showed that the urinary content of DESs was significantly higher in patients than in the age-matched control group and that DESs excretion was significantly correlated with the disease severity. These observations, together with the evidence that fibers were fragmented and distorted, were the proof that destruction and re-synthesis of elastic fibers are a chronic process in patients with CF.

3.2. Immunochemical methods

The application of immunochemical methods for detection of DESs originally started around the 1980s as soon as polyclonal antibodies against these cross-links became available. Although at the beginning RIA and ELISA assays suffered for poor specificity, they have been widely applied for at least 20 years. Their success was attributable to the remarkable increase in sensitivity that made these approaches, at least for a while, a valid alternative to chromatographic ones.

3.2.1. RIA

Initially developed to study the mechanisms of elastogenesis in cell cultures, DESs determination *via* RIA assay was also helpful for a better understanding of in vivo elastogenesis [35]. In this context, the levels of urinary DESs have been used as a measure of elastin catabolism in vivo to study the progression over months of experimental emphysema induced in hamsters by a single intratracheal injection of elastase [36]. DESs excretion was found to increase progressively during the first 24 h after injection, and to normalize after 6 days. This behavior seemed to suggest that the late progression of elastase-induced emphysema was not accompanied by increased elastolysis.

A remarkable improvement in the original RIA assay was the binding of antibodies to magnetic particles. Applied to urine of CF patients, this modified procedure allowed to show that urinary DESs in CF were higher than in controls [37]. Although a few interfering compounds present in the samples competed with DESs for the antibody, thus impoverishing method precision, this was a clear message that DESs measurement could be used to discriminate the two cohorts.

Although with conflicting results, other numerous lung disorders have been investigated by monitoring with RIA the levels of urinary DESs. These assays have been used: (i) to demonstrate that the amount of urinary DESs excreted by healthy non-smokers over 24 h was around 10-fold lower than that of smokers with evidence of COPD [38, 39]; (ii) to compare urinary DESs excretion in homozygous AAT-deficient patients with emphysema; patients with interstitial lung diseases and healthy subjects [40]; (iii) to detect DESs levels in patients with adult respiratory distress syndrome [41] and in patients with acute lung injury [42] and (iv) to assess elastin maturation during the development of human lungs [43].

3.2.2. ELISA

The sake of rapid, specific, safe and sensitive immunochemical assays resulted in the development of ELISA methods. Although the early approaches suffered of poor specificity due to the cross-reactivity toward pyridoline of antibodies against desmosines [44, 45], optimization in the production of antibodies allowed to overcome this problem. The design of an anti-desmosine antiserum characterized by high specificity and sensitivity had a positive impact on the precision of DESs detection in tissue hydrolyzates [46, 47]. A very specific indirect competitive ELISA test was also used to compare the urinary content of DESs in COPD patients with that of healthy controls. The finding that the amount of DESs excreted by the former cohort was significantly higher than that of controls, while not being a novelty, was a confirmation that these cross-links could be potential indicators of lung status [48]. Of great attractiveness was the observation that DESs concentration was higher in patients which showed no evidence of emphysema (or with only mild emphysema) than in those with moderate-to-severe emphysema. This allowed to reason that urinary DESs could be a remarkable tool from a clinical point of view, being potential markers for the identification of subjects at risk of developing emphysema and for assessing the efficacy of therapeutic interventions.

3.3. High-performance liquid chromatography

With the development of high-performance liquid chromatography (HPLC), a marked improvement in terms of resolution and robustness was contributed to the chromatographic platform. This big difference allowed HPLC methods to be used, for the first time, for simultaneous detection of DESs and other cross-links. These methods have been mostly dedicated to the analysis of DESs in tissues, being tailored for specific applications which range from the determination of DESs concentration on hamsters aorta [49] to their detection in hamsters lungs [50] or to the estimate of the amount of tissue elastin in human and dog aorta [30]. A great deal of research was also focused on the age-related changes in the content of elastin and collagen cross-links. To this aim, human aorta [51], human yellow ligament [52] and bovine *Ligamentum nuchae* or rat aorta have been often used as sources of DESs: desmopyridine and isodesmopyridine. These studies were designed to understand the correlation of the elastic properties of this tissue with age [53] and to explore possible defects in elastin metabolism [54–56]. To find a biochemical explanation for the dilatation of vein wall, DESs and 4-L-Hydroxyproline have been quantified by HPLC in specimens of normal and varicose, dilated and non-dilated veins [57]. These determinations showed that the levels of cross-links were reduced in dilated vein, thus proving that dilatation may be related to elastin metabolism. To investigate the biochemical basis of alterations present in upper esophagical sphincter of patients with Zenker's diverticulum, the same cross-links were detected in samples of cricopharyngeal muscle [58].

Despite their above-mentioned features, when HPLC procedures were applied to urine, it appeared immediately evident that the interfering substances present in this matrix spuriously increased DESs levels, thus being a strong limit for their quantification. After exploring a number of routes to implement resolution, RP-HPLC with isotope dilution was suggested as an affordable procedure to overcome this problem. This approach, successfully applied to investigate a variety of disorders, led to the achievement of significant biochemical insights, in particular from detection of DESs in urine of smokers with/without COPD or with/without rapid decline of lung function and of patients with CF [59, 60].

3.4. Electrophoresis and capillary electrophoresis

Pioneering experiments carried out in the 1980s demonstrated that electrophoretic procedures could be helpful for detecting DESs in biological samples [61]. While 1-DE allowed a rapid detection of small amounts of these cross-links in hydrolyzates of elastin from *Lingamentum nuchae*, the complete separation of desmosine, isodesmosine and merodesmosine was achieved by applying a thin-layer methodology. The same procedure was used to measure cross-linked elastin synthesis in hamsters with pulmonary fibrosis induced by bleomycin [62]. The synthesis of cross-linked elastin was found to be significantly elevated in animals at 1–3 weeks after exposure to bleomycin. The message associated with this increase was that this tissue component was, most likely, an important part of the fibrotic response of the pulmonary parenchyma. The 2-D fingerprint (first dimension ascending chromatography, second one electrophoresis) of peptides produced by elastase digestion of elastin was another useful approach for DESs detection.

In the early 1990s, capillary electrophoresis (CE) was developed as a modern approach to obtain high efficiency, fast analysis times and excellent flexibility in changing the selectivity of the separation. As for other techniques described in previous paragraphs, capillary zone electrophoresis (CZE) was initially applied to detect DESs in elastin hydrolyzates but, although analysis was fast, the peaks of the two amino acids could not be completely separated under the conditions used [63]. Although several attempts to improve resolution have been made, it became soon clear that CZE could not become the "gold standard" for the detection of DESs in human matrices for at least two reasons. First, separating the two analytes, whose peaks were largely overlapping, was not possible due to the strict similarity of DESs structures, and second, the sensitivity of the method, although promising for the analysis of hydrolyzates and urine, was not suitable to allow the investigation of other important matrices (plasma, sputum). Nevertheless, the application of this method to study elastin content of abdominal aorta and aortic function in rats exposed to Vitamin D during gestation and in postnatal period led to the finding that the mean content of DESs was higher in control rats than in those treated with high/low Vitamin D doses [64]. By addition of ionic and/or nonionic detergents to the BGE, the separation mode was switched to micellar electrokinetic chromatography (MEKC) in an effort to verify whether the formation of micelles was a suitable tool to obtain the differential migration of DES and IDES. The efficacy of MEKC was explored on urine of healthy controls and COPD patients. The finding that DESs levels were higher in patients than in controls, while not being surprising, was a sort of "proof of the pudding" that the path taken was the right one. Further confirmation of the robustness of this approach came from the analysis of a large number of urine samples from patients with a variety of pulmonary diseases ranging from stable COPD [14] to subjects with acute exacerbation of COPD [13, 14], with AAT deficiency [14]; bronchiectasis [14] and cystic fibrosis [14, 65]. Taken together, the results of these analyses generated a clear picture of DESs excretion in all subjects characterized. They allowed to speculate that the level of these cross-links, being able to report the airway inflammation or to evaluate the efficacy of replacement therapy, could indeed reflect the lung conditions. Unfortunately, to meet the requirements of sensitivity needed for the use of this approach, a preliminary concentration step of urines, with related drawbacks, was mandatory. In an effort to improve sensitivity, a new procedure was developed in which urinary DESs were labeled with a fluorescent probe to be visualized with a LIF detection system [66]. Due to heavy similarity of their final structure, the two labeled desmosines could not be resolved at all by the electrophoretic system. Despite this potential limitation, the method showed to be reliable and allowed DESs to be quantified with precision as the sum of the two isomers. Since its development and for at least a decade, the high sensitivity, the good accuracy and robustness of MECK-LIF have encouraged its utilization for the determination of DESs in a wide variety of human fluids. As a matter of fact, to the best of our knowledge, this is the protocol applied for screening the largest number of real samples ever investigated [4]. The scheme of **Figure 2** indicates the number of samples analyzed per year and all different matrices in which DESs have been detected by MEKC and CE-LIF.

3.5. LC-MS

Despite being very sensitive, MEKC-LIF suffered for a limit in the quantitation of DESs. The procedure, based on the integration of peak corresponding to the two amino acids, was not free from errors which derived from the possible presence, under this peak, of small amounts of interfering substances. The advent of mass spectrometry-based approaches completely circumvented these technical limitation. In fact, the possibility of monitoring only selected ions and their fragments may lead to a higher chance for better analyses in terms of specificity and sensitivity. These technological advances have strongly increased the popularity of this approach, allowing liquid chromatography-mass spectrometry (LC-MS) to merge as the most popular protocol to date applied in the DESs field. In addition, only the extreme sensitivity and specificity of these techniques allowed, for the first time, very minute amounts of free DESs to be observed in urine after a simple clean up of the specimens. An example is the accurate measurement of DESs in urine and sputum of healthy volunteers and patients with

	Urine CE-UV 1998/1999 *n* = 281	Urine CE-UV 2000/2001 *n* = 40	Urine CE-UV 2002/2003 n = 50	Urine-Plasma/Serum CE-LIF 2004/2005 *n* = 323	Urine-Plasma/Serum CE-LIF 2006 *n* = 289	Sputum CE-LIF 2007 *n* = 34	Urine CE-LIF 2010 *n* = 224	Urine-Plasma CE-LIF 2011 *n* = 77
Healthy controls	94	16	16	125	140	20	90	51
Smokers with normal lung functions	78	14	15	—	—	—	—	—
Cystic fibrosis	27	—	5	—	—	—	52	—
AATD	37	6	5	91	58	—	32	—
COPD	45	9	9	107	91	14	50	26

Figure 2. The scheme indicates the number of samples analyzed per year in all different matrices in which DESs have been detected by MEKC and CE-LIF.

previously diagnosed COPD [67]. A remarkable implementation of the analytical procedures, in terms of sensitivity and specificity, was indeed the coupling of ESI-MS to the HPLC system. This platform drove the researchers to look forward and explore its potential also for the analysis of matrices other than urine. Once applied to sputum and plasma of AATD- and non-AATD-related COPD patients, this approach resulted in the production of very precise data which demonstrated that the amount of DESs in these matrices was significantly different between the two cohorts [68]. In addition, the data allowed to hypothesize that it could be possible to differentiate patients with COPD of various phenotypes based on the levels of excreted DESs. A LC-MS/MS method with selected reaction monitoring (SRM) of transition ions was also standardized to obtain an accurate measurement, in all body fluids, of DESs as biomarkers for in vivo measurement of tissue elastin degradation in man and animals [69]. The data showed an increase of total DESs in sputum and plasma of COPD patients over normal controls along with an increase of free DESs in urine of these patients. The suggestion that the total/free DESs ratio could be a possible parameter useful for studying the course of COPD and the response to therapy was also an interesting speculation. The measurement of DESs in plasma, urine and sputum of COPD patients was also used as a useful tool to demonstrate therapeutic effects of different pharmacological interventions aimed at reducing elastin degradation in this disorder [70, 71]. The precision of LC-MS data seemed to indicate that DESs might indeed have a role as potential biomarkers for evaluating therapeutic effects of any treatment. The mentioned accuracy of the method was even increased by the use of deuterated DES (DES-d_4) as internal standard in the LC-MS/MS analysis [72] and with the advent of ultra-performance liquid chromatography coupled to tandem mass spectrometry (UPLC-MS/MS) with selected reaction monitoring [73]. This was a further methodological improvement that allowed to measure the level of DESs in small amounts of urine from patients with lymphangioleiomyomatosis, an elastin degrading disorder that, similarly to COPD, affects lung tissues. The LC-MS/MS analysis of DESs with the multiple reaction monitoring (MRM) acquisition modes for monitoring the transitions of interest was applied to urine of COPD patients, and to urine and blood of patients with COPD and asthma. This approach evidenced that while the elevation of urinary DESs levels was associated with the exacerbation status in COPD patients, blood DESs levels were strongly associated with age and were negatively correlated with lung diffusing capacity for carbon monoxide [12, 74, 75]. From among the recently introduced high-sensitive techniques, a prominent position is also held by nano-LC-MS/MS. It has been applied to detect DESs in urine of: (i) COPD rapid decliners; (ii) COPD slow decliners; (iii) healthy smokers and healthy nonsmokers and also to detect hydroxyl-lysyl-pyridinoline and lysyl-pyridinoline as biomarkers for Chronic Graft-versus-Host disease [76, 77]. These latter studies got the conclusion that the chemotherapy treatment had significant effect on the turnover of elastin and collagen.

Investigations performed to check the agreement between LC-MS and MEKC-LIF demonstrated the compatibility of the two methods although the latter showed the tendency to overestimate DESs levels (likely due to the presence of interferents co-eluting with DESs peak) compared to the former [78]. Based on these results, the conclusion was drawn that, despite the advent of very sophisticated LC-MS/MS techniques, MEKC-LIF may still be considered a valuable method for assessment of DESs concentration in clinical investigations.

4. Clinical validity of desmosines as biomarkers of lung disorders

Although the recent technological advancements have made the measurement of DESs more common practice, the question of whether these cross-links are ready to be introduced into the biomarker "hall of fame" is still unanswered. In fact, it remains unclear whether these surrogate markers can predict effects or clinically relevant endpoints of pulmonary disorders. This is mainly due to a number of critical questions that, by making the clinical validity and utility of this assay controversial, need to be addressed. One of the major questions is the fact that elastic fibers, being present in many tissues including large elastic arteries and the dermis, are obviously not unique to lung interstitium. As a consequence, not necessarily an increase in elastin turnover could be primarily related to pulmonary diseases. Thus, the question arises whether increased DESs levels might be associated, for example, with accelerated elastin turn-over in the skin or major vessels. In this case, the use of DESs as markers of lung diseases will be inappropriate. However, the well-documented finding of DESs in sputum of individuals with pulmonary disorders and their good correlation with the lung conditions strongly supported the hypothesis that lung would be the major source of elastin cross-links in body fluids of these subjects. This being said, given that the analytical validation of a method is mandatory to guarantee the reliability of results, an equally important question is whether fully validated methods for DESs quantification are currently available. The rationale is that data of inadequate quality may lead to inaccurate patient monitoring and incorrect conclusions in clinical studies or over-/under-estimation of new drug effects. The immunochemical tools for testing DESs, while being able to differentiate patients with lung disorders from healthy controls, evidenced a number of drawbacks. First of all, the matrix analyzed for their determination was mostly urine, a fluid not devoid of potential limits. These included: (i) the use, in most cases, of single-point urine samples which are less representative than urine collected over 24 h; (ii) the variability over time of urinary excretion of DESs in individuals with pulmonary disorders and (iii) the effect (never taken into account) that decreased renal function might have on DESs excretion into urine. Moreover, based on the absolute values determined (which were much higher than those observed with amino acid analysis or HPLC), immunochemical methods appeared to overestimate DESs, most likely because of the presence in urine of cross-reacting substances. The obvious consequence of such spurious elevation was the masking of important differences between controls and patients. Last but not least, the low size of individuals analyzed for each set of experiments was another limitation to be reckoned with. Thus, in spite of the efforts made to refine the data, it appeared clear that these methods, while being a tool for the measurement of urinary DESs, were not sufficiently powerful for pointing to these cross-links as biomarkers helpful for clinical use. The horizon was widened by the methodological progresses achieved with the advent of HPLC and CE that allowed DESs to be detected in a larger variety of human fluids. In particular, a tremendous improvement in data accuracy and reliability came from the combination of direct DESs analysis by CE-LIF, with an increase in sample size. This method of analysis had apparently the potential to provide important information in the understanding of the pathogenesis of pulmonary disorders in which degradation of lung elastin is believed to be an ongoing part of the disease process. In addition, at least in part, it fulfilled its goal. The improved consistency of results allowed incontrovertible proofs of DESs differences existing among different groups of subjects or different clinical phenotypes within the group of patients affected by the same disease to be evidenced. However, since the effect of confounding factors (including gender, age and body mass index) which influence urinary DESs excretion was not considered, the clinical utility and validity of urinary DESs measurements remained unproven.

The application of new experimental strategies promoted by the rapid state of flux of this field, while allowing to dig deeper into the catabolism of elastin through the quantification of more and more minute amounts of DESs, did not answer the question of whether these cross-links could be considered surrogate markers of pulmonary diseases, in particular of COPD. Moreover, the lack of consensus on what should be quantified (free or total DESs, or DES + IDES) and in which biological fluid (plasma, sputum and urine) might obscure the view of these cross-links as reliable biomarkers for COPD diagnosis or prognosis. Nevertheless, the standardized methodologies already developed, together with the implementation of sample size and the taking into account of possible confounding factors, seem to indicate that DESs are mature to be addressed as candidate for becoming in the near future the "gold standard" for the study of COPD. Another important aspect of DESs validation is the longitudinal behavior and the relationship with progression and severity of the disease. Large longitudinal studies are necessary to confirm their predictive power for patients' clinical outcome. Indeed, these studies would add to the understanding of whether, besides their association with COPD in cross-sectional studies, DESs could be related to FEV1 decline and to the worsening of diffusing capacity in longitudinal cases, and perhaps to changes in lung CT scan densitometry. This would certainly confirm their capacity for monitoring progression of disease severity and response in effective interventional trials. In this context, something has already been done. After adjustment for age, sex, height, body mass index, and smoking status convincing evidence has been gained that, while urinary DESs had a significant association with several lung function parameters (FEV1, FVC, RV, RV/TLC and DL, CO), plasma DESs correlated with FEV1, DL, and CO only. These correlations were much more pronounced in COPD subjects than in individuals without COPD. Of great interest was the finding that DESs can be independently influenced by a number of factors after adequately correcting for risk factors to avoid confounding results.

Thus, what has emerged from the scientific literature over the course of these years is that we are on the right path to utilizing these cross-links as valid tools or "biomarkers" in the differential diagnosis and clinical management of pulmonary diseases, COPD in particular. It remains to be seen whether DESs measurement could have an evidence-based role in stratifying patients for specific treatment or prognosis.

Acknowledgements

The authors are deeply grateful to Dr. Maurizio Luisetti, a great scientist and a good friend, for introducing them into the interesting world of pulmonary pathology.

Dr. Luisetti passed away on 2014, October the 20th, and all of us have missed him a lot.

http://dx.doi.org/10.5772/intechopen.68570

Appendices and nomenclatures

COPD	Chronic obstructive pulmonary disease
DESs	Desmosines
IDES	Isodesmosines
HNE	Elastase
AAT	α1-Antitrypsin
CF	Cystic fibrosis
HPLC	High-performance liquid chromatography
CE	Capillary electrophoresis
CZE	Capillary zone electrophoresis
MEKC	Micellar electrokinectic chromatography
LC-MS	Liquid chromatography-mass spectrometry
UPLC-MS/MS	Ultra-performance liquid chromatography coupled to tandem mass spectrometry

Author details

Simona Viglio[1], Monica Di Venere[1], Maddalena Cagnone[1], Marco Fumagalli[2] and Paolo Iadarola[2]*

*Address all correspondence to: piadarol@unipv.it

1 Department of Molecular Medicine, Biochemistry Unit, University of Pavia, Pavia, Italy

2 Department of Biology and Biotechnologies, Biochemistry Unit, University of Pavia, Pavia, Italy

References

[1] Foster JA, Curtiss SW. The mechanism of lung elastin synthesis. American Journal of Physiology. 1990;**259**:13–23

[2] Akagawa M, Suyama K. Mechanism of formation of elastin crosslinks. Connective Tissue Research. 2000;**41**(2):131–141

[3] Viglio S, Zanaboni G, Luisetti M, Trisolini R, Grimm R, Cetta G, Iadarola P. Micellar electrokinetic chromatography for the determination of urinary desmosine and isodesmosine in patients affected by chronic obstructive pulmonary disease. Journal of Chromatography B. 1998;**714**:87–89

[4] Viglio S, Annovazzi L, Luisetti M, Stolk J, Casado B, Iadarola P. Progress in the methodological strategies for the detection in real samples of desmosine and isodesmosine, two biological markers of elastin degradation. Journal of Separation Science. 2007;**30**(2):202–213

[5] Starcher BC. Elastin and the lung. Thorax. 1986;**41**(8):577–585

[6] Luisetti M, Ma S, Iadarola P, Stone PJ, Viglio S, Casado B, Lin YY, Snider GL, Turino GM. Desmosine as a biomarker of elastin degradation in COPD: Current status and future directions. European Respiratory Society. 2008;**32**(5):1–12. DOI: 10.1183/09031936.00174807

[7] Barnes PJ. Chronic obstuctive pulmonary disease. The New England Journal of Medicine. 2000;**343**(4):269–280. DOI: 10.1056/NEJM200007273430407

[8] Shapiro SD, Ingenito EP. The pathogenesis of chronic obstructive pulmonary disease: Advances in the past 100 years. American Journal of Respiratory Cell and Molecular Biology. 2005;**32**(5):367–372. DOI: 10.1165/rcmb.F296

[9] Snider GL. Emphysema: The first two centuries—and beyond. A historical overview, with suggestions of future research: Part 2. American Review of Respiratory Disease. 1992;**146**(6):1615–1622. DOI: 10.1164/ajrccm/146.6.1615

[10] Barnes PJ, Chowdhury B, Kharitonov SA, Magnussen H, Page CP, Postma D, Saetta M. Pulmonary biomarkers in chronic obstructive pulmonary disease. American Journal of Respiratory and Critical Care Medicine. 2006;**174**(1):6–14. DOI: 10.1164/rccm.200510-1659PP

[11] Luisetti M, Stolk J, Iadarola P. Desmosine, a biomarker for COPD: Old and in the way. European Respiratory Society. 2012;**39**(4):797–798. DOI: 10.1183/09031936.00172911

[12] Huang JT, Chaudhuri R, Albarbarawi O, Barton A, Grierson C, Rauchhaus P, Weir CJ, Messow M, Stevens N, McSharry C, Feuerstein G, Mukhopadhyay S, Brady J, Palmer CN, Miller D, Thomson NC. Clinical validity of plasma and urinary desmosine as biomarkers for chronic obstructive pulmonary disease. Thorax. 2012;**67**(6):502–508. DOI: 10.1136/thoraxjnl-2011-200279

[13] Fiorenza D, Viglio S, Lupi A, Baccheschi J, Tinelli C, Trisolini R, Iadarola P, Luisetti M, Snider GL. Urinary desomosine excretion in acute exacerbations of COPD: A preliminary report. Respiratory Medicine. 2002;**96**(2):110–114

[14] Viglio S, Iadarola P, Lupi A, Trisolini R, Tinelli C, Balbi B, Grassi V, Worlitzsch D, Doring G, Meloni F, Meyer KC, Dowson L, Hill SL, Stockley RA, Luisetti M. MECK of desmosine and isodesmosine in urine of chronic destructive lung disease patients. European Respiratory Society. 2000;**15**(6):1039–1045

[15] Lindberg CA, Engstrom G, De Verdier MG, Nihlén U, Anderson M, Forsman-Semb K, Svartengren M. Total desmosines in plasma and urine correlate with lung function. European Respiratory Society. 2012;**39**(10):839–845. DOI: 10.1183/09031936.00064611

[16] Fregonese L, Ferrari F, Fumagalli M, Luisetti M, Stolk J, Iadarola P. Long-term variability of desmosine/isodesmosine as biomarker in alpha-1-antitrypsin deficiency-related COPD. Journal of Chronic Obstructive Pulmonary Disease. 2011;**8**(5):329–333. DOI: 10.3109/15412555.2011.589871

[17] Stolk J, Veldhuisen B, Annovazzi L, Zanone C, Versteeg EM, van Kuppevelt TH, Berden JH, Nieuwenhuzen W, Iadarola P, Luisetti M. Short-term variability of biomarkers of proteinase activity in patients with emphysema associated with type Z alpha-1-antitrypsin deficiency. Respiratory Research. 2005;**6**:47. DOI: 10.1186/1465-9921-6-47

[18] Viglio S, Stolk J, Luisetti M, Ferrari F, Piccinini P, Iadarola P. From micellar electrokinetic chromatography to liquid chromatography-mass spectrometry: Revisiting the way of analyzing human fluids for the search of desmosines, putative biomarkers of chronic obstructive pulmonary disease. Electrophoresis. 2014;**51**(1):109–118. DOI: 10.1002/elps.201300159

[19] Iadarola P, Luisetti M. The role of desmosines as biomarkers for chronic as biomarkers for chronic obstructive pulmonary disease. Expert Review of Respiratory Medicine. 2013;**7**(2):137–144. DOI: 10.1586/ers

[20] Francis G, Jhon R, Thomas J. Biosynthetic pathway of desmosines in elastin. Biochemical Journal. 1973;**136**(1):45–55

[21] Partridge SM, Elsden DF, Thomas J. Constitution of cross-linkages in elastin. Nature. 1963;**197**:1297–1298

[22] Partridge SM, Elsden DF, Thomas J, Dorfman A, Tesler A, Ho PL. Biosynthesis of the desmosine and isodesmosine cross-bridges in elastin. Biochemical Journal. 1964;**93**(3):30–33

[23] Miller EJ, Martin GR, Piez KA. The utilization of lysine in the biosynthetic of elastin crosslinks. Biochemical and Biophysical Research Communications. 1964;**17**(3):248–253

[24] Anwar RA. Comparison of elastins from various sources. Canadian Journal of Biochemistry. 1966;**44**(6):725–734

[25] Corbin KW. Ion-exchange column chromatography of desmosine and isodesmosine. Analytical Biochemistry. 1969;**32**(1):118–121

[26] Starcher BC. Determination of elastin content of tissues by measuring desmosine and isodesmosine. Analytical Biochemistry. 1977;**79**:11–15

[27] Goldstein RA, Starcher BC. Urinary excretion of elastin peptides containing desmosine after intratracheal injection of elastase in hamsters. Journal of Clinical Investigation. 1978;**61**(5):1286–1290. DOI: 10.1172/JC*109045*

[28] Mecham RP, Foster JA. A structural model for desmosine cross-linked peptides. Biochemical Journal. 1978;**173**(2):617–625

[29] Lonky SA, Gochman N, Smith S, Bergeron-Lynn G, Jacobs K. Amino acid analysis of elastin hydrolysates using a lithium citrate gradient: Quantification of elastin from whole lung. Clinica Chimica Acta. 1981;**110**(2–3):227–233

[30] Covault HP, Lubrano T, Dietz AA, Rubinstein HM. Liquid-chromatographic measurement of elastin. Clinical Chemistry. 1982;**28**(7):1465–1468

[31] Velebny V, Kasafirek E, Kanta J. Desmosine and isodesmosine contents and elastese activity in normal and cirrhotic rat liver. Biochemical Journal. 1983;**214**(3):1023–1025

[32] Zarkadas CG, Zarkadas GC, Karatzas CN, Khalili AD, Nguyen Q. Rapid method for determining desmosine, isodesmosine, 5-hydroxylysine, tryptophan, lysinoalanine and the amino sugars in proteins and tissues. Journal of Chromatography. 1986;**378**(1): 67–76

[33] Gunja-Smith Z, Boucek RJ. Collagen cross-linking compounds in human urine. Biochemical Journal. 1981;**197**(3):759–762

[34] Bruce MC, Poncz L, Klinger JD, Stern RC, Tomashefski JF Jr, Dearborn DG. Biochemical and pathologic evidence for proteolytic destruction of lung connective tissue in cystic fribrosis. American Review of Respiratory Disease. 1985;**132**(3):529–535. DOI: 10.1164/arrd.1985.132.3.529

[35] Starcher BC, Mecham RP. Desmosine radioimmunoassay as a means of studying elastogenesis in cell culture. Connective Tissue Research. 1981;**8**(3–4):255–258

[36] Kuhn C, Engleman W, Chraplyvy M, Starcher BC. Degradation of elastin in experimental elastase-induced emphysema measured by a radioimmuneassay for desmosine. Experimental Lung Research. 1983;**5**(2):115–123

[37] Starcher BC, Green M, Scott M. Measurement of urinary desmosine as an indicator of acute pulmonary disease. Connective Tissue Research. 1995;**62**(2):252–257

[38] Harel S, Janoff A, Yu SY, Hurewitz A, Bergofsky EH. Desmosine radioimmunoassay for measuring elastin degradation in vivo. American Review of Respiratory Disease. 1980;**122**(5):769–773. DOI: 10.1164/arrd.1980.122.5.769

[39] Harel S, Yu SY, Janoff A, Hurewitz A, Bergofsky EH. Measurement of elastin degradation in vivo by desmosine radioimmunoassay. Bulletin Europeen de Physiopathologie Respiratoire. 1980;**16**:75–82

[40] Pelham F, Wewers M, Crystal R, Buist AS, Janoff A. Urinary excration of desmosine (elastin cross-links) in subjects with PiZZ alpha-1-antitrypsin deficiency, a phenotype associated with hereditary predisposition to pulmonary emphysema. American Review of Respiratory Disease. 1985;**132**(1):821–823. DOI: 10.1164/arrd.1985.132.4.821

[41] Tenholder MF, Rajagopal KR, Phillips YY, Dillard TA, Bennett LL, Mundie TG, Tellis CJ. Urinary desmosine excretion as a marker of lung injury in the adult respiratory distress syndrome. Chest. 1991;**100**(5):1385–1390

[42] Fill JA, Brandt JT, Wiedemann HP, Rinehart BL, Lindemann CF, Komara JJ, Bowsher RR, Spence MC, Zeiher BG. Urinary desmosine as a biomarker in acute lung injury. Biomarkers. 2006;**11**(1):85–96. DOI: 10.1080/13547500500343225

[43] Desai R, Wigglesworth JS, Aber V. Assessment of elastin maturation by radioimmunoassay of desmosine in the developing human lung. Early Human Development. 1988;**16**(1):61–71

[44] Matsumoto T, Mizusaki S, Nishimura K, Oshima S. An enzyme-linked immunosorbent assay for desmosine. Bulletin of the Chest Disease Research Institute, Kyoto University. 1986;**19**(1–2):9–22

[45] Verplanke AJ, Watanabe T, Ishimori K, Matsuki H, Kasuga H. Experience with an enzyme-linked immuno sorbent assay for the quantitation of urinary desmosine. Tokai Journal of Experimental and Clinical Medicine. 1988;**13**(3):159–163

[46] Laurent P, Magne L, De Palmas J, Bignon J, Jaurand MC. Quantitation of elastin in human urine and rat pleural mesothelial cell matrix by a sensitive avidin-biotin ELISA for desmosine. Journal of Immunological Methods. 1988;**107**(1):1–11

[47] Osakabe T, Seyama Y, Yamashita S. Comparison of ELISA and HPLC for the determination of desmosine or isodesmosine in aortic tissue elastin. Journal of Clinical Laboratory Analysis. 1995;**9**(5):293–296

[48] Cocci F, Miniati M, Monti S, Cavarra E, Gambelli F, Battolla L, Lucattelli M, Lungarella G. Urinary desmosine excretion is inversely correlated with the extent of emphysema in patients with chronic obstructine pulmonary disease. The International Journal of Biochemistry & Cell Biology. 2002;**34**(6):594–604

[49] Faris B, Ferrera R, Glembourt T, Mogayzel PJ Jr, Crombie G, Franzblau C. Rapid quantitation of desmosine content in tissue hydrolysates by high-performance liquid chromatography. Analytical Biochemistry. 1981;**114**(1):71–74. DOI: 10.1016/0003-2697(81)90453-X

[50] Soskel NT. High-performance liquid chromatographic quantitation of desmosine plus isodesmosine in elastin and whole tissue hydrolysates. Analytical Biochemistry. 1987;**160**(1):98–104

[51] Fujimoto D. Aging and cross-linking in human aorta. Biochemical and Biophysical Research Communications. 1982;**109**(4):1264–1269

[52] Chen JR, Takahashi M, Kushida K, Suzuki M, Suzuki K, Horiuchi K, Nagano A. Direct detection of crosslinks of collagen and elastin in the hydrolysates of human yellow ligament using single-column high performance liquid chromatography. Analytical Biochemistry. 2002;**278**(2):99–105. DOI: 10.1006/abio.1999.4412

[53] Watanabe M, Sawai T, Nagura H, Suyama K. Age-related alteration of cross-linking amino acids of elastin in human aorta. The Tohoku Journal of Experimental Medicine. 1996;**180**(2):115–130

[54] Hanis TL, Deyl Z, Struzinsky R, Miksik I. Separation of elastin cross-links as phenylisothiocyanate derivatives. Journal of Chromatography. 1991;**553**(1–2):93–99

[55] Nakamura F, Suyama K. Silica gel high-performance liquid chromatography for the determination of cross-links in elastin. Journal of Chromatography Science. 1991;**29**(5): 217–220

[56] Umeda H, Aikawa M, Libby P. Liberation of desmosine and isodesmosine as amino acids from insoluble elastin by elastolytic proteases. Biochemical and Biophysical Research Communications. 2011;**411**(2):281–286. DOI: 10.1016/j.bbrc.2011.06.124

[57] Venturi M, Bonavina L, Annoni F, Colombo L, Butera C, Peracchia A, Mussini E. Biochemical assay of collagen and elastin in the normal and varicose vein wall. Journal of Surgical Research. 1996;**60**(1):245–248. DOI: 10.1006/jsre.1996.0038

[58] Venturi M, Bonavina L, Colombo L, Antoniazzi L, Bruno A, Mussini E, Peracchia A. Biochemical markers of upper esophageal sphincter compliance in patietns with Zenker's diverticulum. Journal of Surgical Research. 1997;**70**(1):46–48. DOI: 10.1006/jsre.1997.5049

[59] Gottlieb DJ, Stone PJ, Sparrow D, Gale ME, Weiss ST, Snider GL, O'Connor GT. Urinary desmosine excration in smokers with and without rapid decline of lung function: The Normative Aging Study. American Journal of Respiratory and Critical Care Medicine. 1996;**154**(2):1290–1295. DOI: 10.1164/ajrcem.154.5.8912738

[60] Stone PJ, Beiser A, Gottlieb DJ. Circadian variation of urinary excretion of elastin and collagen crosslinks. Proceedings of the Society for Experimental Biology and Medicine. 1998;**218**(3):229–233

[61] Keller S, Turino GM, Mandl L. Separation of elastin components by thin layer chromatography and electrophoresis. Connective Tissue Research. 1981;**8**(3–4):251–254

[62] Cantor JO, Osman M, Keller S, Cerreta JM, Mandl I, Turino GM. Measurement of cross-linked elastin syntesis in bleomycin-induced pulmonary fibrosis using a highly sensitive assay for desmosine and isodesmosine. Journal of Laboratory and Clinical Medicine. 1984;**103**(3):389–392

[63] Giummelly P, Botton B, Friot R, Prima-Putra D, Atkinson JJ. Measurement of desmosine and isodesmosine by capillary zone electrophoresis. Journal of Chromatography A. 1995;**710**(2):357–360

[64] Norman P, Moss I, Sian M, Gosling M, Powell J. Maternal and postnatal vitamin D ingestion influences rat aortic structure, function and elastin content. Cardiovascular Research. 2002;**55**(2):369–374. DOI: 10.1016-50008-6363(02)00444-3

[65] Urlich M, Worlitzsch D, Viglio S, Siegmann N, Iadarola P, Shute JK, Geiser M, Pier GB, Friedel G, Barr ML, Schuster A, Meyer KC, Ratjen F, Bjarnsholt T, Gulbins E, Doring G. Alveolar inflammation in cystic fibrosis. Journal of Cystic Fibrosis. 2010;**9**(3):217–227. DOI: 10.1016/j.jfc.2010.03.001

[66] Annovazzi L, Viglio S, Perani E, Luisetti M, Baraniuk J, Casado B, Cetta G, Iadarola P. Capillary electrophoresis with laser-induced fluorescence detection as a novel sensitive approach for the analysis of desmosines in real samples. Electrophoresis. 2004;**25**(4–5):683–691. DOI: 10.1002/elps.200305607

[67] Ma S, Lieberman S, Turino GM, Lin YY. The detection and quantitation of free desmosine and isodesmosine in human urine and their peptide-bound forms in sputum. Proceedings of the National Academy of Sciences USA. 2003;**100**(22):12941–12943. DOI: 10.1073/pnas.2235344100

[68] Ma S, Lin YY, Turino GM. Measurements of Desmosines and Isodesmosines by Mass Spectrometry in COPD. Chest. 2007;**13**(5):1363–1371. DOI: 10.1378/chest.06-2251

[69] Ma S, Turino GM, Lin YY. Quantitation of desmosine and isodesmosine in urine, plasma and sputum by LC-MS/MS as biomarkers for elastin degradation. Journal of Chromatography B, Analytical Technologies in the Biomedical and Life Sciences. 2011;**879**(21):1893–1898. DOI: 10.1016/j.jchromb

[70] Ma S, Lin YY, Tartell L, Turino GM. The effect of titotopium therapy on markers of elastin degradation in COPD. Respiratory Research. 2009;**10**:12. DOI: 10.1186/1465-9921-10-12

[71] Ma S, Lin YY, J He J, Rouhani FN, Brantly M, Turino GM. Alpha-1-antitrypsin augmentation therapy and biomarkers of elastin degradation. Journal of Chronic Obstructive Pulmonary Disease. 2013;**10**(4):1–9. DOI: 10.3109/15412555.2013.771163

[72] Ma S, Turino GM, Hayashi T, Yanuma H, Usuki T, Lin YY. Stable deuterium internal standard for the isotope-diluition LC-MS/MS analysis of elastin degradation. Analytical Biochemistry. 2013;**440**(2):158–165. DOI: 10.1016/j.ab.2013.05.014

[73] Shiraishi K, Matsuzaki K, Matsumoto A, Hashimoto Y, Iba K. Development of a robust LC-MS/MS method for determination of desmosine and isodesmosine in human urine. Journal of Oleo Science. 2010;**59**(8):431–439

[74] Albarbarawi O, Barton A, Lin Z, Takahashi E, Buddharaju A, Brady J, Miller D, Palmer CN, Huang JT. Measurement of urinary total desosine and isodesmosine using isotope-dilution liquid chromagraphy-tandem mass spectrometry. Analytical Biochemistry. 2010;**82**(9):3745–3750. DOI: 10.1021/ac100152f

[75] Ohgay S, Sikma M, Horvatovitch P, Heimans J, Miller BE, ten Hacken MHT, Bischoff R. Free urinary desmosine and isodesmosine as COPD markers: The revelance of counfonding factors. Chronic Obstructive Pulmonary Disease. 2016;**3**:560–569. DOI: 10.15326/jcopf.3.2.2015.019

[76] Boutin M, Berthelette C, Gervais FG, Scholand MB, Hoidal J, Leppert MF, Bateman KP, Thibault P. High-sensitivity nanoLC-MS/MS Analysis of urinary Desmosine and Isodesmosine. Analytical Biochemistry. 2009;**81**(5):1881–1887. DOI: 10.1021/ac801745d

[77] Boutin M, Ahmad I, Jauhiainen M, Lachapelle N, Rondeau C, Roy J, Thibault P. NanoLC-MS/MS analysis of urinary desmosine, hydroxylysylpyridinoline and lysylpyridinoline as biomarkers for chronic graft-versus-host disease. Analytical Biochemistry. 2009;**81**(22):9454–9461. DOI: 10.1021/ac9018796

[78] Ferrari F, Fumagalli M, Piccinini P, Stolk J, Luisetti M, Viglio S, Tinelli C, Iadarola P. Micellar electrokinetic chromatography with laser induced detection and liquid chromatography tandem mass-spectrometry-based desmosine assays in urine of patients with Chronic Obstructive Pulmonary Disease: A comparative analysis. Journal of Chromatography A. 2012;**1266**:103–109. DOI: 10.1016/j.chroma.2012.10.014

Stereoselective Synthesis of α-Aminophosphonic Acids through Pudovik and Kabachnik-Fields Reaction

Mario Ordóñez, José Luis Viveros-Ceballos and
Iván Romero-Estudillo

Additional information is available at the end of the chapter

http://dx.doi.org/10.5772/intechopen.68707

Abstract

Representative examples concerning the Pudovik and Kabachnik-Fields reactions as the main strategies for the stereoselective synthesis of α-aminophosphonic acids are discussed, classifying these reactions according to the chiral auxiliary and chiral catalyst.

Keywords: α-aminophosphonic acids, α-aminophosphonates, stereoselective synthesis, Pudovik and Kabachnik-Fields

1. Introduction

Optically active α-aminophosphonic acids are the most important analogs of α-amino acids, which are obtained by isosteric substitution of the planar and less bulky carboxylic acid (CO_2H) by a sterically more demanding tetrahedral phosphonic acid functionality (PO_3H_2). Several α-aminophosphonic acids have been isolated from natural sources, either as free amino acids or as constituents of more complex molecules [1], such as the phosphonotripeptide K-26 (**Figure 1**) [2].

The α-aminophosphonic acids, α-aminophosphonates, and phosphonopeptides are currently receiving significant attention in organic synthesis and medicinal chemistry as well as in agriculture, due to their biological and pharmacological properties. Additionally, the α-aminophosphonic acids are used as key synthetic intermediates in the synthesis of phosphonic acids, phosphonamides, and phosphinates, which not only play an important role as protease inhibitors but also in the wide range of biochemical pathways (**Scheme 1**) [3].

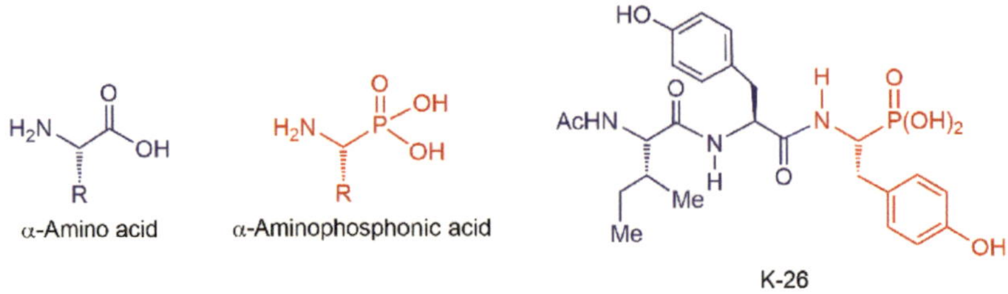

Figure 1. α-Aminophosphonic acid analogs of α-amino acids.

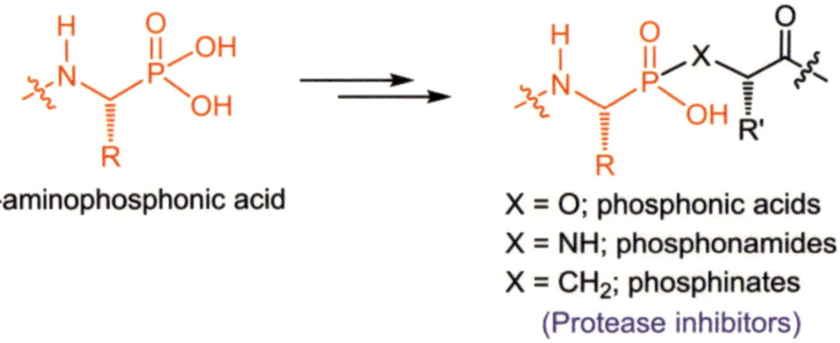

Scheme 1.

The inhibitory activity of the α-aminophosphonic acids and their derivatives has been attributed to the tetrahedral geometry of the substituents around the phosphonic moiety mimicking the tetrahedral high-energy transition state of the peptide bond hydrolysis, favoring the inhibition of a broad spectrum of proteases and ligases (**Scheme 2**) [4].

Furthermore, it is well known that the biological activity of the α-aminophosphonic acids and derivatives depends on the absolute configuration of the stereogenic α-carbon to phosphorous [5]. For example, (*R*)-phospholeucine is a more potent inhibitor of leucine aminopeptidase than the (*S*)-phospholeucine [6], and (*S*,*R*)-alaphosphalin shows higher antibacterial activity against both Gram-positive and Gram-negative microorganisms than the other three diastereoisomers [7]. Additionally, the L-Pro-L-Leu-L-Trp^{P} tripeptide acts as an MMP-8 enzyme inhibitor, wherein the peptide responsible for the biological activity is that in which the three amino acids have L configuration (**Figure 2**) [8].

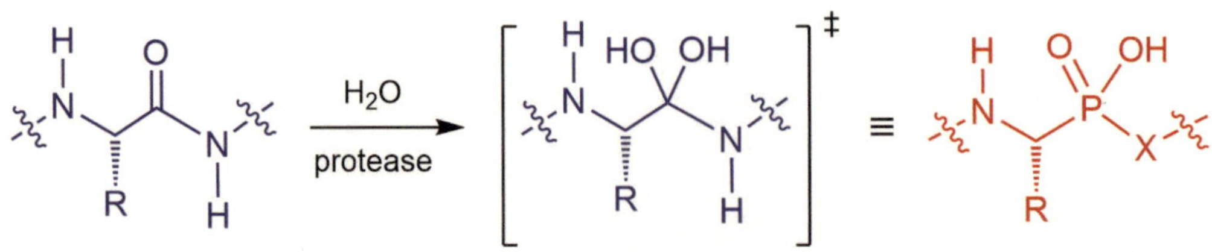

Scheme 2.

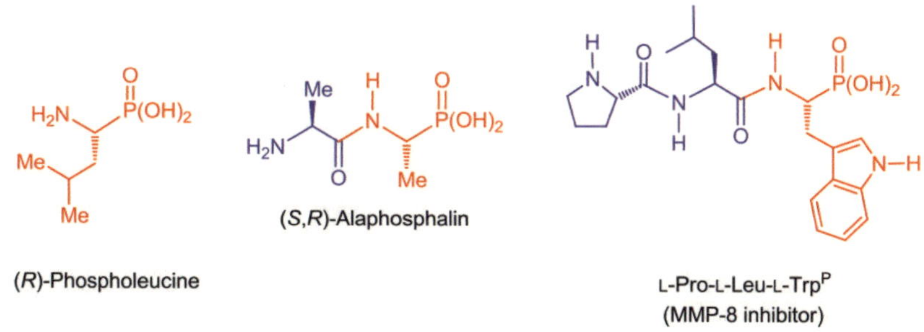

Figure 2. Importance of the chirality of the α-aminophosphonic acids.

In view of the different biological and chemical applications of the α-aminophosphonic acids, nowadays the development of suitable synthetic methodologies for their preparation in optically pure form is a topic of great interest and many reviews have been recently published concerning their stereoselective synthesis [9]. In this context, Pudovik and Kabachnik-Fields reactions the main synthetic strategies for the stereoselective synthesis of α-aminophosphonic acids will be described in this chapter.

2. Stereoselective C-P bond formation (Pudovik methodology)

The diastereoselective and enantioselective hydrophosphonylation of aldimines and ketimines, called as the Pudovik reaction, involves the addition of a phosphorus nucleophile agent over the corresponding imine, in such a way that one or both of the reactants can incorporate a chiral auxiliary or nonchiral reagents may be reacted in the presence of a chiral catalyst (**Scheme 3**).

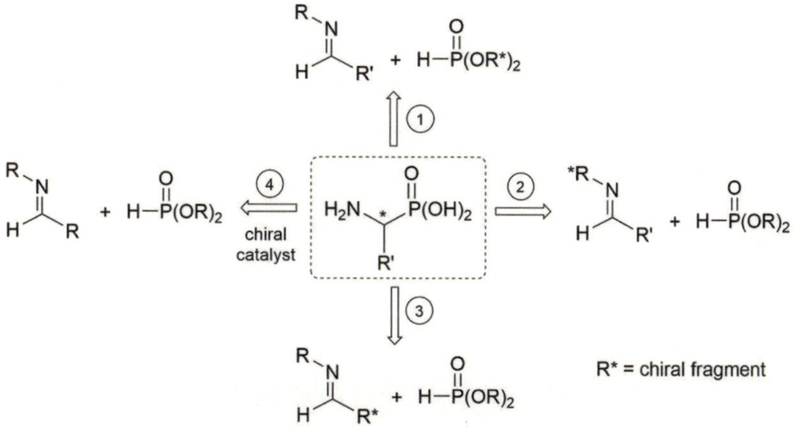

Scheme 3. Diastereo- and enantio-selective synthesis of α-aminophosphonic acids by Pudovik methodology.

2.1. Chiral phosphorus compounds

One of the general methods for the synthesis of α-aminophosphonic acids involves the diastereoselective hydrophosphonylation of achiral imines with chiral phosphites to introduce the phosphonate function, which by hydrolysis afforded the optically enriched α-aminophosphonic acids. For example, the nucleophilic addition of chiral C_3-symmetric trialkyl phosphite **2**, obtained from the naturally occurring (1*R*,2*S*,5*R*)-(–)-menthol to the aldimine **1** in the presence of trimethylsilyl chloride (TMSCl) as an activator, provided the α-aminophosphonate **3** in 60% yield and moderate induction at the α-carbon atom (50% diastereoisomeric excess), which by hydrolysis with HCl in dioxane, followed by catalytic hydrogenolysis using Pd/C, produced the (*R*)-phosphophenyl glycine **4** in 70% yield and with 95% enantiomeric excess (**Scheme 4**) [10].

Palacios et al. [11] proposed also the chiral cyclic (R,R)- α,α,α′,α′-tetraphenyl-2,2-disubstituted 1,3-dioxolane-4,5-dimethanol (TADDOL) phosphite **5**, derived from natural tartaric acid, as a suitable phosphorus nucleophile in the stereoselective synthesis of α-aminophosphonic acids. In this context, the diastereoselective hydrophosphonylation reaction of *N*-diphenylphosphinoyl aldimines **6a,b** with (*R,R*)-TADDOL-derived phosphite **5** in the presence of $ZnEt_2$ and *N,N,N′,N′*-tetramethylethylenediamine (TMEDA) in tetrahydrofuran (THF) at −80°C afforded the α-aminophosphonates **7a,b** in good yields and diastereoselectivities. Finally, the simultaneous hydrolysis of (*R,R*)-TADDOL phosphonate and diphenylphosphinoyl groups in the diastereoisomerically pure **7a,b** with 4 *N* HCl, led to the optically pure (*R*)-α-aminophosphonic acids hydrochlorides (*R*)-**8a,b** in 77 and 82% yield, respectively (**Scheme 5**).

Additionally, the (*R,R*)-TADDOL framework has also proved its usefulness as a chiral auxiliary in the diastereoselective addition of Grignard reagents to chiral α-aminophosphonates. Thus, nucleophilic addition of chiral phosphite (*R,R*)-**5** to *N*-tosylbenzaldimine **9** in the presence of Et_3N in toluene, afforded the α-aminophosphonate **10** in 93% and 77:23 diastereoisomeric ratio, which by oxidation and by treatment with trichloroisocyanuric acid (TCCA) and poly(4-vinylpyridine), gave the α-ketiminophoshonate **11** in 82% yield. Addition of methylmagnesium bromide to **11**, furnished the quaternary α-aminophoshonate **12** in good yield and 94:6 diastereoisomeric ratio, which by hydrolysis with 10 M HCl, produced the optically enriched (*S*)-α-aminophosphonic acid **13** in 80% yield (**Scheme 6**) [12].

The Pudovik reaction has also been reported incorporating the chiral auxiliary attached not only to the phosphite residue, but also to the imine fragment. As a proof of concept, Olszewski and

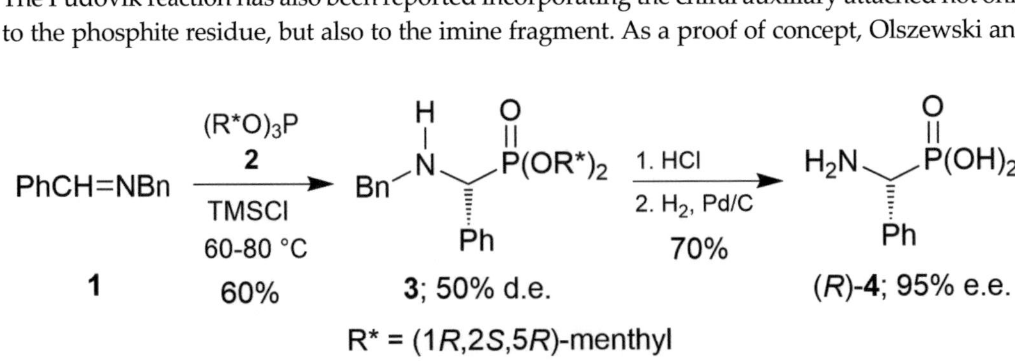

Scheme 4.

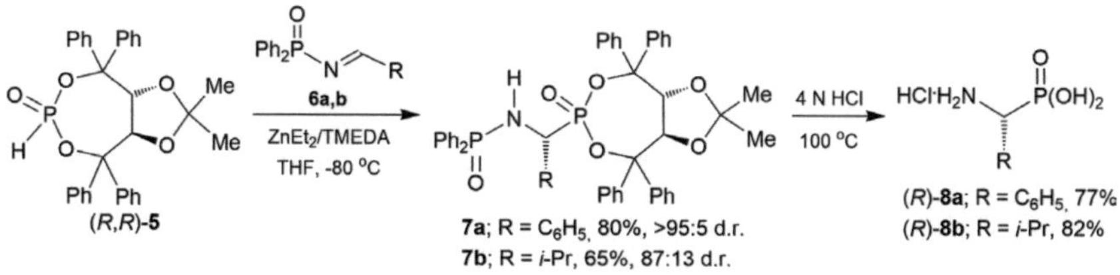

Scheme 5.

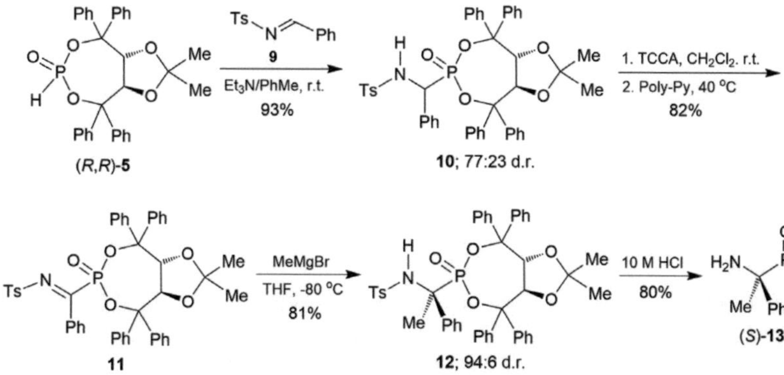

Scheme 6.

Majewski [13] reported the hydrophosphonylation reaction of (*S*)-*N*-*tert*-butylsulfinylaldimines **14a-d** with readily available chiral (*R*,*R*)-TADDOL phosphite **5** in the presence of potassium carbonate in CH_2Cl_2 at room temperature, obtaining the α-aminophosphonates **15a-d** in 80–87% yield and diastereoisomeric ratio (>95:5 d.r.). Simultaneous removal of both chiral auxiliaries in **15a-d** by hydrolysis with 4 M HCl at 100°C, produced the (*R*)-α-aminophosphonic acids **4, 16a-c** in 78–92% yield (**Scheme 7**).

2.2. Imines from chiral carbonyl compounds

The hydrophosphonylation of chiral Schiff bases is another general method for the synthesis of optically enriched α-aminophosphonates, which can be performed by addition of alkyl phosphites to chiral imines readily obtained by condensation of chiral aldehydes with nonchiral

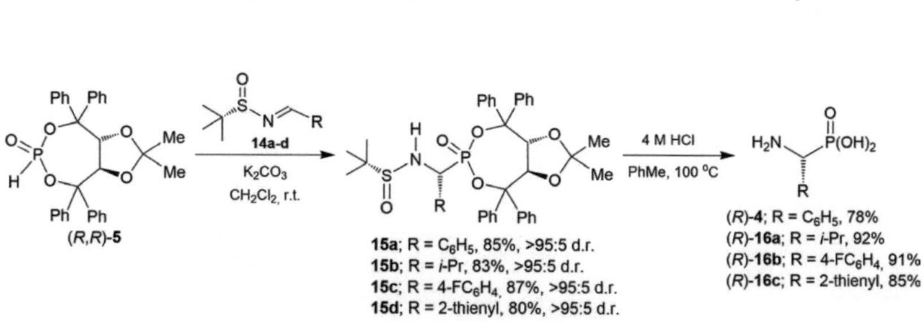

Scheme 7.

amines. In this context, Bongini et al. [14] carried out the synthesis of (*S*,*S*)-phosphothreonine **19** through nucleophilic addition of trimethylsilyl diethyl phosphite to the chiral imine (*S*)-**17**, obtained by condensation of (*S*)-2-triisopropylsilyloxy lactaldehyde and *N*-trimethylsilyl amine. The reaction proceed in excellent way to give the β-silyloxy-α-aminophosphonate (1*S*,2*S*)-**18** in 85% yield and >98:2 *syn*/*anti* diastereoisomeric ratio. Cleavage of the *O*-$SiMe_3$ bond and hydrolysis of the diethyl phosphonate in (*S*,*S*)-**18** with 6 N HCl provided the (1*S*,2*S*)-phosphothreonine **19** in quantitative yield. Under identical conditions, the (1*R*,2*R*)-phosphothreonine **19** was obtained starting from the enantiopure aldimine (*R*)-**17** (**Scheme 8**).

2.3. Imines from chiral amino compounds

On the other hand, the stereoselective hydrophosphonylation of chiral Schiff bases can also be conducted by addition of alkyl phosphites to chiral imines readily obtained by condensation of nonchiral aldehydes with chiral amines. For example, the nucleophilic addition of dimethyl phosphite to the imine (*S*)-**20**, readily obtained from the condensation of isobutyraldehyde and (*S*)-α-methylbenzylamine at 140°C, under solvent-free conditions, afforded the α-aminophosphonates (*R*,*S*)-**21** and (*S*,*S*)-**22** with a 85:15 diastereoisomeric ratio. Hydrolysis of the phosphonates in (*R*,*S*)-**21** and (*S*,*S*)-**22** followed by separation and hydrogenolysis using $Pd(OH)_2/C$ afforded the (*R*)-Val^P **16a** in 65–70% yield. The (*S*)-Val^P **16a** was obtained also from (*R*)-α-methylbenzylamine-derived imine (**Scheme 9**) [15].

On the other hand, Vovk et al. [16] carried out the addition of sodium diethyl phosphite to the imine (*S*)-**23**, obtaining the α-aminophosphonate (*S*,*R*)-**24** in 98% yield and 95% diastereoisomeric excess. Hydrogenolysis of the chiral auxiliary in (*S*,*R*)-**24** and hydrolysis of the diethyl phosphonate with trimethylsilyl bromide (TMSBr) in chloroform followed by the treatment with methanol gave the enantiomerically pure (*R*)-α-aminophosphonic acid **25** (**Scheme 10**).

Nucleophilic addition of triethyl phosphite to the chiral base imines (*S*)-**26a-c** bearing (*S*)-1-(α-aminobenzyl)-2-naphthol, promoted by trifluoroacetic acid (TFA) in toluene at room

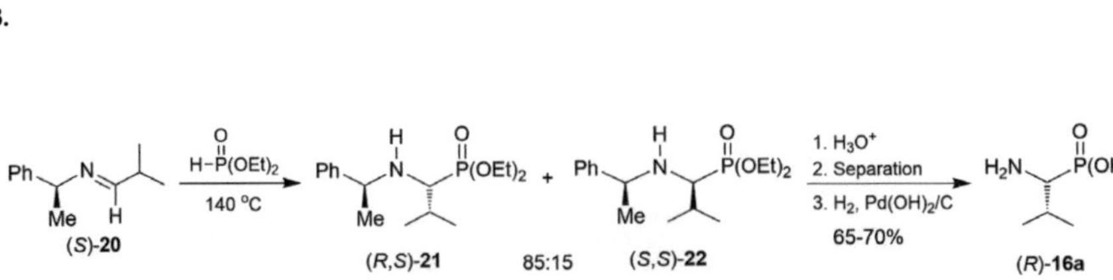

Scheme 8.

Scheme 9.

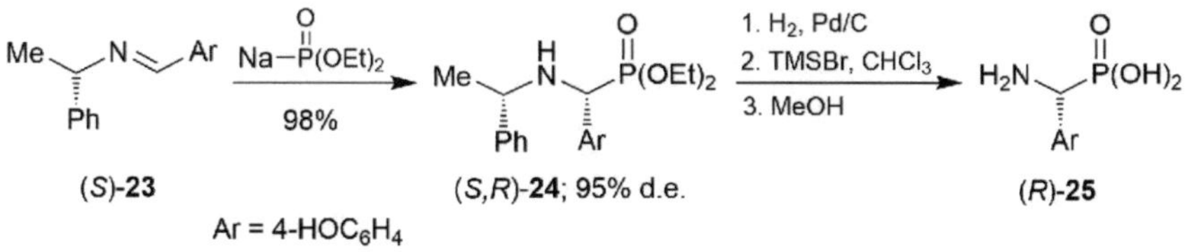

Scheme 10.

temperature and subsequent crystallization, provided the α-aminophosphonates (*S*,*S*)-**27a-c** in 44–57% yield and with excellent diastereoisomeric excess (>98%), which by cleavage of the chiral auxiliary and hydrolysis of the diethyl phosphonate with HCl in 1,4-dioxane at 80°C, afforded the (*S*)-α-aminophosphonic acids **4, 28a,b** in 76–88% yield (**Scheme 11**) [17]. Additionally, the (*R*)-α-aminophosphonic acid **4** was obtained also starting from the aldimine (*R*)-**26c**.

Smith et al. explored the generality of the diastereoselective addition of the lithium salt of diethyl phosphite to a variety of imines. Thus, addition of $LiPO_3Et_2$ to aldimines (*R*)-**29a-c** bearing the methyl ether of (*R*)-phenylglycinol as chiral auxiliary, furnished the α-aminophosphonates (*R*,*R*)-**30a-c** in 37–81% yield and 96 to >99% diastereoisomeric excess. Cleavage of the chiral fragment in (*R*,*R*)-**30a-c** by hydrogenolysis using $Pd(OH)_2$/C followed by hydrolysis of the diethyl phosphonate with concentrated HCl at 100°C gave the enantiomerically pure (*R*)-Glu^P **31a,** (*R*)-Leu^P **31b,** and (*R*)-Met^P**31c** in 55–74% yield (**Scheme 12**) [18].

The readily available chiral sulfinimides [19] containing an aryl- or *tert*-butylsulfinyl moiety represent valuable chiral auxiliaries in stereoselective synthesis [20]. In this regard [21], the nucleophilic addition of the lithium salt of the diethyl phosphite to the enantiopure imine

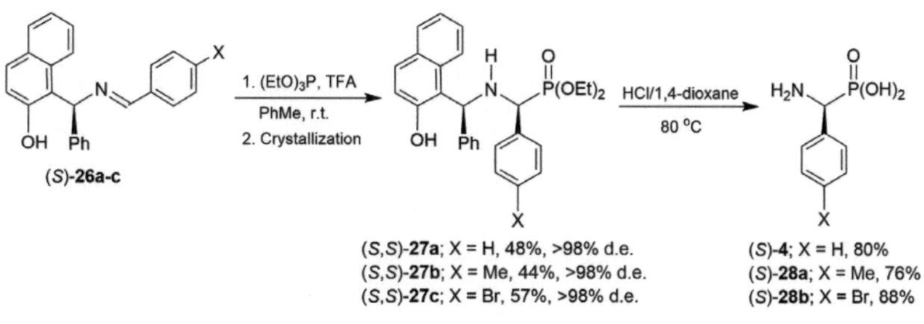

Scheme 11.

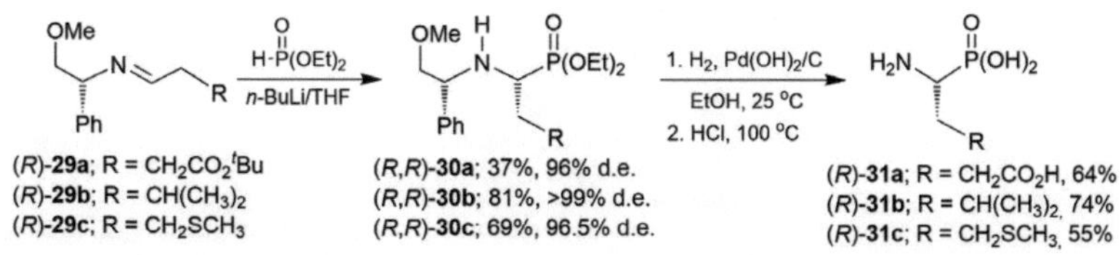

Scheme 12.

(*S*)-**32a** [22, 23], readily obtained by condensation of (*S*)-*p*-toluenesulfinamide with benzaldehyde gave the α-aminophosphonate (S_S,R_C)-**33a** in 85% yield and 92:08 diastereoisomeric ratio. When the lithium salt of bis(diethylamido) phosphite was reacted with (*S*)-**32a**, the α-aminophosphonate (S_S,S_C)-**33b** was obtained in good yield and diastereoselectivity [24]. Cleavage of the chiral auxiliary and hydrolysis of the diethyl phosphonate and diamidophosphite in (S_S,R_C)-**33a** and (S_S,S_C)-**33b** with hydrochloric acid in acetic acid at 100°C led to the enantiomerically pure (*R*)- and (*S*)-phosphophenyl glycine **4** (**Scheme 13**).

Mikolajczyk et al. [25] reported the addition of the lithium salt of the bis(diethylamido)phosphine borane complex to the *p*-toluenesulfinyl imines (*S*)-**32a-e** in THF at −78°C, obtaining mainly the (S_S,S_C)-**34a-e** diastereoisomers in 72–100% yield. Finally, cleavage of the *N*-sulfinyl auxiliary and hydrolysis of the bis(diethylamido)phosphine borane function with hydrochloric acid in acetic acid at reflux gave the (*S*)-α-aminophosphonic acids **4, 16b,c, 28b, 35** in 75–93% yield and 76 to >98% enantiomeric excess. Under identical conditions, the (*R*)-α-aminophosphonic acids (*R*)-**4, 16b,c, 28b, 35** were obtained from the imines (*R*)-**32a-e** (**Scheme 14**).

On the other hand, the addition of the lithium salt of diethyl phosphite to the enantiopure *p*-toluenesulfinyl imines (*S*)-**36a-c**, readily obtained by the $Ti(OEt)_4$ catalyzed condensation of (*S*)-*p*-toluenesulfinamide with the corresponding ketones [26], furnished the α-aminophosphonates (S_S,R_C)-**37a-c** in 73–97% yield and excellent diastereoisomeric ratio (>99:1 d.r.). Cleavage of the chiral auxiliary and hydrolysis of the diethyl phosphonate in (S_S,R_C)-**37a-c** with 10 N HCl at reflux followed by the treatment with propylene oxide led to the (*R*)-α-aminophosphonic acids **38a-c** in 68–84% yield (**Scheme 15**) [23].

With the aim of obtaining the phosphonic analog of aspartic acid (*R*)-**42**, Mikołajczyk et al. [27] reported the nucleophilic addition of the lithium salt of diethyl phosphite to the enantiopure sulfinylaldimine (*S*)-**39** at −78°C in THF, obtaining the α-aminophosphonate (R_C,S_S)-**40** in 62%

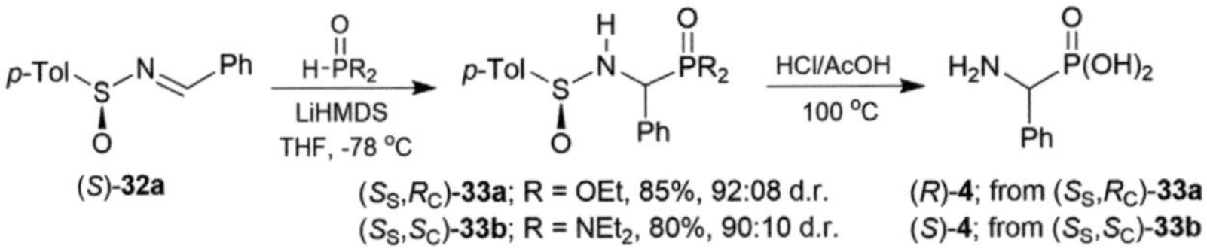

Scheme 13.

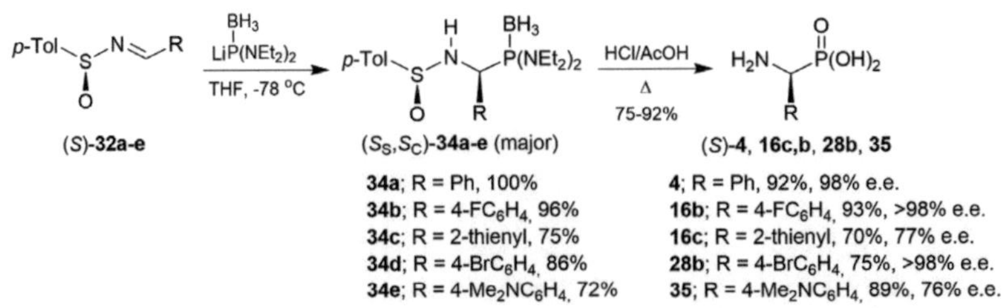

Scheme 14.

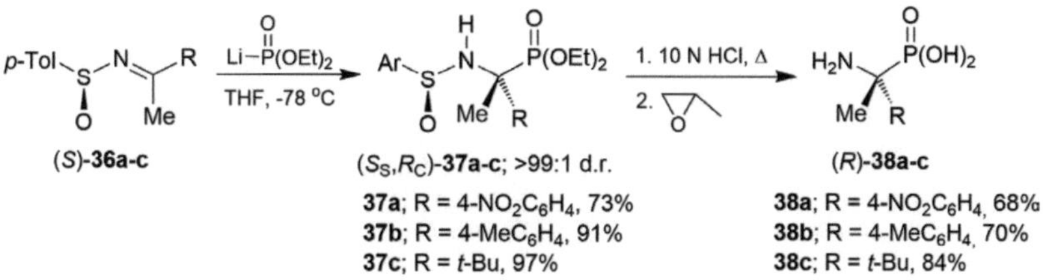

Scheme 15.

yield and 16:1 diastereoisomeric ratio. Ozonolysis of diastereoisomerically pure (R_C,S_S)-**40** followed by $NaBH_4$ reduction, provided the α-aminophosphonate (R_C,S_S)-**41** in 99% yield, which under Mitsunobu reaction conditions led to the cyanide, that by hydrolysis with HCl in AcOH gave the phosphoaspartic acid (*R*)-**42** in 53% yield (**Scheme 16**).

The *N-tert*-butylsulfinyl group activates the imines for the nucleophilic addition, serves as a powerful chiral directing group and after the addition reaction is readily cleaved upon treatment of the product with acid. Competitive nucleophilic attack at the sulfur atom is minimized in the addition to *N-tert*-butylsulfinyl imines versus *N-p*-tolylsulfinyl imines, due to the greater steric hindrance and reduced electronegativity of the *tert*-butyl group relative to the *p*-tolyl moiety [28]. Under this context, reaction of the chiral *N-tert*-butylsulfinyl imines (*S*)-**43a-e** with dimethyl phosphite in the presence of K_2CO_3 in Et_2O at room temperature gave the α-aminophosphonates (S_S,R_C)-**44a-e** in 80–85% yield and with >95% diastereoisomeric excess, which by simultaneous cleavage of the sulfinyl group and hydrolysis of the diethyl phosphonate with 10 N HCl at reflux, followed by treatment with propylene oxide, produced the (*R*)-α-aminophosphonic acids **13, 38a, 45a-c** in 83–88% yield (**Scheme 17**) [29].

On the other hand, the addition of diethyl trimethylsilyl phosphite to chiral *N-tert*-butyl-sulfinylaldimine (*S*)-**46** afforded the α-aminophosphonate (S_C,S_S)-**47** in 69% yield and 84% diastereoisomeric excess. Cleavage of the *N-tert*-butylsulfinyl group in (S_C,S_S)-**47** with 4 N HCl in methanol, produced the α-aminophosphonate (*S*)-**48** in 89% yield. Finally, the hydrolysis of the diethyl phosphonate in (*S*)-**48** with 10N HCl at reflux followed by the treatment with propylene oxide gave the enantiomerically pure (*S*)-phosphonotrifluoroalanine **49** in 96% yield (**Scheme 18**) [30].

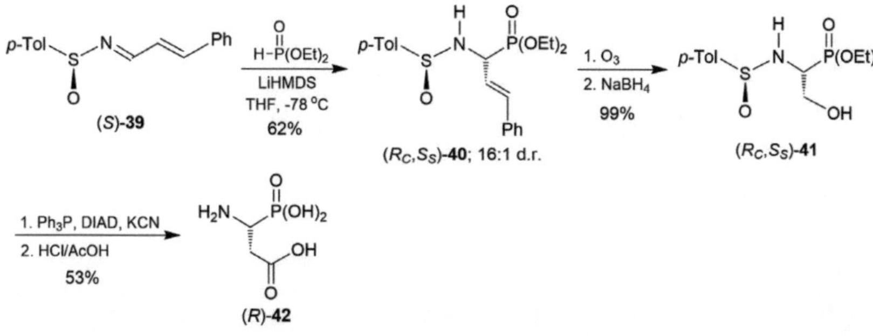

Scheme 16.

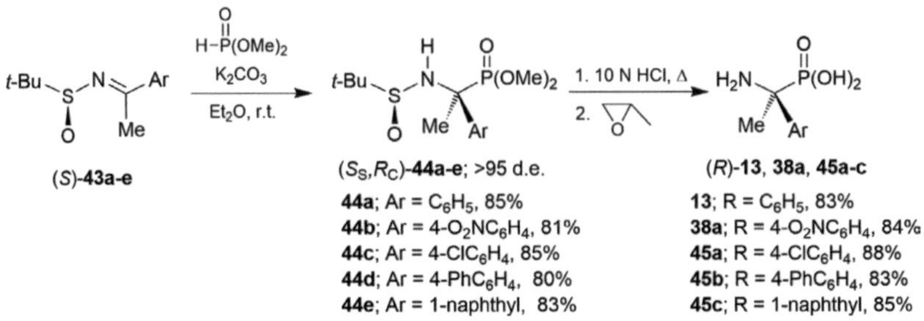

Scheme 17.

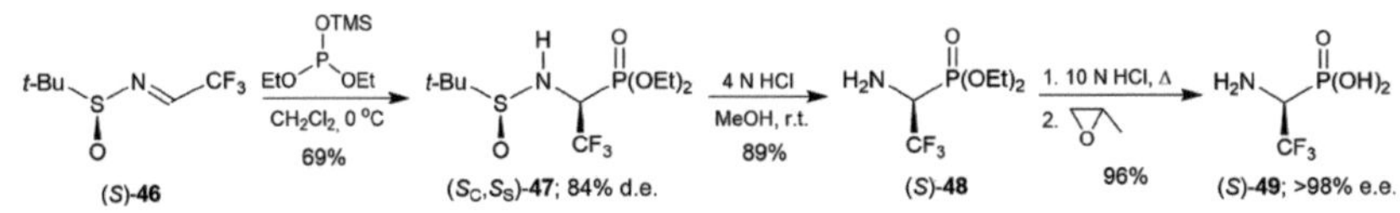

Scheme 18.

Lu et al. [31] reported the addition of diethyl phosphite to the enantiopure sulfinylketimines (*R*)-**50a-c**, to obtain the quaternary α-aminophosphonates (S_C,R_S)-**51a-c** in 73–84% yield and 8:1–43:1 diastereoisomeric ratio. Cleavage of the *N*-*tert*-butylsulfinyl group and hydrolysis of the diethyl phosphonate in (S_C,R_S)-**51a-c** with 6 N HCl at reflux, led to the enantiomerically pure α-aminophosphonic acids (*S*)-**52a-c** in excellent yield (**Scheme 19**).

On the other hand, the sugar-derived nitrones have also emerged as valuable synthetic intermediates in the stereoselective synthesis of α-aminophosphonic acids. For example, the hydrophosphonylation reaction of the nitrones **53a-c** with the lithium salt of diethyl or dibenzyl phosphite, provided the *N*-glycosyl-α-aminophosphonates **54a-c** in 41–63% yield and 90–98.7% diastereoisomeric excess, which by hydrolysis of the sugar fragment and the phosphonate with concentrated HCl and subsequent cleavage of the N-OH bond by hydrogenation using Pd/C, afforded the optically enriched α-aminophosphonic acids (*S*)-**16a, 55a,b** in 36–80% yield. Additionally, the nucleophilic addition of *tris*(trimethylsilyl) phosphite to the enantiomerically pure nitrone **53c** in the presence of $HClO_4$ followed by hydrolysis of the sugar fragment and the phosphonate, led to the *N*-hydroxyphosphovaline (*R*)-**56** in 78%

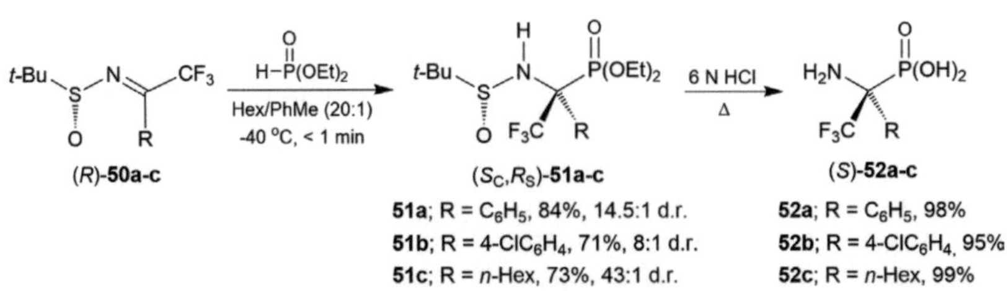

Scheme 19.

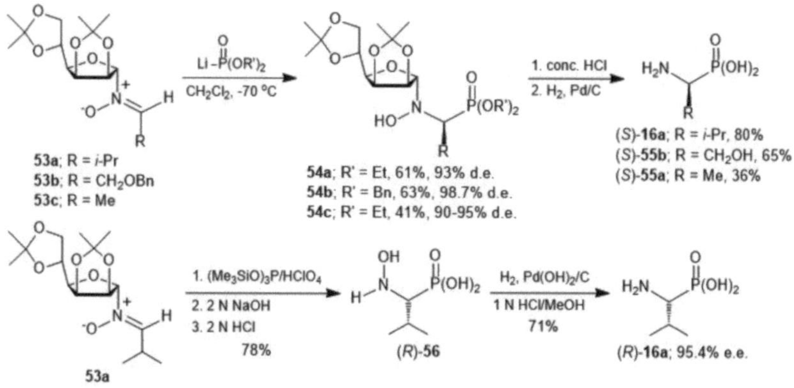

Scheme 20.

yield, which by hydrogenation of the N–OH bond and treatment with 1 N HCl, gave the (*R*)-Val^P **16a** in 71% yield and 95.4% enantiomeric excess (**Scheme 20**) [32].

Huber and Vasella [33] reported the synthesis of optically enriched (*S*)-Val^P **16a** and (*S*)-Ser^P **55b** from the enantiopure nitrones **53a,b** with a slight modification of the reaction conditions. Thus, the nucleophilic addition of *tris*(trimethylsilyl) phosphite to the sugar-derived nitrones **53a,b** catalyzed by $ZnCl_2$/HCl afforded directly the corresponding α-aminophosphonic acids (*S*)-**16a**, **55b** in good yield and with 43.8 and 87.7% enantiomeric excess, respectively (**Scheme 21**).

Similarly, the addition of *tris*(trimethylsilyl) phosphite to the enantiopure nitrone **57** in the presence of $Zn(OTf)_2$ at −40°C and subsequent treatment with 1 N HCl in MeOH, led to the *N*-hydroxy-α-aminophosphonic acid (*R*)-**58** in 71% yield, which by cleavage of the N-OH bond by hydrogenation using $Pd(OH)_2$/C, provided the (*R*)-Met^P **31c** in 88% yield and 76.8% enantiomeric excess (**Scheme 22**) [33].

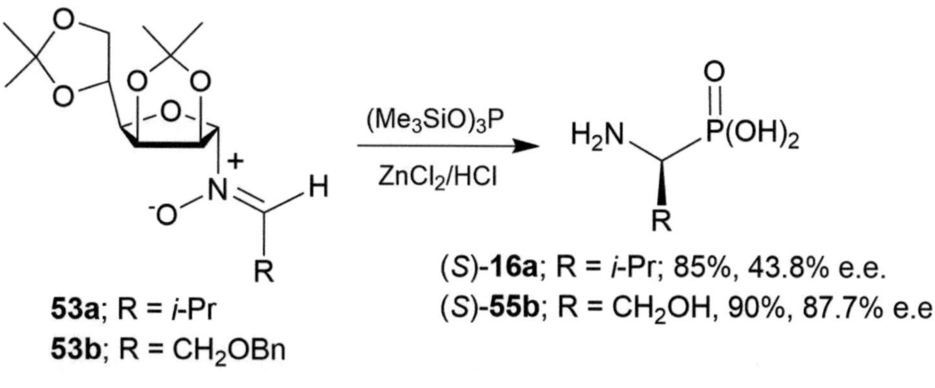

Scheme 21.

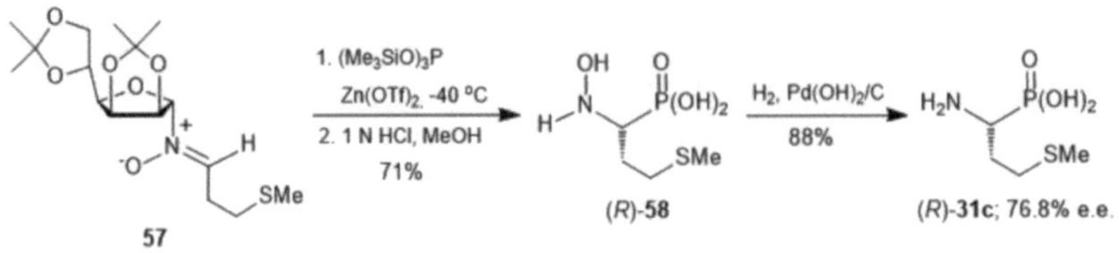

Scheme 22.

2.4. Chiral catalyst

Catalytic asymmetric synthesis is one of the most important topics in modern synthetic chemistry and is considered the most efficient methodology to bring about the synthesis of enantiomerically pure compounds [34]. For example, the hydrophosphonylation reaction of *N*-sulfonylaldimine **59** with diphenyl phosphite in the presence of catalytic amounts of hydroquinine gave the (*S*)-α-aminophosphonate **60** in quantitative yield and excellent enantiomeric excess (>99%). Cleavage of the *N*-sulfonyl group in (*S*)-**60** by treatment with Mg in AcOH/AcONa and *N,N*-dimethylformamide (DMF) afforded the (S)-α-aminophosphonate **61** in 86% yield, which by hydrolysis of the diphenyl phosphonate with HBr in acetic acid followed by treatment with propylene oxide, produced the (*S*)-phosphophenyl glycine **4** in 83% yield and 98% enantiomeric excess (**Scheme 23**) [35].

In order to obtain the optically enriched (*R*)-phosphophenyl glycine **4**, Wang et al. [36] carried out the nucleophilic addition of diethyl phosphite to the *N*-benzoylimine **62** in the presence of catalytic amounts of (*S,S*)-**63** and $ZnMe_2$, obtaining the (*R*)-α-aminophosphonate **64** in 91% yield and >99% enantiomeric excess. Simultaneous hydrolysis of the diethyl phosphonate and *N*-benzoyl group in (*R*)-**64** with concentrated HCl at reflux, produced the optically enriched (*R*)-phosphophenyl glycine **4** in 96% yield (**Scheme 24**).

On the other hand, Joly and Jacobsen [37] reported that the addition of di(*o*-nitrobenzyl) phosphite to the achiral *N*-benzyl aldimines **1**, **65a,b** in the presence of catalytic amounts of the chiral urea **66**, produced the (*R*)-α-aminophosphonates **67a-c** in 87–93% yield and 90–98% enantiomeric excess. Finally, the simultaneous cleavage of the di(*o*-nitrobenzyl) phosphonate

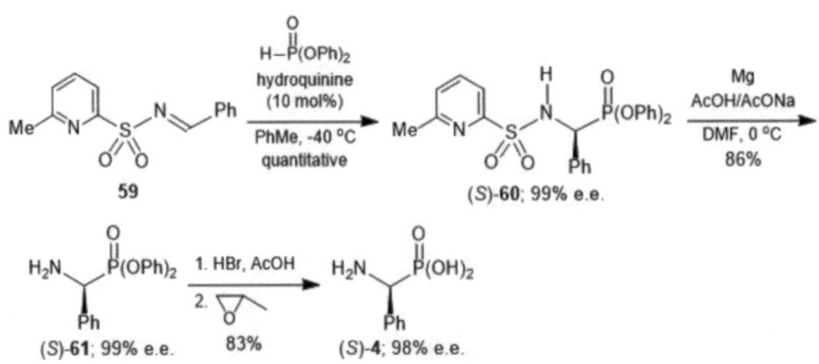

Scheme 23.

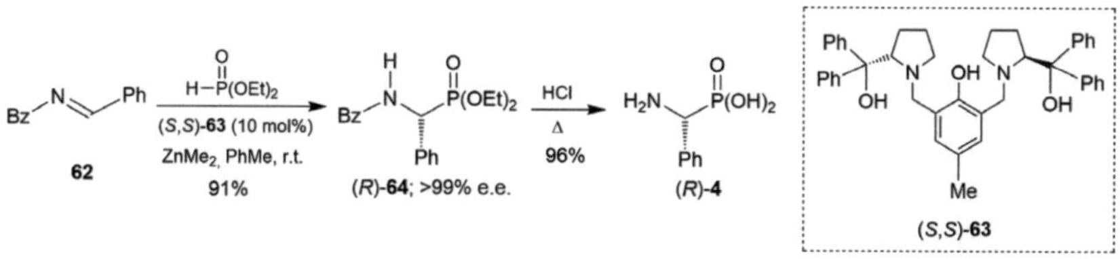

Scheme 24.

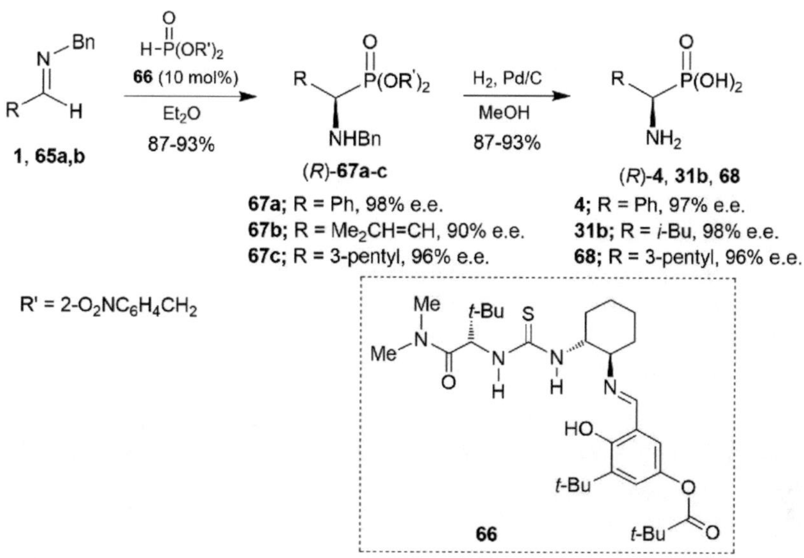

Scheme 25.

and *N*-Bn bond by hydrogenolysis in (*R*)-**67a-c** using Pd/C in MeOH afforded the enantiomerically enriched (*R*)-*α*-aminophosphonic acids **4**, **31b**, **68** in 87–96% yield and excellent enantioselectivity (**Scheme 25**).

Another exceptional example of the chiral catalyst approach is reported by Shibasaki et al. [38] who found that the catalytic hydrophosphonylation of the aldimine **69** in the presence of the lanthanoid-potassium-1,1′-bi-2-naphthol (BINOL) complex [(*R*)-LPB] afforded the (*R*)-*α*-aminophosphonate **70** in 70% yield and 96% enantiomeric excess. Cleavage of *p*-anisylmethyl fragment and simultaneous hydrolysis of the dimethyl phosphonate in (*R*)-**70** with concentrated HCl at reflux, produced the enantiomerically enriched (*R*)-Val^{P} **16a** (**Scheme 26**).

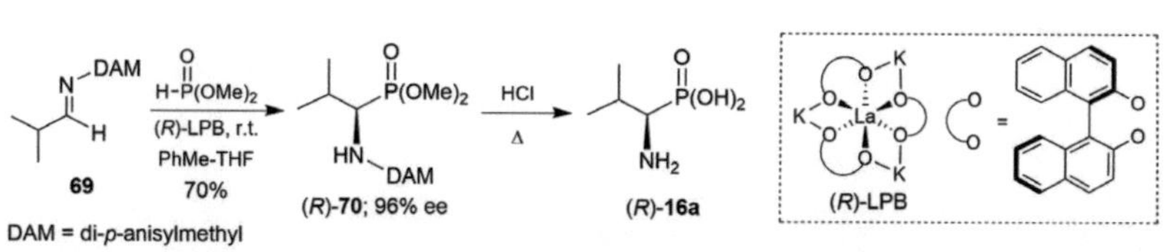

Scheme 26.

3. Stereoselective C-P bond formation (Kabachnik-Fields methodology)

Another important method for the stereoselective synthesis of α-aminophosphonic acids is the "one-pot" three-component reaction, known as the Kabachnik-Fields reaction. In this process, the reactants (carbonyl compound, amine and the phosphorus nucleophile agent) are placed all together to give the diastereo or enantiomerically pure α-aminophosphonates, which are easily transformed into the corresponding α-aminophosphonic acids. To induce the stereochemistry in the α-aminophosphonates, the chirality inducer may be at the source of phosphorus, in the amine, in the aldehyde or ketone, or in a chiral catalyst. Additionally, the reactions are carried out in solvent or under solvent free conditions (**Scheme 27**).

3.1. Chiral phosphorus compounds

The "one-pot" three-component reaction of benzyl carbamate, benzaldehyde, and diethyl (*R,R*)-2-chloro-1,3,2-dioxaphospholane-4,5-dicarboxylate **71**, readily obtained from the reaction of diethyl L-tartrate with phosphorous trichloride, followed by dioxaphospholane ring opening with H_2O, led to the α-aminophosphonates (*R,R,R*)-**72** and (*S,R,R*)-**73** in 40% yield and 1.9:1.0 diastereoisomeric ratio. Saponification of diastereoisomer (*R,R,R*)-**72** gave the (*R*)-*N*-Cbz-phosphophenyl glycine **74** in 53% yield (**Scheme 28**) [39].

On the other hand, Xu and Gao [40] carried out the stereoselective synthesis of the depsiphosphonopeptides **76** and **77**, as key intermediates in the synthesis of α-aminophosphonic acids. Thus, the three-component reaction of (*R*)-1-carboethoxy phosphorodichloridite **75** with benzyl carbamate and benzaldehyde in benzene at room temperature and subsequent treatment with H_2O, produced the depsiphosphonopeptides (*S,R*)-**76** and (*R,R*)-**77** in 86% yield and

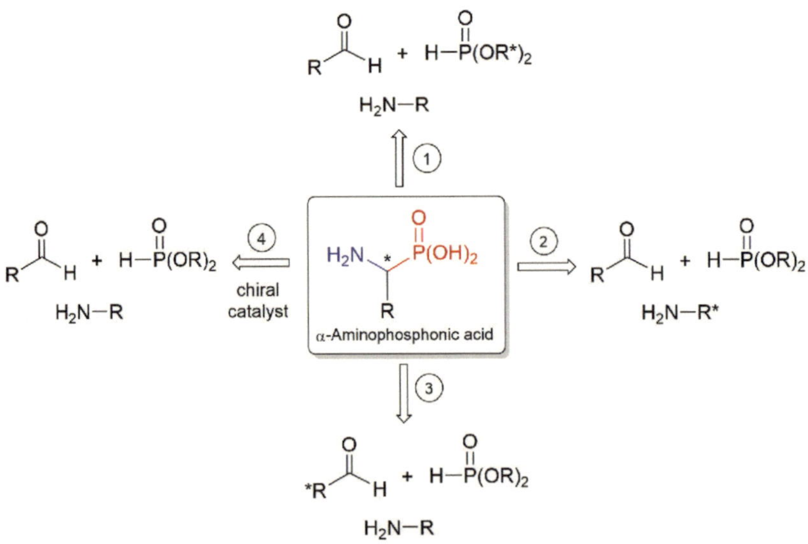

Scheme 27. Diastereo and enantioselective synthesis of α-aminophosphonic acids by Kabachnik-Fields methodology.

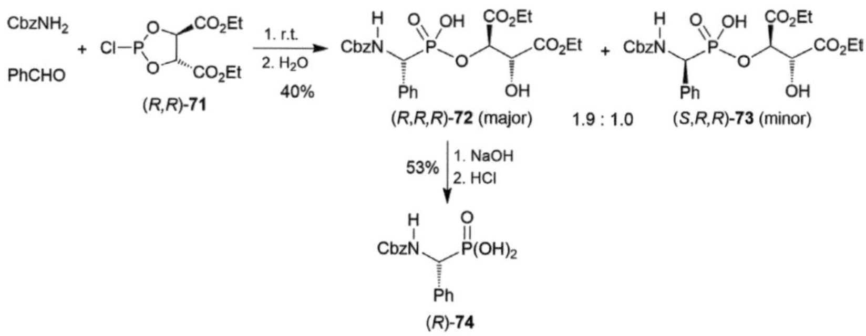

Scheme 28.

85:15 diastereoisomeric ratio. Saponification of the phosphonic ester in the diastereoisomerically pure (*S*,*R*)-**76** followed by hydrogenolysis of *N*-Cbz bond using Pd/C in AcOH gave the enantiomerically pure (*S*)-phosphophenyl glycine **4** in 57% yield (**Scheme 29**).

3.2. Chiral carbonyl compounds

In order to prepare conformationally restricted α-aminophosphonic acids, Fadel et al. [41] carried out the TMSCl promoted three-component reaction of the chiral ketal (2*S*)-**78** with (*S*)-α-methylbenzylamine hydrochloride and triethyl phosphite in EtOH at 55°C, obtaining the α-aminophosphonates (1*S*,2*S*)-**79** and (1*R*,2*S*)-**80** in 80% yield and 87:13 diastereoisomeric ratio. Cleavage of the methylbenzyl fragment by hydrogenolysis in the major diastereoisomer (1*S*,2*S*)-**79** using $Pd(OH)_2$/C in EtOH at room temperature, provided the α-aminophosphonate (1*S*,2*S*)-**81** in 82% yield, which by hydrolysis of the diethyl phosphonate with trimethylsilyl iodide (TMSI) followed by treatment with propylene oxide, produced the enantiomerically pure (1*S*,2*S*)-1-amino-2-methylcyclopropane phosphonic acid **82** in 86% yield (**Scheme 30**).

Similarly, the one-pot reaction of chiral ketal (2S)-**78**, (*R*)-phenylglycinol and triethyl phosphite catalyzed by TMSCl in ethanol at 55°C, led to the α-aminophosphonates (1*S*,2*S*)-**83** and (1*R*,2*S*)-**84** in 71% yield and 89:11 diastereoisomeric ratio. Hydrogenolysis of diastereoisomerically pure (1*S*,2*S*)-**83** over Pearlman's catalyst in EtOH, provided the α-aminophosphonate

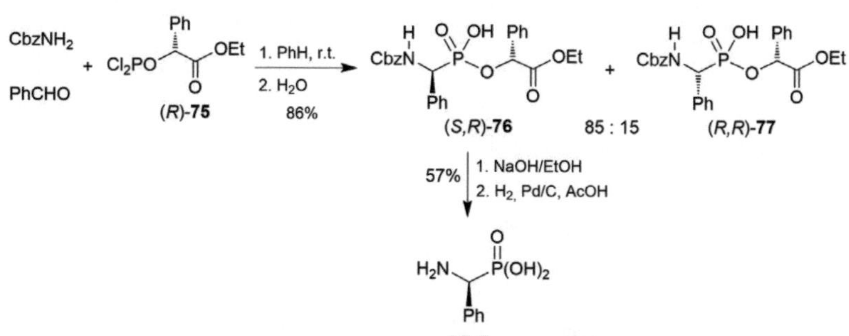

Scheme 29.

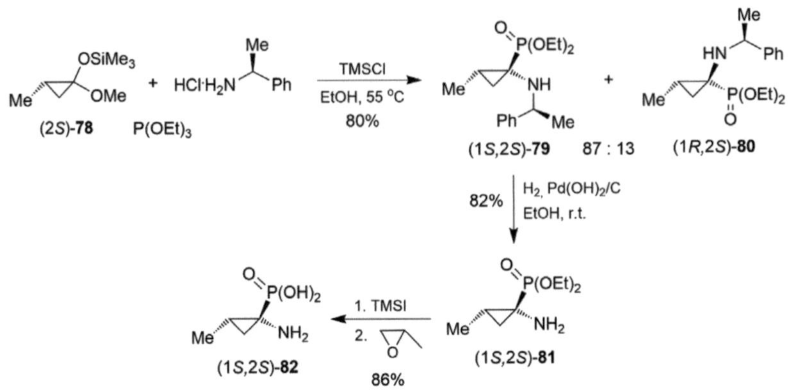

Scheme 30.

monoester (1*S*,2*S*)-**85** in 79% yield. Finally, hydrolysis of (1*S*,2*S*)-**85** with TMSI followed by treatment with propylene oxide afforded the enantiomerically pure α-aminophosphonic acid (1*S*,2*S*)-**82** in 87% yield. In a similar way, the α-aminophosphonate (1*R*,2*S*)-**84** was transformed into α-aminophosphonic acid (1*R*,2*S*)-**86** (**Scheme 31**) [42].

3.3. Chiral amino compounds

The "one-pot" three-component reaction of (*S*)-α-methylbenzylamine, anhydrous hypophosphorous acid and different aldehydes in EtOH at reflux, furnished the corresponding α-aminophosphonous acids (*S*,*R*)-**88a-e** as a single diastereoisomers in 19–50% yield, which by treatment with bromine water solution at 70°C and subsequent treatment with propylene oxide, gave the enantiomerically pure (*R*)-α-aminophosphonic acids **4, 16a, 31b, 89a,b** in 65–88% yield (**Scheme 32**). Using (*R*)-α-methylbenzylamine as starting material, the (*S*)-α-aminophosphonic acids **4, 16a, 31b, 89a,b** were obtained in good yields [43].

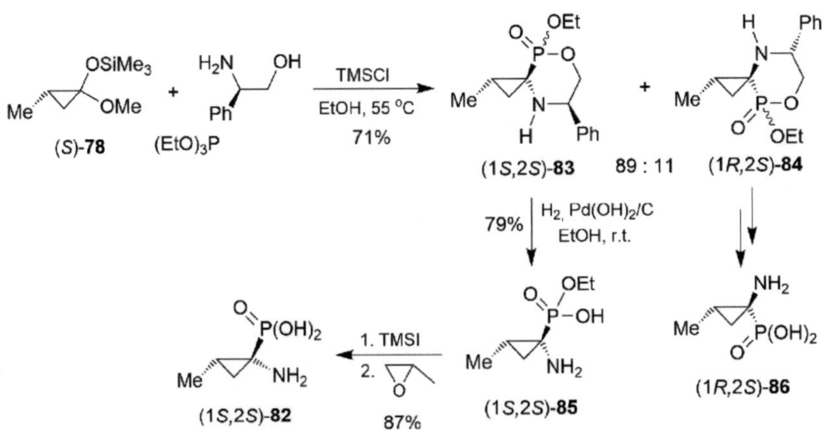

Scheme 31.

http://dx.doi.org/10.5772/intechopen.68707

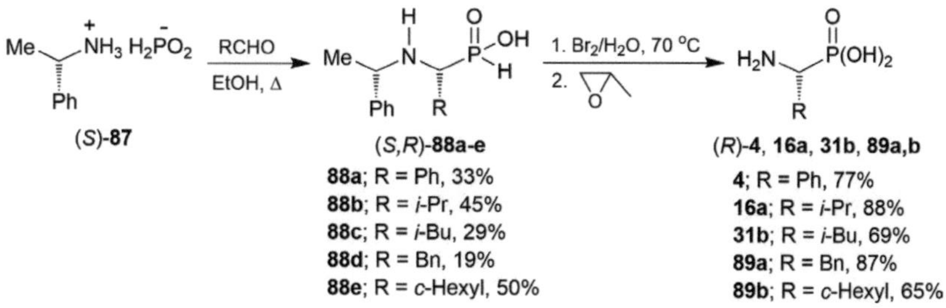

Scheme 32.

On the other hand, Fadel et al. [44] carried out the "one-pot" reaction of *N*-Boc-piperidin-3-one, (*S*)-α-methylbenzylamine (*X* = *H*), triethyl phosphite and AcOH as catalyst in ethanol at 50°C, to obtain the quaternary α-aminophosphonates (3*R*,1'*S*)-**90** and (3*S*,1'*S*)-**91** in 75% yield and 60:40 diastereoisomeric ratio. The use of (*S*)-α-methoxymethylbenzylamine (X = OMe) as chiral amine in this three-component reaction afforded the α-aminophosphonates (3*R*,1'*R*)-**92** and (3*S*,1'*R*)-**93** in 55% yield and with the same diastereoisomeric ratio (60:40). Cleavage of *N*-Boc bond with TFA at room temperature, chromatographic separation, and removal of the chiral fragment by hydrogenolysis using $Pd(OH)_2/C$ in each pure diastereoisomer, furnished the quaternary (*R*)- and (*S*)-α-aminophosphonates **94** in good yield. Finally, the hydrolysis of diethyl phosphonate in (*R*)- and (*S*)-**94** with 6 M HCl at reflux followed by treatment with propylene oxide gave the enantiomerically pure α-aminophosphonic acids (*R*)- and (*S*)-**95** in 98% yield (**Scheme 33**).

Enantiomerically pure carbamates and urea have also shown a potential as chiral auxiliaries in the stereoselective synthesis of α-aminophosphonic acids. For example, the "one-pot" reaction of carbamate **96**, readily obtained from naturally occurring (–)-menthol or the urea **97** derived from (*S*)-α-methylbenzylamine, with acetaldehyde or propionaldehyde and triphenyl phosphite in the presence of acetic acid as catalyst, provided the α-aminophosphonates (*R*)-**98**, which by hydrolysis with concentrated HCl followed by treatment with propylene oxide, afforded the (*R*)-Ala^P **55a** and (*R*)-Val^P **16a** in moderate yield but low enantiomeric excess (**Scheme 34**) [45].

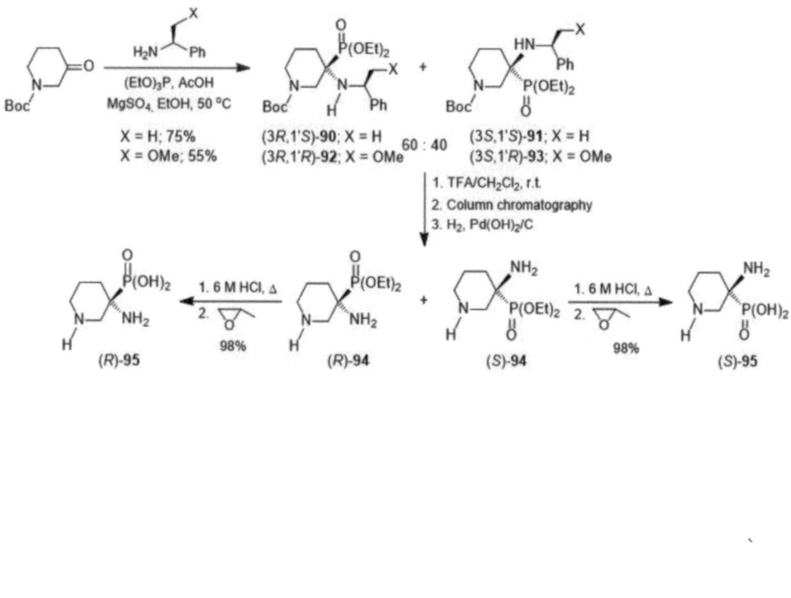

Scheme 33.

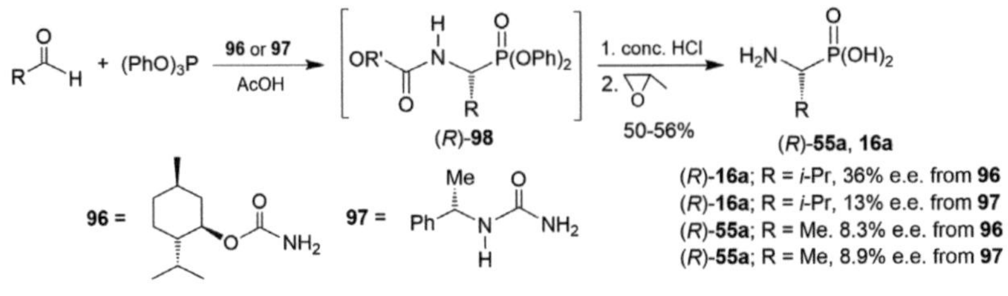

Scheme 34.

3.4. Chiral catalyst

The development of methodologies under chiral catalysis protocols has become one of the most relevant issues in the field of modern synthetic chemistry [46]. In this respect, List et al. [47] described the Kabachnik-Fields reaction of 2-cyclopentyl-2-phenylacetaldehyde, *p*-anisidine and di-(pent-3-yl) phosphite in the presence of catalytic amounts of the chiral phosphoric acid (*S*)-**99** in cyclohexane at 50°C, obtaining the (*R*,*R*)-α-aminophosphonate **100** in 86% yield with both high diastereoisomeric and enantiomeric ratio. Removal of *p*-methoxyphenyl fragment with cerium ammonium nitrate (CAN) followed by the hydrolysis of the diethyl phosphonate in (*R*,*R*)-**100** with TMSBr, produced the optically enriched (*R*,*R*)-α-aminophosphonic acid **101** in 54% yield (**Scheme 35**).

In another example, Shibata et al. [48] reported that the reaction of benzaldehyde, *p*-anisidine, and di-(*o*-methoxyphenyl) phosphite in the presence of catalytic amounts of $Zn(NTf_2)_2$ and **102** as chiral ligand in CH_2Cl_2 at −50°C gave the (*S*)-α-aminophosphonate **103** in 99% yield and 90% enantiomeric excess. Removal of *p*-methoxyphenyl group in (*S*)-**103** was accomplished

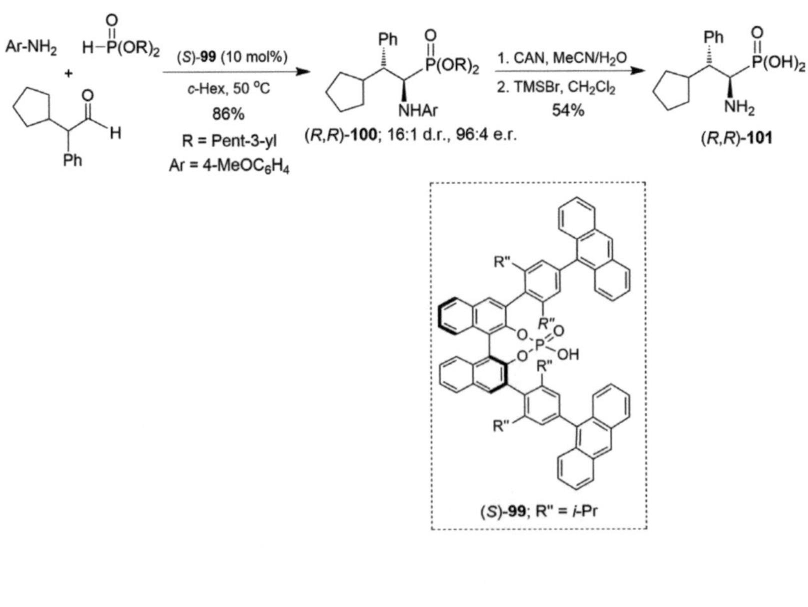

Scheme 35.

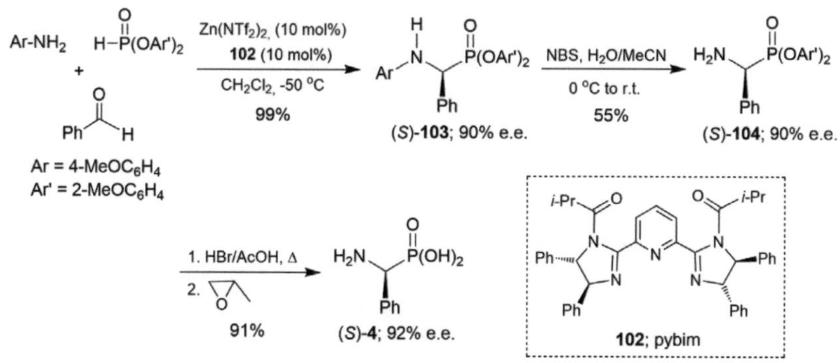

Scheme 36.

by treatment with *N*-bromosuccinimide (NBS), obtaining the (*S*)-α-aminophosphonate **104** in 55% yield without racemization, which by hydrolysis of the phosphonate with HBr/AcOH followed by treatment with propylene oxide, led to the optically enriched (*S*)-phosphophenyl glycine **4** in 91% yield and 92% enantiomeric excess (**Scheme 36**).

Author details

Mario Ordóñez*, José Luis Viveros-Ceballos and Iván Romero-Estudillo

*Address all correspondence to: palacios@uaem.mx

Center of Chemical Research CIQ-IICBA, Autonomous University of Morelos State, Cuernavaca, Morelos, Mexico

References

[1] (a) Horiguchi M, Kandatsu M. Isolation of 2-aminoethane phosphonic acid from rumen protozoa. Nature. 1959;**184**;901-902; (b) Bayer E, Gugel KH, Hägele K, Hagenmaier H, Jessipow S, Konig WA, Zähner H. Phosphinothricin and phosphinothricyl-alanin-alanine. Helvetica Chimica Acta. 1972;**55**:224-239; (c) Kamiya T, Hemmi K, Takeno H, Hashimoto M. Studies on phosphonic acid antibiotics. I. Structure and synthesis of 3-(n-acetyl-n-hydroxyamino)propylphosphonic acid (FR-900098) and its n-formyl analogue (FR-31564). Tetrahedron Letters. 1980;**21**:95-98

[2] (a) Yamato M, Koguchi T, Okachi R, Yamada K, Nakayama K, Kase H, Karasawa A, Shuto K. K-26, a novel inhibitor of angiotensin I converting enzyme produced by an actinomycete K-26. The Journal of Antibiotics 1986;**39**:44-52; (b) Ntai I, Bachmann BO. Identification of ACE pharmacophore in the phosphonopeptide metabolite K-26. Bioorganic & Medicinal Chemistry Letters. 2008;**18**;3068-3071

[3] (a) Kukhar VP, Hudson HR, editors. Aminophosphonic and Aminophosphinic Acids: Chemistry and Biological Activity. Chichester: Wiley; 2000; (b) Orsini F, Sello G, Sisti M. Aminophosphonic acids and derivatives. Synthesis and biological applications. Current Medicinal Chemistry. 2010;**17**:264-289; (c) Lejczak B, Kafarski P. Biological activity of aminophosphonic acids and their short peptides. Topics in Heterocyclic Chemistry. 2009;**20**:31-63; (d) Sienczyk M, Oleksyszyn J. Irreversible inhibition of serine proteases-design and in vivo activity of diaryl α-aminophosphonate derivatives. Current Medicinal Chemistry. 2009;**16**:1673-1687; (e) Moonen K, Laureyn I, Stevens CV. Synthetic methods for azaheterocyclic phosphonates and their biological activity. Chemical Reviews. 2004;**104**:6177-6215; (f) Hiratake J, Oda J. Aminophosphonic and aminoboronic acids as key elements of a transition state analogue inhibitor of enzymes. Bioscience, Biotechnology, and Biochemistry. 1997;**61**:211-218; (g) Kafarski P, Lejczak B. Biological activity of aminophosphonic acids. Phosphorus, Sulfur, and Silicon and the Related Elements. 1991;**63**:193-215

[4] Mucha A, Kafarski P, Berlicki L. Remarkable potential of the α-aminophosphonate/phosphinate structural motif in medicinal chemistry. Journal of Medicinal Chemistry. 2011;**54**:5955-5980

[5] (a) Mikolajczyk M. Acyclic and cyclic aminophosphonic acids: Asymmetric syntheses mediated by chiral sulfinyl auxiliarity. Journal of Organometallic Chemistry. 2005;**690**:2488-2486; (b) Huang J, Chen R. An overview of recent advances on the synthesis and biological activity of α-aminophosphonic acid derivatives. Heteroatom Chemistry. 2000;**11**:480-492; (c) Patel DV, Rielly-Gauvin K, Ryono DE, Free CA, Rogers WL, Smith SA, DeForrest JM, Oehl RS, Petrillo Jr EW. α-Hydroxy phosphinyl-based inhibitors of human renin. Journal of Medicinal Chemistry. 1995;**38**:4557-4569; (d) Allen JG, Atherton FR, Hall MJ, Hassall CH, Holmes SW, Lambert RW, Nisbet LJ, Ringrose PS. Phosphonopeptides as antibacterial agents: alaphosphin and related phosphonopeptides. Antimicrobial Agents and Chemotherapy. 1979, 15, 684-695; (e) Atherton FR, Hall MJ, Hassall CH, Lambert RW, Lloyd WJ, Ringrose PS. Phosphonopeptides as antibacterial agents: Mechanism of action of alaphosphin. Antimicrobial Agents and Chemotherapy. 1979;**15**:696-705

[6] (a) Stamper C, Bennett B, Edwards T, Holz RC, Ringe D, Petsko G. Inhibition of the aminopeptidase from aeromonas proteolytic by L-leucine phosphonic acid. Spectroscopic and crystallographic characterization of the transition state of peptide hydrolysis. Biochemistry. 2001;**40**:7035-7046; (b) Kafarzki P, Lejczak B, Szewczyk J. Optically active 1-aminoalkanephosphonic acids. Dibenzoyl-L-tartaric anhydride as an effective agent for the resolution of racemic diphenyl 1-aminoalkanephosphonates. Canadian Journal of Chemistry. 1983;**61**:2425-2430

[7] (a) Solodenko VA, Kukhar VP. Stereoselective papain-catalyzed synthesis of alafosfalin. Tetrahedron Letters. 1989;**30**:6917-6918; (b) Polyak MS. Antibiotics of the phosphonic acid group. Antibiotiki I Meditsinskaia Biotekhnologiia. 1987;**32**:66-75

[8] Agamennone M, Campestre C, Preziuso S, Consalvi V, Crucianelli M, Mazza F, Politi V, Ragno R, Tortorella P, Gallina C. Synthesis and evaluation of new tripeptide phosphonate inhibitors of MMP-8 and MMP-2. European Journal of Medicinal Chemistry. 2005;**40**:271-279

[9] (a) Ordóñez M, Viveros-Ceballos JL, Cativiela C, Sayago FJ. An update on the stereoselective synthesis of α-aminophosphonic acids and derivatives. Tetrahedron. 2015;**71**:1745-1784; (b) Ordóñez M, Sayago FJ, Cativiela C. Synthesis of quaternary α-aminophosphonic acids. Tetrahedron. 2012;**68**:6369-6412; (c) Ordóñez M, Viveros-Ceballos JL, Cativiela C, Arizpe A. Stereoselective synthesis of α-aminophosphonic acids analogs of the 20 proteinogenic α-amino acids. Current Organic Synthesis. 2012;**9**:310-341; (d) Kudzin ZH, Kudzin MH, Drabowicz J, Stevens CV. Aminophosphonic acids-phosphorus analogues of natural amino acids. Part 1: Syntheses of α-aminophosphonic acids. Current Organic Chemistry. 2011;**15**:2015-2071; (e) Ordóñez M, Rojas-Cabrera H, Cativiela C. An overview of stereoselective synthesis of α-aminophosphonic acids and derivatives. Tetrahedron. 2009;**65**:17-49

[10] Kolodiazhnyi OI, Sheiko S, Grishkun EV. C_3-Symmetric trialkyl phosphites as starting compounds of asymmetric synthesis. Heteroatom Chemistry. 2000;**11**:138-143

[11] Palacios F, Olszewski TK, Vicario J. Diastereoselective hydrophosphonylation of imines using (R,R)-TADDOL phosphite. Asymmetric synthesis of α-aminophosphonic acids derivatives. Organic & Biomolecular Chemistry. 2010;**8**:4255-4258

[12] Vicario J, Ortiz P, Palacios F. Synthesis of tetrasubstituted α-aminophosphonic acids derivatives from trisubstituted α-aminophosphonates. European Journal of Organic Chemistry. 2013:7095-7100

[13] Olszewski TK, Majewski M. Highly diastereoselective addition of chiral H-phosphonate to tert-butylsulfinyl aldimines: A convenient approach to (R)-α-aminophosphonic acids. Tetrahedron: Asymmetry. 2015;**26**:846-852

[14] (a) Bongini A, Camerini R, Panunzio M. Efficient synthesis of the four diastereomers of phosphothreonine from lactaldehyde. Tetrahedron: Asymmetry. 1996;**7**:1467-1476; (b) Bongini A, Camerini R, Hofman S, Panunzio M. Synthesis of (1S,2S)-phosphothreonine via N-trimethylsilylimine of (S)-lactic aldehyde. Tetrahedron Letters. 1994;**35**:8045-8048

[15] (a) Gilmore WF, McBride HN. Synthesis and optically active α-aminophosphonic acid. Journal of the American Chemical Society. 1972;**94**:4361-4361; (b) Glowiak T, Sawka-Dobrowolska W, Kowalik J, Mastalerz P, Soroka M, Zon J. Absolute configuration of optically active aminophosphonic acids. Tetrahedron Letters. 1977;**18**:3965-3968

[16] Cherenok S, Vovk A, Muravyova I, Shivanyuk A, Kukhar V, Lipkowski J, Kalchenko V. Calix[4]arene α-aminophosphonic acids: Asymmetric synthesis and enantioselective inhibition of an alkaline ahosphatase. Organic Letters. 2006;**8**:549-552

[17] (a) Metlushka KE, Kashemirov BA, Zheltukhin VF, Sadkova DN, Büchner B, Hess C, Kataeva ON, McKenna CE, Alfonsov VA. 1-(α-Aminobenzyl)-2-naphthol: A new chiral auxiliary for the synthesis of enantiopure α-aminophosphonic acids. Chemistry: A European Journal. 2009;**15**:6718-6722; (b) Metlushka KE, Sadkova DN, Shaimardanova LN, Kataeva ON, Alfonsov VA. Stereoselective synthesis of α-aminophosphonic acids using the Betti base as chiral auxiliary. Phosphorus, Sulfur, and Silicon. 2011;**186**:712-717

[18] Smith AB, Yager KM, Taylor CM. Enantioselective synthesis of diverse α-aminophosphonate diesters. Journal of the American Chemical Society. 1995;**117**:10879-10888

[19] Davis FA, Reddy RE, Szewczyk JM, Portonovo PS. Asymmetric synthesis of sulfinimines: Chiral ammonia imine synthons. Tetrahedron Letters. 1993;**34**:6229-6232

[20] For representative publications, see: (a) Feng X, Wang Y, Wei B, Yang J, Du H. Simple N-sulfinyl-based chiral sulfur-olefin ligands for rhodium-catalyzed asymmetric 1,4-additions. Organic Letters. 2011;**13**:3300-3303; (b) Steurer M, Bolm C. Synthesis of amino-functionalized sulfonimidamides and their application in the enantioselective Henry reaction. Journal of Organic Chemistry. 2010;**75**:3301-3310; (c) Solà J, Revés M, Riera A, Verdaguer X. N-phosphino sulfonamide ligands: An efficient manner to combine sulfur chirality and phosphorus coordination behavior. Angewandte Chemie International Edition. 2007;**46**:5020-5023; (d) Huang Z, Lai H, Qin Y. Syntheses of novel chiral sulfinamido ligands and their applications in diethylzinc additions to aldehydes. Journal of Organic Chemistry. 2007;**72**:1373-1378; (e) Robak MT, Trincado M, Ellman JA. Enantioselective aza-Henry reaction with an N-sulfinyl urea organocatalyst. Journal of the American Chemical Society. 2007;**129**:15110-15111; (f) Gnas Y, Glorius F. Chiral auxiliaries—Principles and recent applications. Synthesis. 2006;1899-1930; (g) Schenkel LB, Ellman JA. Application of P,N-sulfinyl imine ligands to Iridium-catalyzed asymmetric hydrogenation of olefins. Journal of Organic Chemistry. 2004;**69**:1800-1802; (h) Fernández I, Khiar N. Recent developments in the synthesis and utilization of chiral sulfoxides. Chemical Reviews. 2003;**103**:3651-3705

[21] (a) Lefebvre IM, Evans, Jr SA. Studies toward the asymmetric synthesis of α-aminophosphonic acids via the addition of phosphites to enantiopure sulfinimines. Journal of Organic Chemistry. 1997**62**:7532-7533; (b) Lefebvre IM, Evans, Jr SA. Asymmetric synthesis of α-aminophosphonic acids employing versatile sulfinimines and sulfonimines. Phosphorus, Sulfur, and Silicon and the Related Elements. 1999;**144-146**:397-400

[22] For the synthesis of chiral sulfinimines, see: Davis FA, Reddy RE, Szewczyk JM, Reddy GV, Portonovo PS, Zhang H, Fanelli D, Reddy RT, Zhou P, Carroll PJ. Asymmetric synthesis and properties of sulfinimines (thiooxime S-oxides). Journal of Organic Chemistry. 1997;**62**:2555-2563 and references therein

[23] Davis FA, Lee S, Yan H, Titus DD. Asymmetric synthesis of quaternary α-aminophosphonates using sulfinimines. Organic Letters. 2001;**3**:1757-1760

[24] Mikolajczyk M, Lyzwa P, Drabowicz J. Asymmetric adition of dialkyl phosphite and diamino phosphite anions to chiral, enantiopure sulfinimines: A new, convenient route to enantiomeric α-aminophosphonic acids. Tetrahedron: Asymmetry. 1997;**8**:3991-3994

[25] Mikolajczyk M, Lyzwa P, Drabowicz J. A new efficient procedure for asymmetric synthesis of α-aminophosphonic acids via addition of lithiated bis(diethylamino) phosphine borane complex to enantiopure sulfinimines. Tetrahedron: Asymmetry. 2002;**13**:2571-2576

[26] (a) Davis FA, Lee S, Zhang H, Fanelli DL. Applications of the sulfinimine-mediated asymmetric Strecker synthesis to the synthesis of α-alkyl α-amino acids. Journal of Organic Chemistry. 2000;**65**:8704-8708; (b) Davis FA, Zhang Y, Andemichael Y, Fang T, Fanelli DL, Zhang H. Improved synthesis of enantiopure sulfinimines (thiooxime S-oxides) from p-toluenesulfinamide and aldehydes and ketones. Journal of Organic Chemistry. 1999;**64**:1403-1406; (c) Fanelli DL, Szewczyk JM, Zhang Y, Reddy GV, Burns DM, Davis FA. Sulfinimines (thiooximine *S*-oxides): Asymmetric synthesis of methyl (*R*)-(+)-β-phenylalanate from (*S*)-(+)-*N*-(benzylidene)-*p*-toluenesulfinamide. Organic Syntheses. 1999;**77**:50-56

[27] Łyżwa P, Blaszczyk J, Sieron L, Mikołajczyk M. Asymmetric synthesis of structurally diverse aminophosphonic acids by using enantiopure *N*-(*p*-tolylsulfinyl)cinnamaldimines as reagents. European Journal of Organic Chemistry. 2013;2106-2115

[28] Ellman JA, Owens TD, Tang TP. *N-tert*-butanesulfinyl imines: versatile intermediates for the asymmetric synthesis of amines. Accounts of Chemical Research. 2002;**35**:984-995

[29] Chen Q, Yuan C. A new and convenient asymmetric synthesis of α-amino- and α-alkyl-α-aminophosphonic acids using *N-tert*-butylsulfinyl imines as chiral auxiliaries. Synthesis. 2007;3779-3786

[30] Röschenthaler G-V, Kukhar VP, Kulik IB, Belik MY, Sorochinsky AE, Rusanov EB, Soloshonok VA. Asymmetric synthesis of phosphonotrifluoroalanine and its derivatives using *N-tert*-butanesulfinyl imine derived from fluoral. Tetrahedron Letters. 2012;**53**:539-542

[31] Wang L, Shen Q, Lu L. A general and highly selective method for the asymmetric synthesis of trifluoromethyl-substituted α- and β-aminophosphonates. Chinese Journal of Chemistry. 2013;**31**:892-900

[32] Huber R, Knierzinger A, Obrecht J-P, Vasella A. Nucleophilic additions to *N*-glycosylnitrones. Asymmetric synthesis of α-aminophosphonic acids. Helvetica Chimica Acta. 1985;**68**:1730-1747

[33] Huber R, Vasella A. Nucleophilic additions to *N*-glycosylnitrones. Part IV. Asymmetric synthesis of *N*-hydroxy-α-aminophosphonic and α-aminophosphonic acids. Helvetica Chimica Acta. 1987;**70**:1461-1476

[34] For representative publications on the catalytic asymmetric hydrophosphonylation of imines, see: (a) Yin L, Bao Y, Kumagai N, Shibasaki M. Catalytic asymmetric hydrophosphonylation of ketimines. Journal of the American Chemical Society. 2013;**135**:10338-10341; (b) Xie H, Song A, Zhang X, Chen X, Li H, Sheng C, Wang W. Quinine-thiourea catalyzed enantioselective hydrophosphonylation of trifluoromethyl 2(1*H*)-quinazolinones. Chemical Communications. 2013;**49**:928-930; (c) Zhao D, Wang R. Recent devel-

opments in metal catalyzed asymmetric addition of phosphorus nucleophiles. Chemical Society Reviews. 2012;**41**:2095-2108; (d) Bhadury PS, Li H. Organocatalytic asymmetric hydrophosphonylation/Mannich reactions using thiourea, cinchona and Brønsted acid catalysts. Synlett. 2012;**23**:1108-1131; (e) Merino P, Marquéz-López E, Herrera RP. Catalytic enantioselective hydrophosphonylation of aldehydes and imines. Advanced Synthesis & Catalysis. 2008;**350**:1195-1208; (f) Bhadury PS, Song B-A, Yang S, Zhang Y, Zhang S. Some potential chiral catalysts for preparation of asymmetric α-aminophosphonates. Current Organic Synthesis. 2008;**5**:134-150

[35] Nakamura S, Nakashima H, Yamamura A, Shibata N, Toru T. Organocatalytic enantioselective hydrophosphonylation of sulfonylimines having a heteroarenesulfonyl group as a novel stereocontroller. Advanced Synthesis & Catalysis. 2008;**350**:1209-1212

[36] Zhao D, Wang Y, Mao L, Wang R. Highly enantioselective conjugate additions of phosphites to α,β-unsaturated *N*-acylpyrroles and imines: A practical approach to enantiomerically enriched amino phosphonates. Chemistry: A European Journal. 2009;**15**:10983-10987

[37] Joly GD, Jacobsen EN. Thiourea-catalyzed enantioselective hydrophosphonylation of imines: practical access to enantiomerically enriched α-aminophosphonic acids. Journal of the American Chemical Society. 2004;**126**:4102-4103

[38] Sasai H, Arai S, Tahara Y, Shibasaki M. Catalytic asymmetric synthesis of α-aminophosphonates using lanthanoid-potassium-BINOL complexes. Journal of Organic Chemistry. 1995;**60**:6656-6657

[39] Liu H, Cai S, Xu J. Asymmetric synthesis of *N*-protected chiral 1-aminoalkylphosphonic acids and synthesis of side chain-functionalized depsiphosphonopeptides. Journal of Peptide Science. 2006;**12**:337-340

[40] Xu J, Gao Y. Straightforward synthesis of depsiphosphonopeptides via Mannich-type multicomponent condensation. Synthesis. 2006 ;783-788

[41] Fadel A, Tesson N. Synthesis of enantiomerically pure (1*S*,2*S*)-1-aminocyclopropanephosphonic acids from (2*S*)-methylcyclopropanone acetal. European Journal of Organic Chemistry. 2000;2153-2159

[42] Fadel A, Tesson N. Preparation of enantiomerically pure (1*S*,2*S*)-1-aminocyclopropanephosphonic acid from methylcyclopropanone acetal via spirophosphonate intermediates. Tetrahedron: Asymmetry. 2000;**11**:2023-2031

[43] (a) Hamilton R, Walker B, Walker BJ. A highly convenient route to optically pure α-aminophosphonic acids. Tetrahedron Letters. 1995;**36**:4451-4454. Also, see: (b) Drag M, Grembecka J, Pawelczak M, Kafarski P. α-Aminoalkylphosphonates as a tool in experimental optimisation of P1 side chain shape of potential inhibitors in S1 pocket of leucine- and neutral aminopeptidases. European Journal of Medicinal Chemistry. 2005;**40**:764-771; (c) Drag M, Pawelczak M, Kafarski P. Stereoselective synthesis of 1-aminoalkanephosphonic acids with two chiral centers and their activity towards leucine aminopeptidase. Chirality. 2003;**15**:S104-S107; (d) Drag M, Grembecka J, Kafarski P.

The computer-aided design, synthesis, and activity prediction of new leucine aminopeptidase inhibitors. Phosphorus, Sulfur, and Silicon and the Related Elements. 2002; **177**:1591-1595

[44] Louaisil N, Rabasso N, Fadel A. Asymmetric synthesis of *(R)*-and *(S)*-α-amino-3-piperidinylphosphonic acids via phosphite addition to iminium ions. Synthesis. 2007;289-293

[45] Oshikawa T, Yamashita M. Preparation of optically active 1-aminoalkylphosphonic acids from chiral carbamates and chiral ureas. Bulletin of the Chemical Society of Japan. 1989;**62**:3177-3181

[46] For representative publications, see: (a) Thorat PB, Goswami SV, Magar RL, Patil BR, Bhusare SR. An efficient organocatalysis: A one-pot highly enantioselective synthesis of α-aminophosphonates. European Journal of Organic Chemistry. 2013;5509-5516; (b) Dehui W, Mingshu W, Jingya M. Recent progress in asymmetric synthesis of Kabachnik-Fields reactions. Chinese Journal of Organic Chemistry. 2012;**32**:13-18; (c) Wang L, Cui SM, Meng W, Zhang GW, Nie J, Ma JA. Asymmetric synthesis of α-aminophosphonates by means of direct organocatalytic three-component hydrophosphonylation. Chinese Science Bulletin. 2010;**55**:1729-1731; (d) Albrecht Ł, Albrecht A, Krawczyk H, Jørgensen KA. Organocatalytic asymmetric synthesis of organophosphorus compounds. Chemistry: A European Journal. 2010;**16**:28-48; (e) Zhou X, Shang D, Zhang Q, Lin L, Liu X, Feng X. Enantioselective three-component Kabachnik–Fields reaction catalyzed by chiral scandium(III)–*N,N'*-dioxide complexes. Organic Letters. 2009;**11**:1401-1404

[47] Cheng X, Goddard R, Buth G, List B. Direct catalytic asymmetric three-component Kabachnik-Fields reaction. Angewandte Chemie International Edition. 2008;**47**:5079-5081

[48] Ohara M, Nakamura S, Shibata N. Direct enantioselective three-component Kabachnik-Fields reaction catalyzed by chiral bis(imidazoline)-zinc(II) catalysts. Advanced Synthesis & Catalysis. 2011;**353**:3285-3289

Section 2

Role of Amino Acids in Plant

Chapter 7

D-Amino Acids in Plants: New Insights and Aspects, but also More Open Questions

Üner Kolukisaoglu and Juan Suarez

Additional information is available at the end of the chapter

http://dx.doi.org/10.5772/intechopen.68539

Abstract

The relevance of the homochirality of proteinogenic amino acids for life is undisputed, but also to their D-enantiomers a growing number of biological functions could be assigned. When it comes to D-amino acids in plants, information was relatively scarce for a long time. Nowadays, also in this field, knowledge is growing which will be presented and discussed in this review. In this respect, it was shown that D-amino acids are taken up by plants from soil but could also be synthesized de novo. Investigations of plant D-amino acid metabolism as well as other studies revealed a central function of D-Ala in plants, which await further elucidation. Also other D-amino acids are shown to cause physiological effects in plants, ranging from nitrogen utilization over stress adaptation to chloroplast division, and indicate that D-amino acids are responsible for a variety of yet poorly understood or even undiscovered functions in plants.

Keywords: D-amino acids in plants, D-amino acid biochemistry, functions of D-amino acids in plants and bacteria, D-alanine

1. Introduction

L-amino acids (L-AAs) are the basis of life on our planet (and maybe also on other animate ones), mainly due to their property to be the building blocks of all proteins. These proteinogenic amino acids are also one of the fundamentals of the universality of the genetic code. The limitation of protein coding sequences on 20 different L-AAs was one of the key developments in evolution to ensure the compatibility between different life forms, regardless if they belong to the same species or if their genetic material is exchanged between a bacterial pathogen and its plant host, as it is the case in the *Agrobacterium*-plant relationship.

In the course of limitation to 20 proteinogenic amino acids, also the convention of exclusive usage of L-AAs (homochirality) in the primary structure of proteins evolved to ensure the intended structure and functionality of a protein. But since the very beginning of evolution also the enantiomers of L-AAs, the D-amino acids (D-AAs) were existent. These D-AAs are mainly products of abiotic and enzymatic racemization of L-AAs [1] or synthesized by aminotransferases from other D-AAs [2]. One possibility of organisms during evolution to handle D-AAs would have been to develop mechanisms for their elimination. But instead almost all organismal classes in the tree of life learned to live with substantial amounts of D-AAs and even made use of them. One prominent example of such usage is the bacterial cell wall. It contains many layers of peptidoglycan, polysaccharide chains cross linked by oligopeptides. Parts of these oligopeptides are D-AAs, especially D-Ala and D-Glu, which protect the cell due to their resistance to cleavage by conventional proteases [3].

The decay of the bacterial cell wall is also one of the major sources of D-AAs in soil [4] and the reason why plants are specifically surrounded and challenged by D-AAs. In soil samples, almost all D-enantiomers of proteinogenic amino acids could be found in significant amounts [4–7]. For a long time, it was assumed that D-AAs are just inhibitory for plant growth and therefore plants evolved mechanisms to avoid and eliminate them. But recent studies have shown that plants are instead able to import D-AAs and metabolize them. They even synthesize D-AA themselves for physiological reasons, which raised the question about the beneficial effects of D-AAs for plants. In this review, we want to summarize the current knowledge about these processes and highlight different aspects and questions of future research with a focus on *Arabidopsis thaliana* as a model plant to investigate D-AAs in plants.

2. How do D-AAs get into the plant?

It is a widely accepted fact that plants harbour free D-AAs as they could be identified in different plant species and tissues [8–12]. In this regard, the question arose, if all these amino acids are synthesized by plants themselves or also taken up from the soil. By detecting various D-AAs in seedlings of runner and soy beans, garden and water cress, as well as alfalfa, raised on amino acid free media in Ref. [10], first indirect indications were given that these plants are able to synthesize D-AAs de novo. This hypothesis was supported by the discovery, identification and characterization of alanine, serine and isoleucine racemases from different plant species [13–16]. The toxicity of D-AAs on Arabidopsis [12, 17] and the toxic effect of D-Ser on other species [18] were first hints for a general D-AA uptake mechanism in plants. Furthermore, it was shown that almost all D-enantiomers of proteinogenic amino acids could be detected in Arabidopsis plants after their exogenous application [12, 19]. The direct uptake of D-Ala and its utilization could be demonstrated for the first time in wheat [20].

At that point, it was interesting which transporters are involved in the uptake of D-AAs and which properties they have. One of the first hints in this respect was given by the works of Ref. [21]. In an initial screen, they germinated Arabidopsis mutants on 3 mM D-Ala, a toxic D-Ala concentration for wild-type plants and found plants to survive with mutations in the *LYSINE HISTIDINE TRANSPORTER 1 (LHT1)* gene. Furthermore, the uptake for D-Ala in these plants was reduced by more than 90% in these mutants. Also the uptake of many other D-AAs was found to be reduced in *lht1* mutants [12]. This was the first evidence that a broad range specificity L-AA transporter in plants was also able to take up D-AAs from soil. A second example for plant D-AA transporters are the proline transporters of *A. thaliana* (AtProTs); they facilitate the uptake of L- and D-Pro, and mutants of *AtProT2* show reduced sensitivity against D-Pro [22].

These reports implied that transporters involved in the uptake and transport of L-AAs could also be responsible for the same processes of D-AAs. That D-AA transporting proteins are most probably not restricted to the LHT, and ProT families were given by experiments of our group: in toxicity tests, performed as described before [12], an Arabidopsis mutant of *LHT1* was confirmed to be less sensitive against D-Ala than the corresponding wild type (**Figure 1A**). A mutant of *AAP1*, belonging to the *Amino Acid Permease* family and shown to be responsible for root uptake of uncharged L-AAs [23], revealed a higher resistance against D-Met and D-Phe than Col-0 (**Figure 1B** and **C**). This result implies that AAP1 is involved in the import of D-AAs, specifically of D-Met and D-Phe.

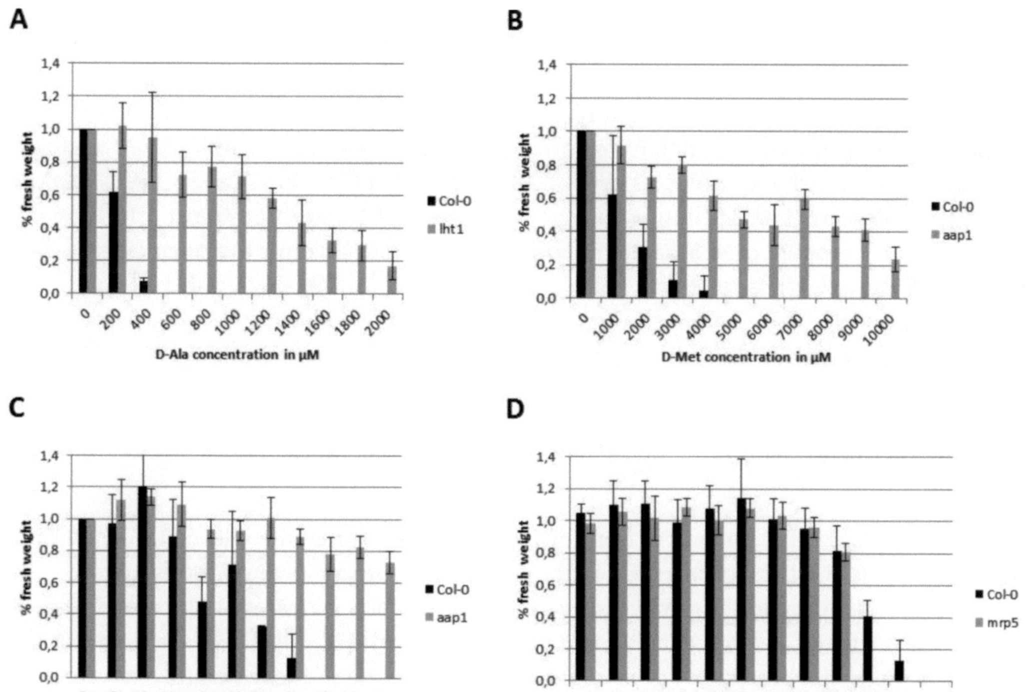

Figure 1. Seed growth inhibition of D-AAs on different transporter mutants compared to their corresponding background accession Col-0.

For these diagrams, different mutants and their corresponding background line Col-0 were germinated for 14 days in ½ MS + 1% sucrose including different D-AAs. Afterward the fresh weight was recorded. For each measurement, three times eight seedlings were measured. In (A), *lht1* was treated with D-Ala, in (B) and (C), *aap1* was treated with D-Met and D-Phe, respectively, and in (D), *mrp5* was treated with D-Phe. Mutant values are always represented by grey blocks, control values in black blocks. All values are calculated and given in relation to the untreated control seedlings. Error bars indicate standard deviation.

But the interpretation of toxicity experiments using transporter mutants should be handled with care as another example shows: in a series of toxicity tests with different D-AAs, an Arabidopsis mutant of *AtMRP5* showed less resistance against D-Phe compared to the corresponding control (**Figure 1D**), instead of increased resistance like in the case of the tested *AAP1* mutant (**Figure 1C**). *AtMRP5* belongs to a gene family of 14 ABC transporters in the Arabidopsis genome [24] and found to transport inositol phosphate for phytate storage [25]. A functionality as amino acid transporter has not been reported for this protein, yet. Surprisingly, at *mrp5* mutant allele showed also drastically reduced root exudation of almost all L-AAs [26]. It is tempting to speculate whether the reduced D-Phe resistance of the *mrp5* mutant in our experiments may be a consequence of reduced exudation of this amino acid, which may lead to accumulation of it to toxic levels.

Altogether the presented studies in this chapter indicate that plants seem to take up D-AAs actively from their rhizosphere. As also shown above candidate transporter proteins for this uptake are found among L-AA transporting proteins like LHT1, ProT2 and AAP1. There is a certain possibility that these three transporters are not the only ones within their families to transport D-AAs. In the Arabidopsis genome, there are 10 LHTs, 3 ProTs and 8 AAPs encoded [22, 27]. This means that at least among the members of these three transporter families, further D-AA transport proteins may be found. Taking into account that members of other amino acid transporter families may also be able to do so, raises the number of candidates in Arabidopsis up to 63. Even more could be found in other plant species, as there are for instance 189 putative amino acid transporter genes encoded in the soybean genome [28].

3. What happens to D-AAs in the plant?

The fact that plants take up D-AAs from their surrounding rhizosphere leads to the question what happens to them in consequence in the plant. An approach to answer this question was given by our group by feeding Arabidopsis mutants and accessions with different D-AAs to measure the D- and L-AA contents in these plants afterwards [12, 29]. These analyses revealed two major metabolic processes which could be observed; one of them was the conversion of particular D-AAs like D-His, D-Met, D-Phe and D-Trp to their L-enantiomers. In this respect, the increase of these L-AAs was about 2–50 times compared to the untreated control plants, depending on the applied D-AA. The other one was the increase of D-Glu and D-Ala contents after treatment with any given D-AA. In this regard, D-Ala was the major compound to be found after D-AA application with concentrations more than 20 times higher than the ones of D-Glu. These observations led to speculations about the metabolic processes responsible for these effects. To explain the outcome of our amino acid profiling three different possibilities of enzymatic reaction have been discussed: racemization, deamination and transamination [4]. Recent studies of our group revealed that as well the D- to L-AA conversion as also the occurrence of D-Glu and D-Ala can be explained by the activity of a single D-AA specific transaminase in the Arabidopsis genome (Suarez et al., unpublished results).

All these studies shifted our focus towards the evolution and metabolic fate of D-Ala in plants. D-Ala appears to be the major product of D-AA metabolism in Arabidopsis, but, at the same time, it is one of the most toxic D-AAs for this species when applied exogenously [12, 17]. This raises the question how plants process D-Ala specifically and why it is the preferred product of D-AA metabolization. Several possibilities for this process are summarized in **Figure 2**.

A common feature of all possible pathways in **Figure 2** is that none of them have been characterized sufficiently to date in plants, especially in Arabidopsis. But there are a series of reports and evidences arguing for the scheme in this figure, which will be discussed in this section.

The ligation of D-Ala to its dipeptide D-Ala-D-Ala is among the best characterized ways to metabolize D-Ala (**Figure 2**). D-Ala-D-Ala could be detected in different grasses and tobacco long before [30–32], indicating the existence of a D-Ala ligating enzyme. Recently, this enzyme, D-Ala-D-Ala ligase (DDL), could be characterized physiologically for the first time from a plant source, PpDDL1 from the moss *Physcomitrella patens* [33]. As it can be seen in **Table 1** also in the Arabidopsis genome, a DDL encoding gene could be found, an orthologue of PpDDL1, which has not been characterized biochemically, yet. The situation is similar for a putative D-amino acid oxidase (DAO) from Arabidopsis: Its homologue from maize has been biochemically characterized and shown to oxidize preferably D-Ala [34], but its Arabidopsis homologue has not been characterized biochemically or physiologically, yet.

When it comes to the alanine racemase in plants (**Figure 2**), knowledge is rather scarce; Alanine racemase enzyme activity and its corresponding enzyme activity could be isolated and measured in *Chlamydomonasreinhardtii* and alfalfa [13, 35], but identification of the

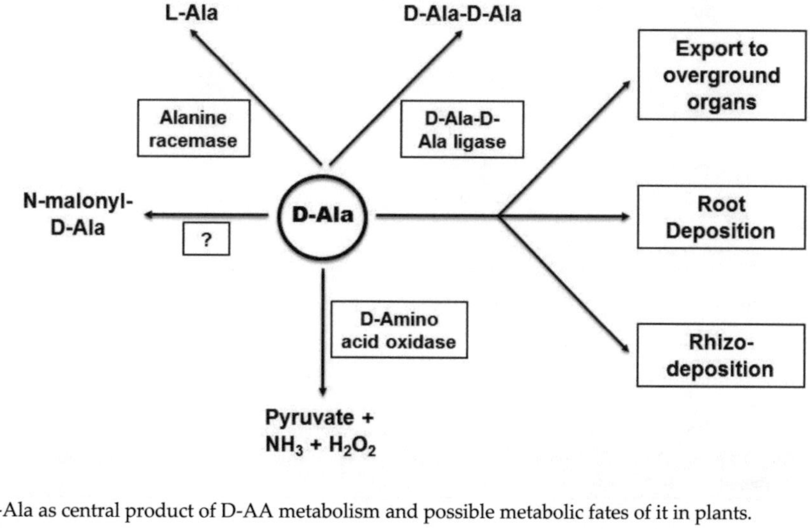

Figure 2. D-Ala as central product of D-AA metabolism and possible metabolic fates of it in plants.

corresponding gene is still pending. In the Arabidopsis genome, two genes with homologies to alanine racemases are annotated (**Table 1**), but characterization is still pending. Even less is known about malonylation of D-Ala, which is given as the fourth major enzymatic way to metabolize D-Ala in **Figure 2**: In pea seedlings, N-malonyl-D-Ala had been detected [36, 37]. Additionally, in mung bean seedlings, D-Ala malonylating activity could be shown [38], but the corresponding enzyme still awaits identification.

Apart from the enzymatic metabolization of D-Ala, other ways of its deposition have to be taken into account as depicted in **Figure 2**: The spatial distribution of D-Ala within the plant, but also of D-AAs in general, would be of interest in this respect. Then, the question could be answered if D-Ala is deposited in the root or if it is transported to other organs, in order to dilute its toxicity. Another possibility would be rhizodeposition, the exudation of metabolites from the root into soil. Rhizodeposition of L-AAs has been shown for plants several times [26, 39, 40]. This process has a strong impact on the microbial community in the rhizosphere, but reports of rhizodeposition of D-AAs are still missing.

Another look into **Table 1** reveals further D-AA processing enzymes in the Arabidopsis genome apart from either synthesizing or metabolizing D-Ala. First of all, there are four D-AA

Function	Name	AGI code	Localization	References
D-amino acid transaminase	AtDAAT1	At5g57850	chloroplast	[41]
		At3g05190	(unknown)	
		At3g54970	(unknown)	
		At5g27410	(unknown)	
Alanine racemase		At1g11930	(chloroplast)	
		At4g26860	(chloroplast)	
D-amino acid racemase	AtDAAR1	At4g02850	(unknown)	[16]
	AtDAAR2	At4g02860	(cytosol)	[16]
Serine racemase	AtSR1	At4g11640	(unknown)	[14]
Asp-Glu racemase		At1g15410	(chloroplast)	
D-aminoacyl-tRNAdeacylase	AtGEK1	At2g03800	cytosol, nucleus	[55]
D-Tyr-tRNAdeacylase		At4g18460	(unknown)	
D-amino acid oxidase		At5g67290	(chloroplast)	
D-Cysdesulfhydrase		At1g48420	mitochondria	[42]
D-Ala-D-Ala ligase		At3g08840	(chloroplast)	

Localization refers to the experimentally determined subcellular localization; localization predictions on the basis of peptide sequencing data by the Plant Proteome Database (PPD; http://ppdb.tc.cornell.edu) and unknown localizations are given in parentheses.

Table 1. Putative D-AA metabolizing genes in the Arabidopsis genome.

specific transaminases, from which just one has been shown to produce D-Ala and preferably D-Glu using various D-AAs as substrates with different affinities [41]. Beside the already mentioned alanine racemase, also three other racemases with specificities for other amino acids can be found in the Arabidopsis genome: First, there is a putative Asp-Glu racemase encoded in the Arabidopsis genome, but currently, there are no reports available about it. The second one is the serine racemase AtSR1, which catalyses the racemization of serine, but also to lesser extent alanine, arginine and glutamine. Beside its racemase activity, it acts also as a dehydratase on D- and L-serine [13]. The third one are the so-called D-amino acid racemases AtDAAR1 and AtDAAR2, which are indeed specific for Ile with D-*allo*-Ile as a product. Leu and Val were just racemized with 1 and 5% relative activity, respectively [16]. The D-Cys desulfhydrase from Arabidopsis is another example for a D-AA metabolizing enzyme with a specificity apart from D-Ala; this specificity to catabolize D-Cys to pyruvate, NH_3 and H_2S has been shown previously [42], but the physiological function of this enzyme and especially of D-Cys still remains unclear and will be discussed in the next chapter. Altogether the collection of D-AA processing enzymes in **Table 1** is a reminder that D-Ala seems to be central product of D-AA metabolism, but that there are far more putative enzyme encoding genes annotated to produce and process also other D-AAs.

4. What are the effects and functions of D-AAs in the plant?

The abundance and fate of D-AAs in plants are indicators that these compounds are actively processed and therefore play a role in the physiology of plants. This leads to the question: Which role(s) are these? In the last years, three different scenarios about the effects of D-AAs on plants were discussed. The first one was that D-AAs have either no effect on plants or even inhibit growth and therefore have to be considered as toxins for plants [19]. In contrast, it could be shown before that at least D-Ile and D-Val promoted seedling growth [17], and that for different D-AAs even the highest tested concentration did not cause growth inhibition [12]. Together with the de novo synthesis of various D-AAs in plants described above a general toxic function of all D-AAs is rather unlikely and depends on dosage, which also applies to many L-AAs.

There are also other arguments speaking against this scenario like the utilization of D-AAs as possible nitrogen source [4, 40], which is the second major postulated function of D-AAs in plants. In this respect, it could be shown that wheat plants are able to assimilate D-Ala as well as D-trialanine, which they took up from the soil [20]. This was the first evidence of direct utilization of D-AAs as a nitrogen source. Additionally, it revealed that plants are able to utilize not just free forms of D-AAs but also as oligomers, as also found as a degradation product of the bacterial cell wall. Nevertheless, more plant species and other D-AAs have to be analysed in this respect to confirm the general utilization of D-AAs as nitrogen sources for plants.

The third major complex of functions of D-AAs in plants is the ones, which have been either just recently discovered, and need to be further analysed and characterized, or which have not been discovered at all. Among these novel functions is, for instance, D-Ala as a stress signal: It has been reported that duckweed seedlings accumulate D-Ala after UV light exposure [43], but the confirmation of this finding by other groups or in other species is still pending. Another, more prominent, example of a novel physiological function of a D-AA in plants is the impact of D-Ser on pollen tube growth in Arabidopsis [44]; In that report, the authors provided evidence that D-Ser affects pollen tube growth via its agonistic action on the glutamate receptor AtGLR1.2. Furthermore, it was shown that the loss of the serine racemase AtSR1 leads to aberrant pollen tube growth. AtGLR1.2 belongs to a protein family of 20 members in Arabidopsis with highest homologies towards the mammalian ionotropic glutamate receptors (GLRs), also known as N-methyl-D-Aspartate (NMDA) receptors [45]. In humans and other mammalians, these receptors, involved in neurotransmission, have been shown to be activated by L-Glu and D-Ser synergistically [46]. The homologous action of D-Ser on GLRs in animals and plants together with the relatively large number of GLRs in the Arabidopsis genome implies further effects of D-Ser on physiological functions in plants, especially on pathogen response, which may be regulated by GLRs, too [45].

Another type of novel functions of D-AAs was unravelled by the analyses of the structure of the chloroplast membrane of mosses [33]. The authors provided evidence that the membranes of chloroplasts from the moss *P. patens* contain the dipeptide D-Ala-D-Ala and therefore possess a major structural component of peptidoglycan, the building block of bacterial cell walls [3]. Another indication of the structural similarity of bacterial cell wall and plastidial envelopes in mosses was given by genetical experiments. Loss-of-function mutants of the Physcomitrella D-Ala-D-Ala ligase, *PpDDL1*, were not able for chloroplast division and therefore showed megachloroplasts in their protonema cells [33]. All these findings fit to the observation made before in the Physcomitrella genome, which harboured all gene homologues from bacteria to synthesize peptidoglycan including *PpDDL1* [47]. The structural similarity between bacterial cell walls and plastidial envelopes seems to be limited to cryptogamic plants, because loss-of-function mutants of *AtDDL1* did not show the megachloroplast phenotype observed in mosses [33]. This observation seems to be in concordance to the situation in the Arabidopsis genome which harbours just four homologues of the ten mentioned genes needed for peptidoglycan synthesis [47]. It is interesting in this respect that homologues of these four genes were found in all higher plant genomes [48].

Nevertheless, D-Ala seems to play a role in chloroplasts of higher plants as well. Many proteins directly involved in D-Ala metabolism in Arabidopsis were either found in the chloroplast or were predicted to be localized there (**Table 1**). Furthermore, we were able to synthesize a fluorescent D-Ala analogue, HADA (7-hydroxycoumarin-3-amino-D-alanine), according to a previously published protocol [49], and fed it to Arabidopsis seedlings. In these experiments, we could trace the HADA fluorescence evenly distributed in the chloroplasts (**Figure 3**). This targeting of the D-Ala analogue to the chloroplasts indicated a central metabolization of this compound in this compartment. The even distribution of it, in contrast to the accumulation, found in moss chloroplast envelopes [33] points to a different function of D-Ala in higher plant chloroplasts, which still awaits to be unravelled.

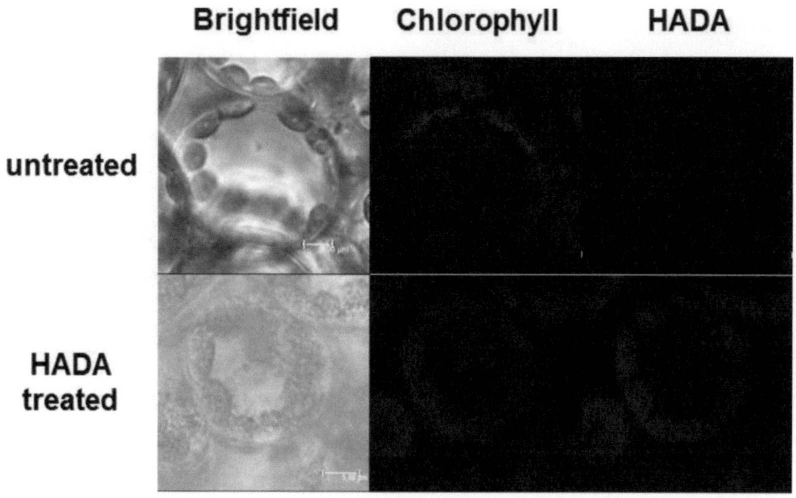

Figure 3. Fluorescent D-Ala analogue HADA accumulates in chloroplasts of Arabidopsis leaves.

Fourteen days old *Col-0* seedlings grown in liquid culture were incubated overnight in 0.1 mM HADA and then analysed microscopically. The pictures in the first column show bright field images of sponge parenchyma cells. The chlorophyll in the chloroplasts was detected by its autofluorescence (Chlorophyll, second column), and fluorescence of HADA was recorded in the DAPI channel (HADA, third column) with a laser-scanning microscope. The upper row shows cells without HADA treatment (untreated) as control, the lower one with HADA treatment (HADA treated). The size bars indicate 5 μm.

Among D-AAs with novel functions in plants, there are D-AAs to be known to affect specific proteins, but not how they cause the associated physiological reactions. One example for such a relationship is the one between D-Cys and drought resistance. As described above, a desulfhydrase specific for D-Cys could be characterized from Arabidopsis, which also produces H_2S [42]. In further experiments, it turned out that increased H_2S production leads to enhanced drought resistance [50], which can be partially assigned to increased D-Cysdesulfhydrase activity [51]. This effect seems to be related to ethylene induced stomatal aperture. Furthermore, H_2S production leads to cross adaptation of plants to several other stress factors [52]. But nevertheless, the source of D-Cys, its significance in stress signalling and adaptation and its detailed way of action still need to be elucidated.

Another example for an enigmatic relation between a D-AA and physiological response was described previously: In this case, an Arabidopsis mutant with hypersensitivity to ethanol, *gek1*, was isolated [53], where the respective mutation could be assigned to a D-aminoacyl-tRNAdeacylase (*AtGEK1*) [54]. Later, it was found that this gene encodes an active enzyme with broad substrate specificity [55]. But its overexpression led neither in *Escherichia coli* nor in yeast to an increase of ethanol tolerance [54]. Therefore, a functional explanation how the loss of *AtGEK1*, and therefore the inability to repair accidental loading of tRNAs with D-AAs, causes ethanol hypersensitivity in plants is still missing.

5. Conclusions

As discussed largely in this text, knowledge gathered in the last decade implies that D-AAs are involved in more plant physiological processes than assumed before. Furthermore, the view of D-AAs as generally toxic molecules needs to be changed to a view of them as physiologically active compounds, which can cause detrimental effects by over dosage. Therefore, the investigation of their uptake and elimination, their metabolic pathways, and their physiological functions in plants will gain more interest and significance in the future.

To understand the transport processes for uptake, distribution (intracellular, intercellular and also at long distances) as well as of possible excretion of D-AAs will be one of the major fields to be investigated. Although there are candidate transport proteins given to be analysed with the classical L-AA transporters as described above, also other, yet unknown, proteins may contribute to D-AA transport processes, as the D-Phe toxicity on *mrp5* mutants demonstrates (**Figure 1**). In this regard, one of the first questions to be solved would be the proof of active uptake of D-AAs by candidate transporters instead of indirect evidences.

When it comes to the metabolization of D-AAs, the formation of D-Ala appeared to be central, which puts also this molecule into the centre of future investigations. As it has been shown in the preceding chapter, D-Ala seems to accumulate in the chloroplast. In this respect, D-Ala may play a double role. On the one hand, it is an intermediate metabolite, which needs to be further metabolized due its toxicity in excess concentration. In this regard, the different putative metabolic pathways await elucidation. On the other hand, D-Ala is a physiologically active compound as it has been shown as a building block of moss chloroplast envelopes. In this context, the function of D-Ala in chloroplasts of higher plant and the concentration of different D-AA-related enzymes in this compartment will be of specific interest. Furthermore, it would be interesting if also other D-AAs take this way over chloroplasts and what functions they fulfil there.

Finally, which physiological role(s) the different D-AAs play in plants are the major questions to be solved. One of these questions will be, to which extent and under which circumstances D-AAs from the rhizosphere are utilized as nitrogen sources. As it was mentioned above, the accumulation of D-Ala and of D-AA-related enzymes in the chloroplasts may point to their involvement in plastid biogenesis, assembly and maintenance. The unresolved functions of D-Cys in stress resistance and the unclear involvement of AtGEK1 in ethanol resistance indicate that there is a high probability that there are still many D-AA-related functions and processes in plants waiting to be discovered.

Acknowledgements

We are grateful to Prof. Klaus Harter (Centre for Plant Molecular Biology, University of Tübingen, Germany) for the provision of research facilities, where the majority of our studies were performed. Studies of Juan Suarez on the effects of D-AAs on Arabidopsis were supported by the DeutscherAkademischerAustauschdienst (DAAD 91567028). Additionally, we would like to thank to Laura Menius for her work on the toxicity of D-AAs on the *mrp5* mutant.

Author details

Üner Kolukisaoglu* and Juan Suarez

*Address all correspondence to: uener.kolukisaoglu@zmbp.uni-tuebingen.de

Department of Plant Physiology, Center for Plant Molecular Biology (ZMBP), Eberhard Karls University of Tübingen, Tübingen, Germany

References

[1] Friedman M. Chemistry, nutrition, and microbiology of D-amino acids. Journal of Agricultural and Food Chemistry. 1999;**47**:3457–3479. DOI: 10.1021/jf990080u

[2] Radkov AD, Moe LA. Bacterial synthesis of D-amino acids. Applied Microbiology and Biotechnology. 2014;**98**:5363–5374. DOI: 10.1007/s00253-014-5726-3

[3] Typas A, Banzhaf M, Gross CA, Vollmer W. From the regulation of peptidoglycan synthesis to bacterial growth and morphology. Nature Reviews Microbiology. 2011;**10**:123–136. DOI: 10.1038/nrmicro2677

[4] Vranova V, Zahradnickova H, Janous D, Skene KR, Matharu KR, Rejsek K, Formanek P. The significance of D-amino acids in soil, fate and utilization by microbes and plants: Review and identification of knowledge gaps. Plant and Soil. 2012;**354**:21–39. DOI: 10.1007/s11104-011-1059-5

[5] Pollock GE, Cheng CN, Cronin SE. Determination of the D and L isomers of some protein amino acids present in soils. Analytical Chemistry. 1977;**49**:2–7. DOI: 10.1021/ac50009a008

[6] Amelung W, Zhang X. Determination of amino acid enantiomers in soil. Soil Biology and Biochemistry. 2001;**33**:553–562. DOI: 10.1016/S0038-0717(00)00195-4

[7] Amelung W, Zhang X, Flach KW. Amino acids in grassland soils: Climatic effects on concentrations and chirality. Geoderma. 2006;**130**:207–217. DOI: 10.1016/j.geoderma.2005.01.017

[8] Robinson T. D-amino acids in higher plants. Life Sciences. 1976;**19**:1097–1102. DOI: 10.1016/0024-3205(76)90244-7

[9] Brückner H, Westhauser T. Chromatographic determination of D-amino acids as native constituents of vegetables and fruits. Chromatographia. 1994;**39**:419–426. DOI: 10.1007/BF02278756

[10] Brückner H, Westhauser T. Chromatographic determination of L- and D-amino acids in plants. Amino Acids. 2003;**24**:43–55. DOI: 10.1007/s00726-002-0322-8

[11] Herrero M, Ibáñez E, Martín-Alvarez PJ, Cifuentes A. Analysis of chiral amino acids in conventional and transgenic maize. Analytical Chemistry. 2007;**79**:5071–5077. DOI: 10.1021/ac070454f

[12] Gördes D, Kolukisaoglu Ü, Thurow K. Uptake and Conversion of D-Amino Acids in *Arabidopsis thaliana*. Amino Acids. 2011;**40**:553–563. DOI: 10.1007/s00726-010-0674-4

[13] Ono K, Yanagida K, Oikawa T, Ogawa T, Soda K. Alanine racemase of alfalfa seedlings (*Medicagosativa* L.): First evidence for the presence of an amino acid racemase in plants. Phytochemistry. 2007;**67**:856–860. DOI: 10.1016/j.phytochem.2006.02.017

[14] Fujitani Y, Nakajima N, Ishihara K, Oikawa T, Ito K, Sugimoto M. Molecular and biochemical characterization of a serine racemase from *Arabidopsis thaliana*. Phytochemistry. 2006;**67**:668–674. DOI: 10.1016/j.phytochem.2006.01.003

[15] Fujitani Y, Horiuchi T, Ito K, Sugimoto M. Serine racemases from barley, *Hordeum vulgare* L., and other plant species represent a distinct eukaryotic group: Gene cloning and recombinant protein characterization. Phytochemistry. 2007;**68**:1530–1536. DOI: 10.1016/j.phytochem.2007.03.040

[16] Strauch RC, Svedin E, Dilkes B, Chapple C, Li X. Discovery of a novel amino acid racemase through exploration of natural variation in *Arabidopsis thaliana*. Proceedings of the National Academy of Sciences U S A. 2015;**112**:11726–11731. DOI: 10.1073/pnas.1503272112

[17] Erikson O, Hertzberg M, Näsholm T. A conditional marker gene allowing both positive and negative selection in plants. Nature Biotechnology. 2004;**22**:455–458. DOI: 10.1038/nbt946

[18] Erikson O, Hertzberg M, Näsholm T. The *dsdA* gene from *Escherichia coli* provides a novel selectable marker for plant transformation. Plant Molecular Biology. 2005;**57**:425–433. DOI: 10.1007/s11103-004-7902-9

[19] Forsum O, Svennerstam H, Ganeteg U, Näsholm T. Capacities and constraints of amino acid utilization in Arabidopsis. New Phytologist. 2008;**179**:1058–1069. DOI: 10.1111/j.1469-8137.2008.02546.x

[20] Hill PW, Quilliam RS, DeLuca TH, Farrar J, Farrell M, Roberts P, Newsham KK, Hopkins DW, Bardgett RD, Jones DL. Acquisition and assimilation of nitrogen as peptide-bound and D-enantiomers of amino acids by wheat.PLoS One. 2011;**6**:e1922. DOI: 10.1371/journal.pone.0019220

[21] Svennerstam H, Ganeteg U, Bellini C, Näsholm T. Comprehensive screening of Arabidopsis mutants suggests the lysine histidine transporter 1 to be involved in plant uptake of amino acids. Plant Physiology. 2007;**143**:1853–1860. DOI: 10.1105/tpc.106.041012

[22] Lehmann S, Gumy C, Blatter E, Boeffel S, Fricke W, Rentsch D. In planta function of compatible solute transporters of the AtProT family. Journal of Experimental Botany. 2011;**62**:787–796. DOI: 10.1093/jxb/erq320

[23] Lee YH, Foster J, Chen J, Voll LM, Weber AP, Tegeder M. AAP1 transports uncharged amino acids into roots of Arabidopsis. The Plant Journal. 2007;**50**:305–319. DOI: 10.1111/j.1365-313X.2007.03045.x

[24] Verrier PJ, Bird D, Burla B, Dassa E, Forestier C, Geisler M, Klein M, Kolukisaoglu U, Lee Y, Martinoia E, Murphy A, Rea PA, Samuels L, Schulz B, Spalding EJ, Yazaki K, Theodoulou FL. Plant ABC proteins – A unified nomenclature and updated inventory. Trends in Plant Sciences. 2008;**13**:151–159. DOI: 10.1016/j.tplants.2008.02.001

[25] Nagy R, Grob H, Weder B, Green P, Klein M, Frelet-Barrand A, Schjoerring JK, Brearley C, Martinoia E. The Arabidopsis ATP-binding cassette protein AtMRP5/AtABCC5 is a high affinity inositol hexakisphosphate transporter involved in guard cell signaling and phytate storage. The Journal of Biological Chemistry. 2009;**284**:33614–33622. DOI: 10.1074/jbc.M109.030247

[26] Zhou D, Huang XF, Chaparro JM, Badri DV, Manter DK, Vivanco JM, Guo J. Root and bacterial secretions regulate the interaction between plants and PGPR leading to distinct plant growth promotion effects. Plant and Soil. 2016;**401**:259. DOI: 10.1007/s11104-015-2743-7

[27] Tegeder M, Ward JM. Molecular Evolution of Plant AAP and LHT Amino Acid Transporters. Frontiers in Plant Science. 2012;**13**:21. DOI: 10.3389/fpls.2012.00021

[28] Cheng L, Yuan HY, Ren R, Zhao SQ, Han YP, Zhou QY, Ke DX, Wang YX, Wang L. Genome-wide identification, classification, and expression analysis of *Amino Acid Transporter* gene family in *Glycine Max*. Frontiers in Plant Science. 2016;**20**;7:515. DOI: 10.3389/fpls.2016.00515

[29] Gördes D, Koch G, Thurow K, Kolukisaoglu U. Analyses of Arabidopsis ecotypes reveal metabolic diversity to convert D-amino acids. Springerplus. 2013;**2**:559. DOI: 10.1186/2193-1801-2-559

[30] Noma M, Noguchi M, Tamaki E. Isolation and characterization of D-alanyl-D-alanine from tobacco leaves. Agricultural and Biological Chemistry. 1973;**37**:2439. DOI: 10.1271/bbb1961.37.2439

[31] Frahn JL, Illman RJ. The occurrence of D-alanine and D-alanyl-D-alanine in *Phalaristube-rosa*. Phytochemistry. 1975;**14**:1464–1465

[32] Manabe H. Occurrence of D-Alanyl-D-alanine in *Oryzaaustraliensis*. Agricultural and Biological Chemistry. 1985;**49**:1203–1204. DOI: 10.1080/00021369.1985.10866877

[33] Hirano T, Tanidokoro K, Shimizu Y, Kawarabayasi Y, Ohshima T, Sato M, Tadano S, Ishikawa H, Takio S, Takechi K, Takano H. Moss chloroplasts are surrounded by a peptidoglycan wall containing D-amino acids. Plant Cell. 2016;**28**:1521–1532. DOI: 10.1105/tpc.16.00104

[34] Gholizadeh A, Kohnehrouz BB. Molecular cloning and expression in *Escherichia coli* of an active fused *Zea mays L.* D-amino acid oxidase. Biochemistry (Moscow). 2009;**74**:137–144. DOI: 10.1134/S0006297909020035

[35] Nishimura K, Tomoda Y, Nakamoto Y, Kawada T, Ishii Y, Nagata Y. Alanine racemase from the green alga *Chlamydomonasreinhardtii*. Amino Acids.2007;**32**:59–62. DOI:10.1007/s00726-006-0352-8

[36] Fukuda M, Tokumura A, Ogawa T. D-Alanine in germinating *Pisumsativum* seedlings. Phytochemistry. 1973;**12**:2893–2595. DOI: 10.1016/0031-9422(73)85061-7

[37] Ogawa T, Fukuda M, Sasaoka K. Occurrence of N-malonyl-D-alanine in pea seedlings. Biochimica et Biophysica Acta. 1973;**297**:60–69. DOI: 10.1016/0304-4165(73)90049-4

[38] Kionka C, Amrhein N. The enzymatic malonylation of 1-aminocyclopropane-1-carboxylic acid in homogenates of mung-bean hypocotyls. Planta. 1984;**162**:226–35. DOI: 10.1007/BF00397444

[39] Chaparro JM, Badri DV, Bakker MG, Sugiyama A, Manter DK, Vivanco JM. Root exudation of phytochemicals in arabidopsis follows specific patterns that are developmentally programmed and correlate with soil microbial functions. PLoS One. 2013;**8**:e55731. DOI:10.1371/journal.pone.0055731

[40] Moe LA. Amino acids in the rhizosphere: From plants to microbes. American Journal of Botany. 2013;**100**:1692–1705. DOI: 10.3732/ajb.1300033

[41] Funakoshi M, Sekine M, Katane M, Furuchi T, Yohda M, Yoshikawa T, Homma H. Cloning and functional characterization of *Arabidopsis thaliana* D-amino acid aminotransferase – D-aspartate behavior during germination. The FEBS Journal. 2008;**275**:1188–1200. DOI: 10.1111/j.1742-4658.2008.06279.x

[42] Riemenschneider A, Wegele R, Schmidt A, Papenbrock J. Isolation and characterization of a D-cysteine desulfhydrase protein from *Arabidopsis thaliana*. The FEBS Journal. 2005;**272**:1291–1304. DOI: 10.1111/j.1742-4658.2005.04567.x

[43] Monselise EB, Levkovitz A, Kost D. Ultraviolet radiation induces stress in etiolated *Landoltiapunctata*, as evidenced by the presence of alanine, a universal stress signal: a ^{15}N NMR study. Plant Biology (Stuttg). 2015;**17**(Suppl 1):101–107. DOI: 10.1111/plb.12198

[44] Michard E, Lima PT, Borges F, Silva AC, Portes MT, Carvalho JE, Gilliham M, Liu LH, Obermeyer G, Feijó JA. Glutamate receptor-like genes form Ca^{2+} channels in pollen tubes and are regulated by pistil D-serine. Science. 2011;**332**:434–437. DOI: 10.1126/science.1201101

[45] Forde BG, Roberts MR. Glutamate receptor-like channels in plants: A role as amino acid sensors in plant defence? F1000Prime Rep. 2014;**6**:37. DOI: 10.12703/P6-37

[46] Wolosker H. NMDA receptor regulation by D-serine: New findings and perspectives. Molecular Neurobiology. 2007;**36**:152–164. DOI: 10.1007/s12035-007-0038-6

[47] Machida M, Takechi K, Sato H, Chung SJ, Kuroiwa H, Takio S, Seki M, Shinozaki K, Fujita T, Hasebe M, Takano H. Genes for the peptidoglycan synthesis pathway are essential for chloroplast division in moss. Proceedings of the National Academy of Sciences U S A. 2006;**103**:6753–6758. DOI: 10.1073/pnas.0510693103

[48] van Baren MJ, Bachy C, Reistetter EN, Purvine SO, Grimwood J, Sudek S, Yu H, Poirier C, Deerinck TJ, Kuo A, Grigoriev IV, Wong CH, Smith RD, Callister SJ, Wei CL, Schmutz J, Worden AZ. Evidence-based green algal genomics reveals marine diversity

and ancestral characteristics of land plants. BMC Genomics. 2016;**17**:267. DOI: 10.1186/s12864-016-2585-6

[49] Kuru E, Tekkam S, Hall E, Brun YV, Van Nieuwenhze MS. Synthesis of fluorescent D-amino acids and their use for probing peptidoglycan synthesis and bacterial growth in situ. Nature Protocols. 2015;**10**:33–52. DOI: 10.1038/nprot.2014.197

[50] Jin Z, Shen J, Qiao Z, Yang G, Wang R, Pei Y. Hydrogen sulfide improves drought resistance in *Arabidopsis thaliana*. Biochemical and Biophysical Research Communications. 2011;**414**:481–486. DOI: 10.1016/j.bbrc.2011.09.090

[51] Hou Z, Wang L, Liu J, Hou L, Liu X. Hydrogen sulfide regulates ethylene-induced stomatal closure in *Arabidopsis thaliana*. Journal of Integrative Plant Biology. 2013;**55**:277–289. DOI: 10.1111/jipb.12004

[52] Li ZG, Min X, Zhou ZH. Hydrogen Sulfide: A Signal Molecule in Plant Cross-Adaptation. Frontiers in Plant Science. 2016;**7**:1621. DOI: 10.3389/fpls.2016.01621

[53] Hirayama T, Fujishige N, Kunii T, Nishimura N, Iuchi S, Shinozaki K. A novel ethanol-hypersensitive mutant of Arabidopsis. Plant and Cell Physiology. 2004;**45**:703–711. DOI: 10.14841/jspp.2008.0.0890.0

[54] Fujishige N, Nishimura N, Iuchi S, Kunii T, Shinozaki K, Hirayama T. A novel Arabidopsis gene required for ethanol tolerance is conserved among plants and archaea. Plant and Cell Physiology. 2004;**45**:659–666. DOI: 10.1093/pcp/pch086

[55] Wydau S, Ferri-Fioni ML, Blanquet S, Plateau P. GEK1, a gene product of *Arabidopsis thaliana* involved in ethanol tolerance, is a D-aminoacyl-tRNAdeacylase. Nucleic Acids Research. 2007;**35**:930–938. DOI: 10.1093/nar/gkl1145

Chapter 8

Amino Acid Metabolism and Transport in Soybean Plants

Takuji Ohyama, Norikuni Ohtake, Kuni Sueyoshi,
Yuki Ono, Kotaro Tsutsumi, Manabu Ueno,
Sayuri Tanabata, Takashi Sato and
Yoshihiko Takahashi

Additional information is available at the end of the chapter

http://dx.doi.org/10.5772/intechopen.68992

Abstract

The ammonium produced by nitrogen fixation in the bacteroid is rapidly excreted to cytosol of infected cell of soybean nodules and then assimilated into glutamine and glutamic acid, by glutamine synthetase/glutamate synthase pathway. Most of the nitrogen is further assimilated into ureides, allantoin, and allantoic acid, via purine synthesis, and they are transported through xylem to the shoots. Nitrate absorbed in the roots is reduced by nitrate reductase and nitrite reductase to ammonia either in the roots or leaves. The ammonia is also assimilated by glutamine synthetase/glutamate synthase pathway, and mainly transported by asparagine, and not ureides. The nitrogen transported into leaves is readily utilized for protein synthesis, and then, some of them are decomposed and retransported to roots, apical shoots, and pods via phloem mainly in the form of asparagine.

Keywords: soybean, amino acid, nitrogen fixation, nitrate absorption, nodule, root, leaf

1. Introduction

1.1. Role of amino acids in plants

Plants are photoautotrophs, and they can synthesize all organic compounds from inorganic materials such as carbon dioxide (CO_2), water (H_2O), and minerals using light energy. Amino acids are the key metabolites in nitrogen (N) metabolism of higher plants. First, the inorganic

N, such as ammonium absorbed in the roots or produced from nitrate reduction, nitrogen fixation, and photorespiration, is initially assimilated into glutamine (Gln) and glutamate (Glu) by the glutamine synthetase (GS)/glutamate synthase (GOGAT) pathway. Second, amino acids are the essential components of proteins. Third, amino acids are used for long-distance transport of nitrogen among organs (roots, nodules, stems, leaves, pod, seeds, and apical buds) through xylem or phloem. Fourth, nonprotein amino acids may play a role in protecting plants from feeding damages by animals, insects, or infection by fungi. In this chapter, we would like to review the amino acid metabolism in soybean nodules, roots, leaves, pods, and seeds. In addition, we will introduce the amino acids transport via xylem and phloem among organs.

Soybean plants absorb inorganic N from the roots, and they can fix atmospheric N_2 in the nodules associated with soil bacteria rhizobia. **Figure 1** shows a model of nutrients and water flow via xylem and phloem in soybean plants. Soybean roots absorb water and nutrients in soil solution, and they are transported to the shoots via xylem vessels by the transpiration and root pressure. The fixed N in nodule is also transported to the shoots via xylem. On the other hand, photoassimilates (mainly sucrose), amino acids (Asn, etc.), and minerals (potassium, etc.) are transported from leaves to the apical buds, roots, nodules, and pods via the phloem by osmotic pressure or protoplasmic streaming.

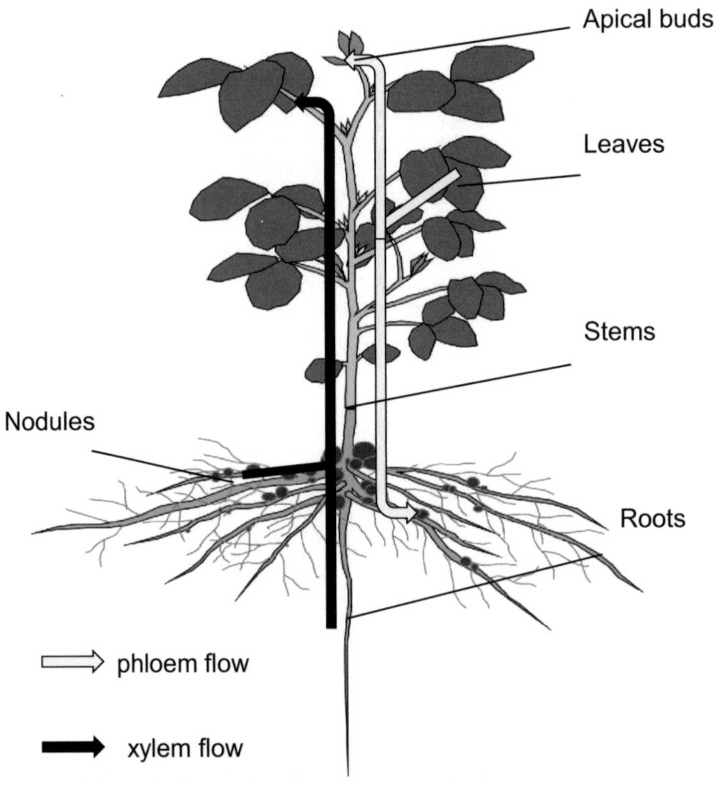

Figure 1. Transport pathways of nutrients and water by xylem and phloem in a soybean plant.

Figure 2 shows the distribution of radioactivity showing xylem flow (**Figure 2A**) [1] or phloem flow (**Figure 2B**) [2]. **Figure 2A** shows the positron imaging of the distribution of radioactivity in nodulated soybean (T202) after 1 hour of $^{13}NO_3^-$ supply to the root medium [1]. All parts of the roots exhibited the highest radioactivity (red), and stems and first trifoliolate leaf were relatively high (yellow). The radioactivity was not observed in the nodules, although they are attached in the roots. **Figure 2B** shows the positron imaging of distribution of radioactivity in nodulated soybean (cv. Williams). After ^{11}C-labeled CO_2 was exposed to the first trifoliolate leaf, and the radioactivity was monitored after 2 hours [2], the highest radioactivity was shown in the $^{11}CO_2$-fed leaf (red) and stems (red) with apical bud (red) and root (yellow) and nodules (red). No radioactivity was observed in the primary leaf and other matured leaves. Nodules showed a higher radioactivity than that in the roots.

1.2. Role of amino acids on inorganic nitrogen assimilation

Ammonium ion (NH_4^+) is first assimilated into glutamine (Gln) combined with glutamic acid (Glu) by the enzyme glutamine synthetase (GS) consuming one molar of ATP (**Figure 3**). Then, the amide group of Gln is transferred to an organic acid, 2-oxoglutarate (2-OG), by glutamate synthase (GOGAT) in plastids using 2 molar of reduced Feredoxin (Fd_{red}). Previously, NH_4^+ was considered to be initially assimilated into Glu by glutamate dehydrogenase (GDH). However, the enzyme GOGAT has been discovered in nitrogen-fixing bacteria *Aerobacter aerogenes* in 1970 [3], and GS/GOGAT cycle has been confirmed as the principal route of ammonium assimilation in plants [4–8].

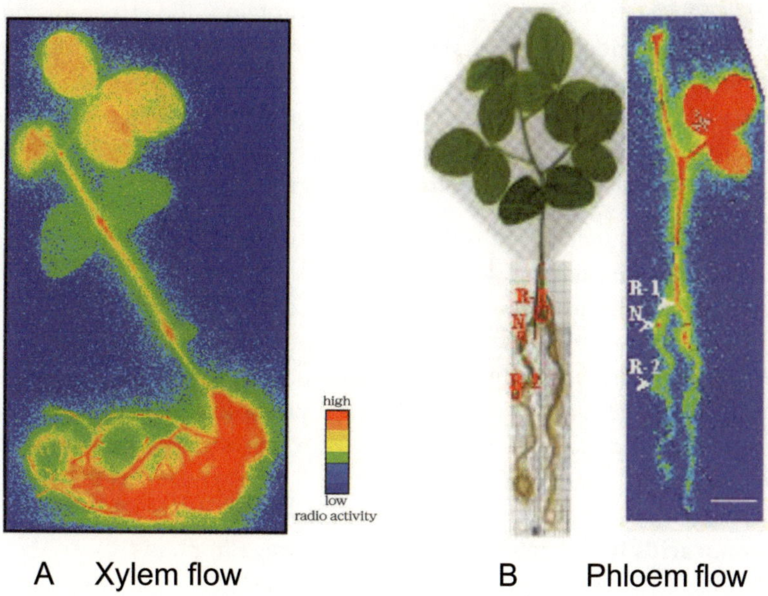

Figure 2. Positron imaging pictures of xylem flow and phloem flow in nodulated soybean plants. (A) Distribution of radioactivity after $^{13}NO_3^-$ was supplied from the roots. (B) Distribution of radioactivity after $^{11}CO_2$ was exposed to the first trifoliolate leaves of a split-rooted soybean.

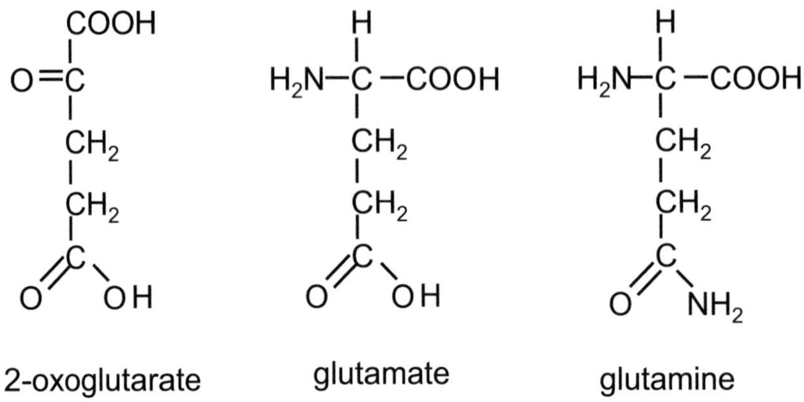

Figure 3. Chemical structures of 2-oxoglutarate, glutamate, and glutamine.

Nitrate is most abundant inorganic nitrogen in upland fields, because NH_4^+ is readily oxidized to NO_3^- by nitrifying bacteria under aerobic conditions. NO_3^- is reduced to nitrite (NO_2^-) by the enzyme nitrate reductase (NR) using one molar of NADH or NADPH as a reductant. The NO_2^- is transported to plastids and further reduced to NH_4^+ by the enzyme nitrite reductase (NiR) using 6 molar of Fd_{red}.

1.3. Role of amino acids for synthesis of proteins and nucleic acids in plants

Protein is a polymer or a complex of polymers of 20 amino acids in higher plants and plays an essential role on metabolism as enzymes, storage proteins, and structure components of the cells. The 20 amino acids consist of alanine, arginine, asparagine, aspartic acid, cysteine, glutamic acid, glutamine, glycine, histidine, isoleucine, leucine, lysine, methionine, phenylalanine, proline, serine, threonine, tyrosine, tryptophan, and valine. Enzyme is a kind of protein that catalyzes a specific chemical reaction in plant cells, and regulation of enzyme synthesis and the activity are essential for maintaining life and growth.

Nucleic acids, deoxyribonucleic acid (DNA), and ribonucleic acid (RNA) are a polymer of purine bases (adenine and guanine) and pyrimidine bases (thiamine, cytosine for DNA, and uracil, cytosine for RNA) with pentose (2-deoxyribose for DNA and ribose for RNA) and phosphate. DNA serves as a template of mRNA, and the mRNA is translated into amino acid sequences of protein. Purine base contains 4 N atoms in a molecule, and they are derived from two glutamines, one aspartic acid, and one glycine. Pyrimidine base contains 2 N atoms in a molecule, and they are derived from one glutamine and one aspartic acid. Amino acids are the precursors of most of N compounds in plants.

1.4. Role of amino acids for nitrogen transport and storage in plants

Amino acids and amides, especially Gln and Asn, are used for N transport through xylem and phloem in many plants. In addition, these amides are used for temporary N storage. Gln and Asn are suitable for N transport and storage, because they have two N atoms in one molecule, and the solubility is high among amino acids.

http://dx.doi.org/10.5772/intechopen.68992

1.5. Role of amino acids for protecting the plants

In addition to the 20 protein amino acids, over 400 nonprotein amino acids are found in various natural sources, and about 240 of them have been found in plants [9]. 4-Methylene glutamine is a nonprotein amino acid found only in groundnut (*Arachis hypogaea*) [10] and tulip (*Tulipa gesneriana*) [11], and this amide is highly accumulated in the leaves and stems of the tulip plants [12], and the tentatively stored N is used for reuse of bulb storage N. Some nonprotein amino acids are toxic such as canavanine and concanavanine A in sword bean (*Canavalia gladiate*), and these toxic amino acids may contribute to protect plant from feeding damages by animals, insects, or fungi [9].

2. Concentrations of free amino acids and soluble N constituents in various parts of soybean plants

In soybean plants, ureides, allantoin, and allantoic acid are mainly used for transport of N in addition to amino acids (**Figure 4**). Ureides have 4 N and 4 C atoms in a molecule, and it is considered to be more efficient to transport of N than asparagine (2N and 4C) by the view of carbon economy.

Table 1 shows the total amino acid-N, ureides-N, nitrate-N, ammonium-N, and others in 80% ethanol soluble fraction of each organ. Both hydrophilic and hydrophobic low-molecular-weight compounds, such as sugars, amino acids, ureides, organic acids, and chlorophylls,

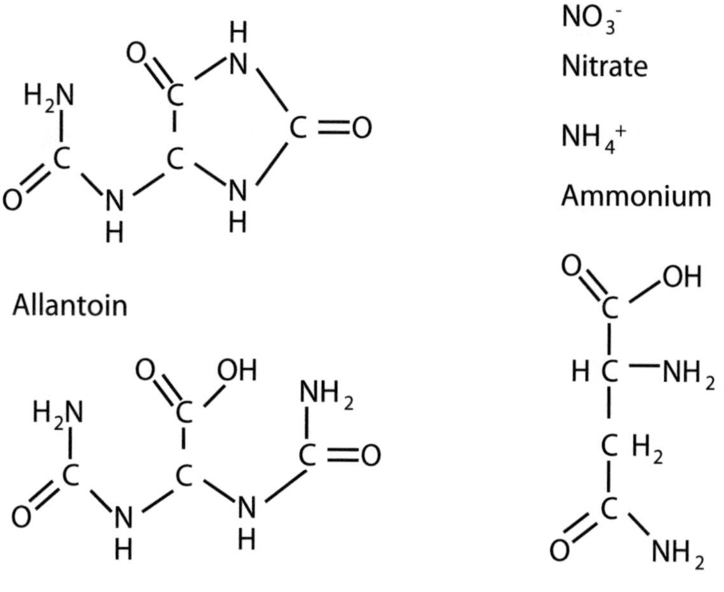

Figure 4. Chemical structures of allantoin, allantoic acid, nitrate, ammonium, and asparagine.

	Nodules	Roots	Stems (UP)	Stems (LP)	Leaves (UP)	Leaves (LP)	Pods (UP)	Pods (LP)	Seeds (UP)	Seeds (LP)
Total amino acid-N	519	147	270	134	96	96	776	415	1243	982
Ureides-N	483	19	143	62	35	36	1529	916	82	82
Nitrate-N	147	32	56	32	33	43	95	32	91	69
Ammonium-N	132	39	36	23	46	46	167	124	119	171
Others-N	4179	477	418	376	622	630	655	1281	8635	6326
Total soluble-N	5460	714	923	627	832	851	3222	2768	10170	7630

Table 1. Concentration of amino acid, ureides, nitrate, ammonium, and other 80% ethanol soluble fraction of soybean organs (μgN/gDW).

can be extracted by 80% ethanol. Macromolecules such as proteins, nucleic acids, and starch remain in the precipitate of 80% ethanol extraction. As shown in **Table 1**, the total soluble N concentration in root nodules (5460 μgN/gDW) was much higher than that in the roots (714 μgN/gDW). In the shoots, the total soluble N concentration was the highest in seeds (10,170 μgN/gDW for upper part and 7630 μgN/gDW for lower part), followed by pods (3222 μgN/gDW for upper part and 2768 μgN/gDW for lower part), and relatively low in stems (923 μgN/gDW for upper part and 627 μgN/gDW for lower part) and leaves (832 μgN/gDW for upper part and 851 μgN/gDW for lower part). The concentration of total amino acid-N was highest in seeds (1243 μgN/gDW in upper part and 982 μgN/gDW in the lower parts), followed by pods (776 μgN/gDW in upper part and 415 μgN/gDW in the lower parts) and nodules (519 μgN/gDW), and lowest in the leaves (96 μgN/gDW in upper part and 96 μgN/gDW in the lower parts). Comparing nodules and roots, amino acid concentration was about 4 times higher in nodules (519 μgN/gDW) than in roots (147 μgN/gDW). The organs in the upper part were relatively high in total amino acid concentrations compared with the lower parts, and this is due to the upper part was younger than lower part. The ureide-N concentration was 25 times higher in the nodules (483 μgN/gDW) than that in the roots (19 μgN/gDW). In the shoots, ureides are highly accumulated in the pods (1529 μgN/gDW for upper part and 916 μgN/gDW for lower part) compared with leaves and seeds. The concentrations of nitrate and ammonium were high in the nodules, but relatively low and constant among other organs.

Table 2 shows the composition of free amino acids in nodules, roots, stems, leaves, pods, and seeds at pod-filling stage [13]. Soybean (cultivar Norin No. 2) seeds were inoculated with *Bradyrhizobium japonicum* (NIAS J-501) and cultivated with hydroponic solution containing nitrate (10 mgN/L). **Table 2** exhibits the amino acids composition at 67 days after planting, and the shoots were separated the upper parts and lower parts. Asparagine was abundant in every part, especially in pods and stems. Asparagine is known to play a major role in transport of nitrogen in legumes [14]. The contents of glutamate were higher in nodules, roots, and leaves, which are the primary nitrogen assimilatory organs, but relatively low in stems, pods, and seeds. Alanine was relatively high in the roots and nodules. The content of γ-amino butylic acid (GABA) was detected at a high level in most organs.

Amino acids	Nodules	Roots	Stems (UP)	Stems (LP)	Leaves (UP)	Leaves (LP)	Pods (UP)	Pods (LP)	Seeds (UP)	Seeds (LP)
Aspartate	22.317	1.764	4.59	3.618	2.688	2.112	10.864	3.735	55.935	34.37
Threonine	8.304	2.205	5.94	3.484	2.016	1.728	14.744	8.3	34.804	35.352
Serine	33.216	8.379	14.04	6.968	10.56	7.392	22.504	14.11	38.533	38.298
Asparagine	120.408	14.7	105.57	56.012	15.456	9.888	401.192	148.57	343.068	169.886
Glutamate	106.395	25.284	3.78	3.886	16.512	10.176	4.656	3.32	17.402	15.712
Glutamine	6.228	2.205	3.78	2.814	0	0	42.68	17.845	4.972	4.91
Proline	2.076	0	4.59	0	0	0	12.416	14.525	32.318	47.136
Glycine	15.57	0	0	0	0	0	2.328	0	27.346	30.442
Alanine	78.369	43.218	9.18	3.082	5.952	7.392	17.848	9.96	52.206	41.244
Citruline	7.785	0	0	0	0	0	25.608	0	31.075	25.532
Valine	8.823	3.234	6.75	3.752	3.648	3.456	20.952	16.185	48.477	47.136
Cysteine	5.19	2.058	2.97	2.948	2.4	2.784	1.552	1.66	2.486	1.964
Isoleucine	7.785	5.145	7.29	6.03	2.496	3.648	17.072	16.185	36.047	36.334
Leucine	6.747	4.263	7.02	5.628	2.4	3.84	11.64	12.45	47.234	52.046
Tyrosine	3.114	0	1.89	1.34	1.44	0	3.88	3.735	13.673	16.694
Phenylalanine	3.114	1.176	3.78	3.082	2.016	2.208	5.432	8.715	19.888	22.586
GABA	46.71	22.785	43.47	25.862	9.888	33.6	126.488	112.05	335.61	233.716
Arginine	11.937	4.557	22.95	5.628	11.904	7.872	22.504	11.62	65.879	43.208

Data from SSPN [13].

Table 2. Concntation of free amino acids in various parts of soybean plants (μgN/gDW).

3. Amino acid metabolism in soybean nodules

Nitrogen is abundant (about 78% in volume) in the atmosphere, but plant itself cannot use the N_2 except for symbiotic association with nitrogen-fixing microorganisms. Symbiotic association by soybean and rhizobia is one of the most efficient nitrogen-fixing system, and it contributes to soybean seed yield [15]. **Figure 5** shows a photograph of nodulated soybean roots cultivated with hydroponics (A) and a model of structure of mature soybean nodule attached to the root (B). Soybean nodules have a spherical form, and they grow up to about 8 mm in diameter. Soybean nodule is classified as a determinate-type nodule, and the cell division and nodule development are completed in early stage of nodule growth. The nodule growth thereafter is mainly due to cell expansion.

Figure 6 shows the scheme of N assimilation in soybean nodules. N_2 gas is diffused into central symbiotic region of the nodule and reduced into ammonia by the enzyme "nitrogenase" in the bacteroid, a symbiotic form of rhizobia. Most of ammonia produced by N_2 fixation

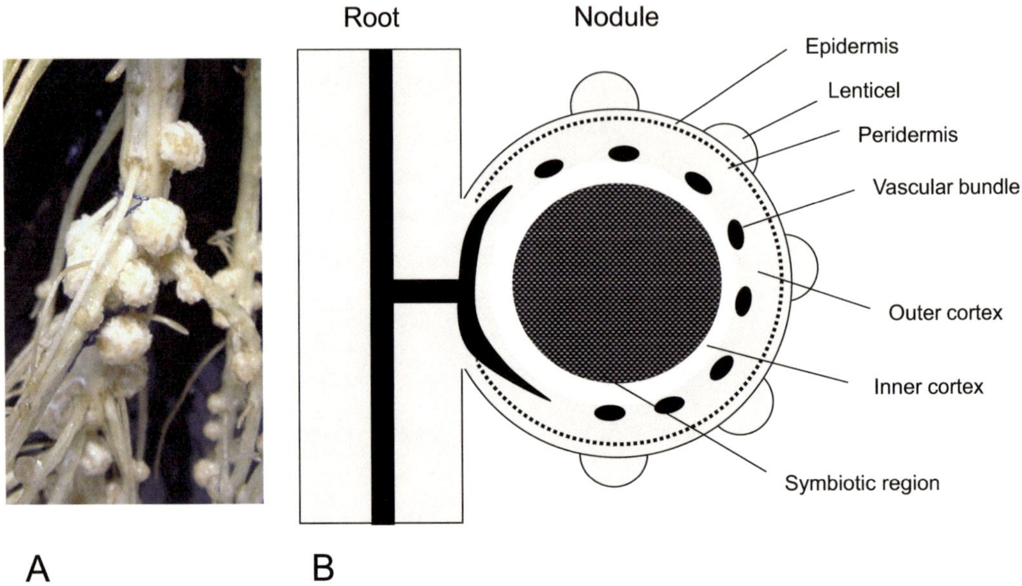

Figure 5. Photograph of nodulated soybean roots and a model of the structure of soybean nodule. (A) Photograph of nodulated soybean roots cultivated in culture solution. (B) A model of the structure of soybean nodule.

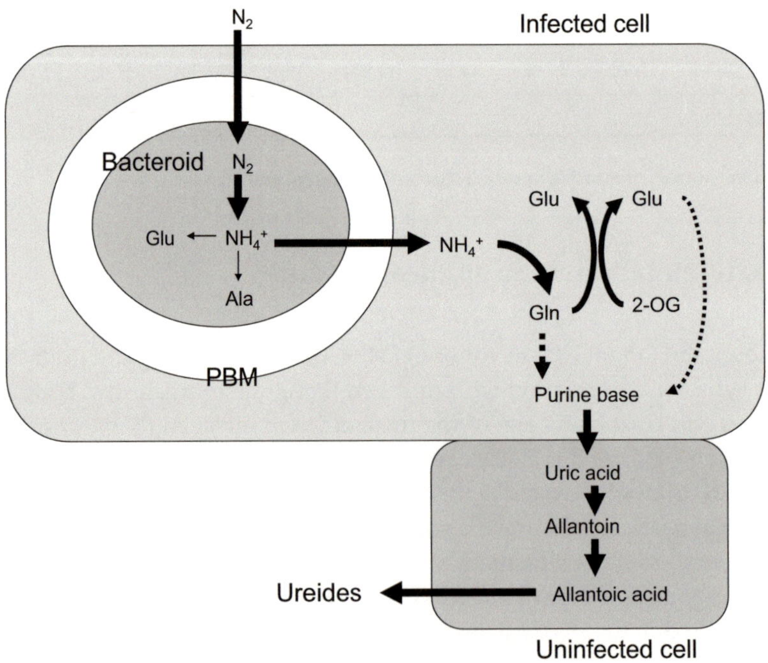

Figure 6. Model of nitrogen flow and metabolism in a soybean nodule.

is rapidly excreted into the cytosol through peribacteroid membrane (PBM) of infected cell. Based on the time course experiment with $^{15}N_2$ feedings in the nodulated intact soybean plants, the ammonia produced by nitrogen fixation is initially assimilated into amide group of Gln with Glu by the enzyme glutamine synthetase (GS) [16, 17]. Then, Gln and 2-OG produce two moles of Glu by the enzyme glutamate synthase (GOGAT). Some part of Gln is used for purine base synthesis, and uric acid is transported from the infected cells to the adjacent uninfected cells in the central symbiotic region of nodule. Uric acid is catabolized into allantoin and allantoic acid in the uninfected cells and then transported to the shoot through xylem vessels in the roots and stems. A small portion of fixed N was assimilated into alanine and Glu in the bacteroids, but it was not by GS/GOGAT pathway [18].

A small portion of fixed N is transported as amino acid-like asparagine (Asn) in addition to ureides, but the percentage is about 10–20% of total fixed N. Small amount of ureides is synthesized from NO_3^- in soybean nodules [19]. Nitrate in culture solution can be absorbed from nodule surface [20] and assimilated in the cortex of the nodules. Nitrate absorbed from lower part of roots has not readily transported to the nodules attached to the upper part of the root system [1].

4. Amino acid metabolism in soybean roots

Higher plants absorb soil inorganic nitrogen such as ammonium or nitrate from roots. **Figure 7** shows the photograph of a nodulated soybean root system (A) and a model of structure of soybean root (B). In a longitudinal direction, there are three regions in root (B): Root cell division occurs in the "apical meristem," and the cells are differentiated and elongated in the upper "region of elongation." Then, the root cells mature in the "region of maturation." There are three parts, epidermis, cortex, and stele, in a cross-section of mature roots. There are two transport pathways, xylem and phloem, in the stele.

Figure 8 shows a model of nutrient flow from soil to xylem vessel in the roots. Nutrients, such as NH_4^+ and NO_3^- in soil solution, are absorbed by epidermal cells or root hairs and are transported cell to cell by the symplastic pathway. Nutrients also enter into the free space of the root cortex by the apoplastic pathway and are absorbed by cortical cells. The apoplastic pathway is blocked by a water proof Casparian strip, so the nutrients should be passed through endodermis by a symplastic pathway into stele. Then, the nutrients are released from parenchima cells of stele to the free space in the stele and loaded into the xylem vessel, which is the vertically connected dead cell walls. The nutrients and water are transported from roots to the shoots by transpiration in leaves and root pressure.

Figure 9 shows a model of ammonium and nitrate assimilation in a plant cell. NH_4^+ is transported into cell cytoplasm by the ammonium transporter, which is located in a plasma membrane. NH_4^+ is first assimilated into glutamine (Gln) combined with glutamic acid (Glu) by the enzyme glutamine synthetase (GS) consuming one molar of ATP. Then, the amide group of Gln is transferred to an organic acid, 2-oxoglutarate (2-OG), by glutamate synthase (GOGAT)

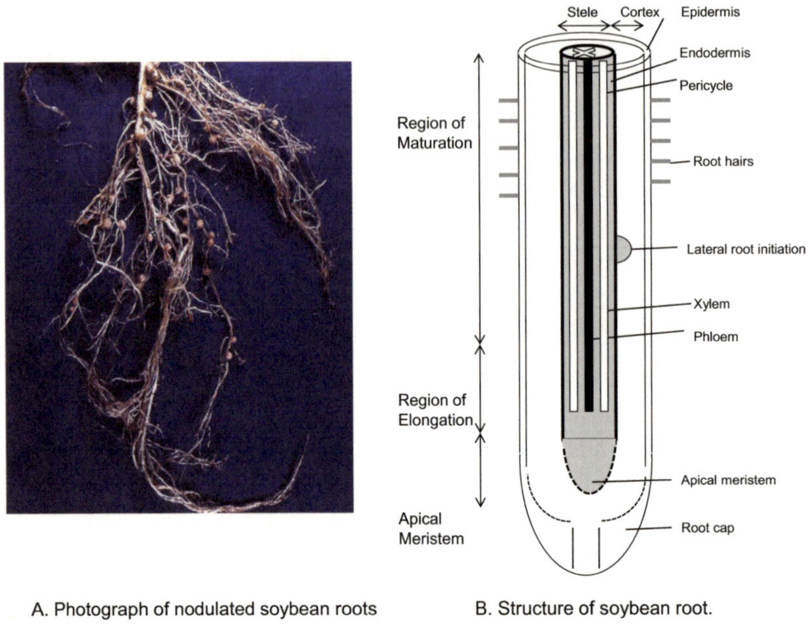

Figure 7. Photograph of nodulated soybean roots and a model of the structure of soybean root. (A) Photograph of nodulated soybean roots cultivated in soil. (B) A model of the structure of soybean root.

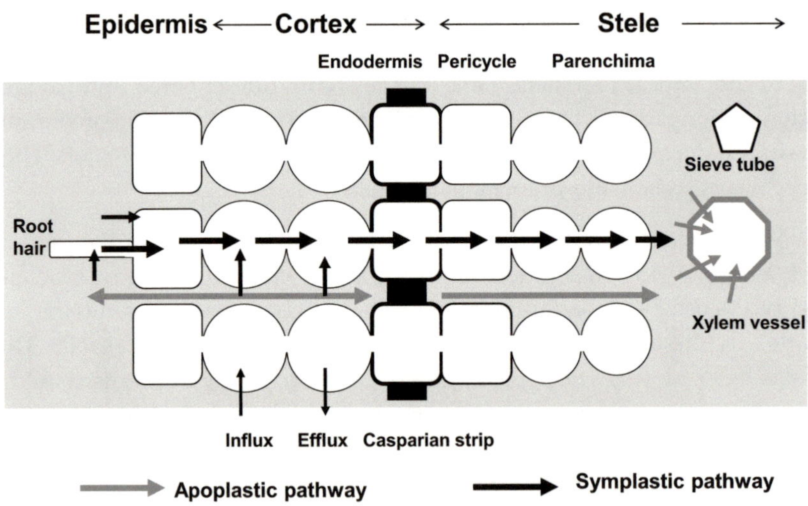

Figure 8. A model of nutrients and water flow in the root.

in plastids using two molar of reduced Feredoxin (Fd_{red}). Various amino acids are synthesized by amino transferases or amino acid metabolism. Amino acids are used for the synthesis of proteins, nucleic acids, etc. Some parts of amino acids are exported outside the cell to transport to the other organs.

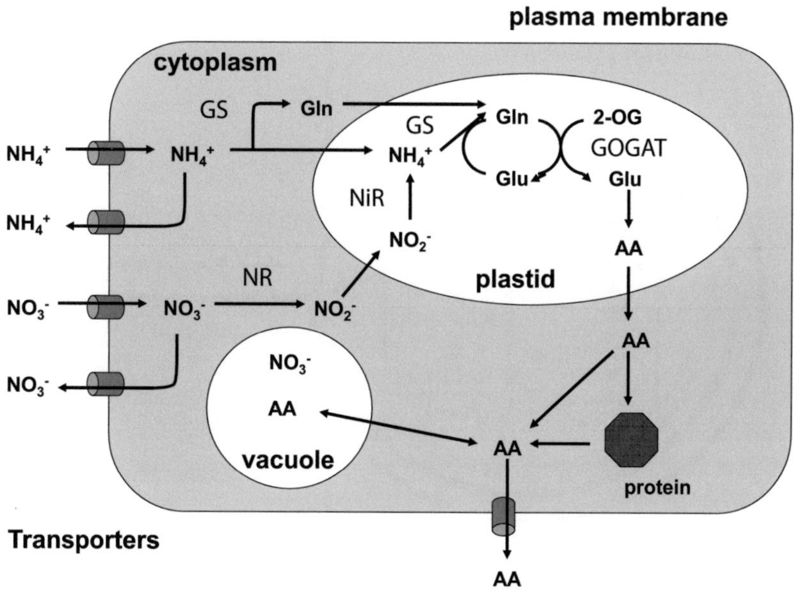

Figure 9. A model of ammonium and nitrate transport and metabolism in root cells. Gln: glutamine, Glu: glutamate, 2-OG: 2-oxo-glutarate, AA: amino acids, GS: glutamine synthetase, GOGAT: glutamate synthase.

Nitrate is most abundant inorganic nitrogen in upland fields under aerobic conditions. NO_3^- is transported into cell cytoplasm by nitrate transporters on plasmamembrane. Two types of nitrate transporters are present in plants, a high-affinity transport system (HATS) and a low-affinity transport system (LATS). Affinity to nitrate is high in HATS (Km is about 10–100 μM) but low in LATS (Km is over 0.5 mM) [21]. The nitrate uptake is driven by a proton motive force and mediated by $2H^+/NO_3^-$symport. By analyzing circulation system of culture solution, the nitrate absorption rate of soybean roots was measured. The kinetics between nitrate concentration versus nitrate absorption rate indicated that soybean root has only one high-affinity nitrate transporter in the roots, which Km was 19 μM. NO_3^- is reduced to nitrite (NO_2^-) by the enzyme nitrate reductase (NR) using one molar of NADH or NADPH as a reductant. The NO_2^- is transported to plastids and further reduced to NH_4^+ by the enzyme nitrite reductase (NiR) using six molars of Fd_{red}. Then, NH_4^+ is assimilated into Gln by GS located in plastids. Finally, Gln and 2-OG are converted to 2 Glu by the enzyme GOGAT. Soybean roots absorb NO_3^- not only during day but also during night. The absorption rate in night was about 2/3 compared with that in day time [22].

In order to compare the assimilation and transport of nitrate, nitrite, and ammonium, $^{15}NO_3^-$, $^{15}NO_2^-$, or $^{15}NH_4^+$ was supplied to the nodulated soybean plants [23]. The N from nitrate and ammonium was rapidly absorbed, and 70% of ^{15}N absorbed was distributed in the shoots at 24 hours after ^{15}N treatments. The partitioning of ^{15}N among the organs was similar between $^{15}NO_3^-$ and $^{15}NH_4^+$ treatments. However, ^{15}N absorption was low, and most of ^{15}N remained in the roots after $^{15}NO_3^-$ treatment.

Figure 10 shows the scheme of N flow from roots and nodules.

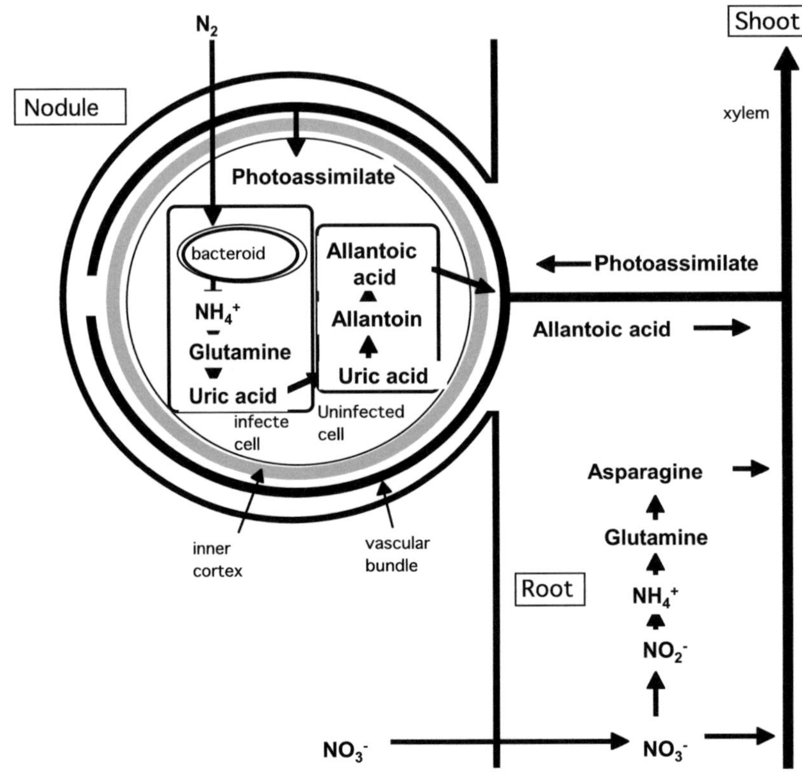

Figure 10. Nitrogen metabolism and transport in nodules and roots of soybean plants.

5. Amino acid metabolism in soybean stems

Stems support the upright structure of shoots, and they connect among roots, leaves, flowers, and fruits in higher plants. Soybean has a main stem and several lateral stems. The structure of the soybean stem was shown in **Figure 11**. In the bark of the stem, there are epidermis and cortex including vascular bundles with phloem. In the central woody part of the stem, there are xylem vessels and pith. Xylem sap comes up through xylem vessels in the stem from root xylem vessels via transpiration stream and root pressure. As shown in **Figure 12**, xylem vessel is not closed pipe, but they have pits on the wall and water and solutes move to the apoplast from xylem vessels and also liquid in apoplast come back to xylem vessels [24].

Stems play a role in storing nutrients temporary or long term in perennial plants. In soybean, ureides and amino acids can be temporarily stored in the stems, and these compounds are eventually transported to the leaves, pods, and seeds. For analyzing concentrations of nitrogen constituents, such as amino acids, ureides, and nitrate, stem extract and stem fluid collected by centrifugation or xylem bleeding sap were used. Stem extract contains all the extractable compounds in the cells including cytoplasm and vacuoles, apoplast fluid outside of the cells, and xylem sap in xylem vessels. Stem extract contains a large amount of cellular

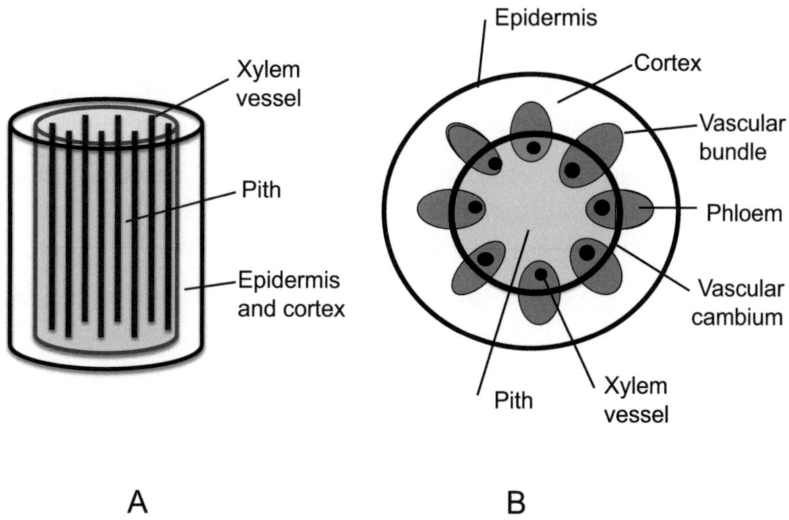

Figure 11. Model of the structure of soybean stem. (A) 3-D image of stem. (B) Cross-section of soybean stem.

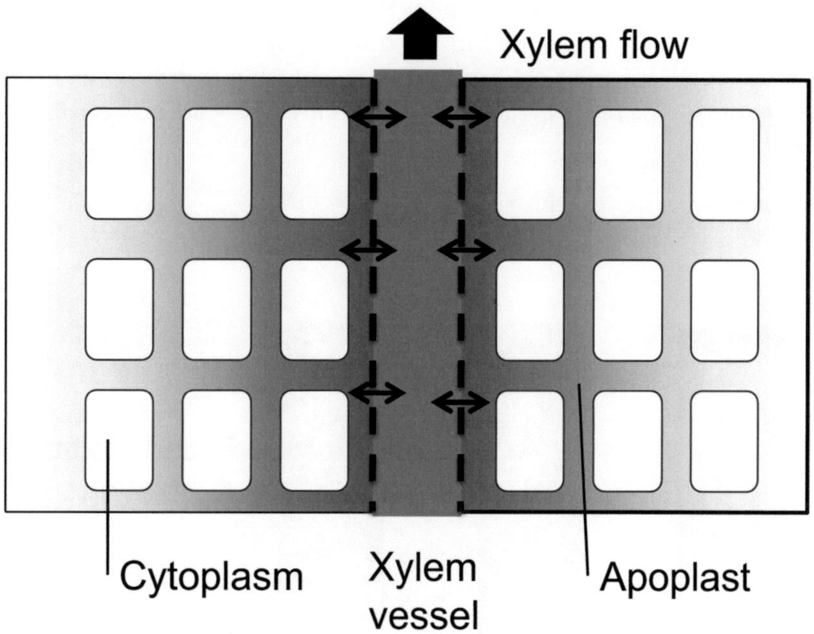

Figure 12. A model of xylem flow in soybean stem.

components such as temporary storage N compounds. When the apoplast fluid collected by centrifugation and xylem bleeding sap were compared, the concentrations of nitrate and ureides are relatively same between two fluids, but the concentrations of amino acids are several times higher in apoplast fluid compared with xylem bleeding sap [25]. Therefore, for the estimation of the percentage dependence of nitrogen fixation in total nitrogen assimilation by relative ureide method, we use the xylem bleeding sap from cut basal stump. Ohtake et al. [26] reported that Asn is the major amino acid in soybean xylem sap, and the average N atoms per an amino acid in xylem sap was about 2.0, irrespective of growth stages.

Sometimes, it is not successful for collecting xylem sap, especially when soil is dry or during late growth stage. Herridge and Peoples [27] used a vacuum extracted stem exudate from lower part of the main stem of the cut shoot or hot-water extraction of the stems.

A relative ureide method is a simple and reliable method for estimating the percentage of nitrogen depending on nitrogen fixation (%Ndfa) in a field-grown soybean [27–29]. **Figure 13** shows the concept of the simple relative ureide method estimating ratio of N_2 fixation and N absorption in the fields. Although a small portion of ureides are transported from non-nodulated soybean grown with nitrate and small portion of amino acids are transported from nodulated soybean without nitrate, we simply estimate the percentage of ureide-N in total N of ureide-N, amide-N plus nitrate-N in root bleeding sap. Takahashi et al. [29] reported the comparison of ureide-N concentration in xylem sap between nodulating and nonnodulating isoline (**Figure 14**), indicating that ureides concentration is much higher in nodulated soybean compared with nonnodulated soybean at any stage [29]. To confirm the origin of ureides in the stems, two treatments have been done for 8 hours. $^{15}N_2$ gas was exposed to a group of soybean plants, and $^{15}NO_3^-$ was added to the culture solution to another group, and then the ratios of ^{15}N abundance from $^{15}N_2/^{15}NO_3^-$ in each compounds in stems and nodules were calculated (**Figure 15**). The ratios in most amino acids in stems were between 0.1 and 0.5, and

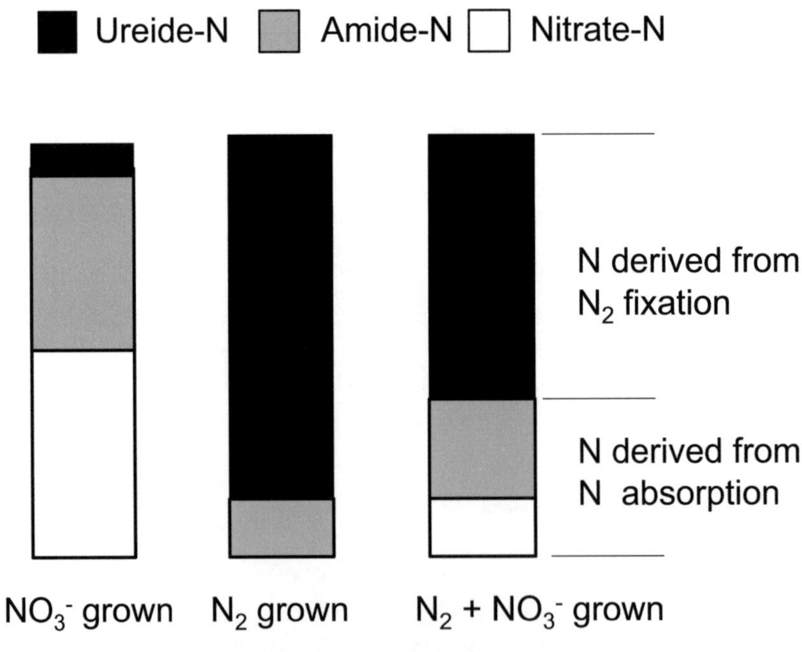

Figure 13. Concept of the simple ureide method for estimating N derived from nitrogen fixation and soil + fertilizer N.

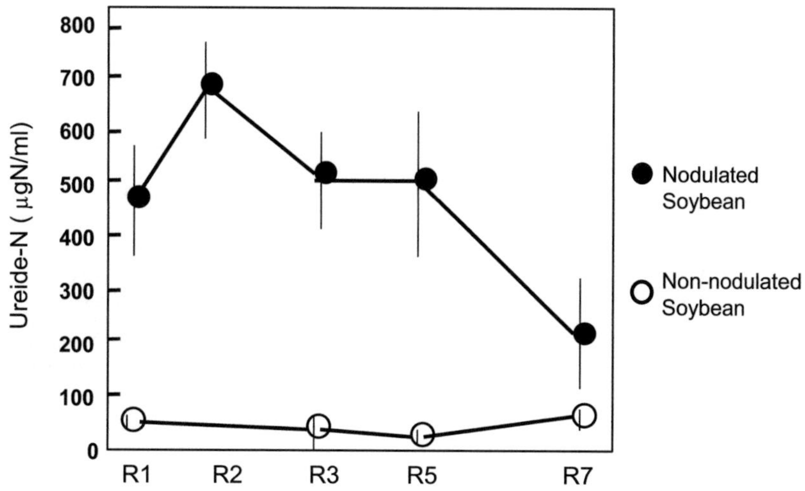

Figure 14. Changes in ureide-N concentration in xylem sap collected from nodulated and nonnodulated soybean.

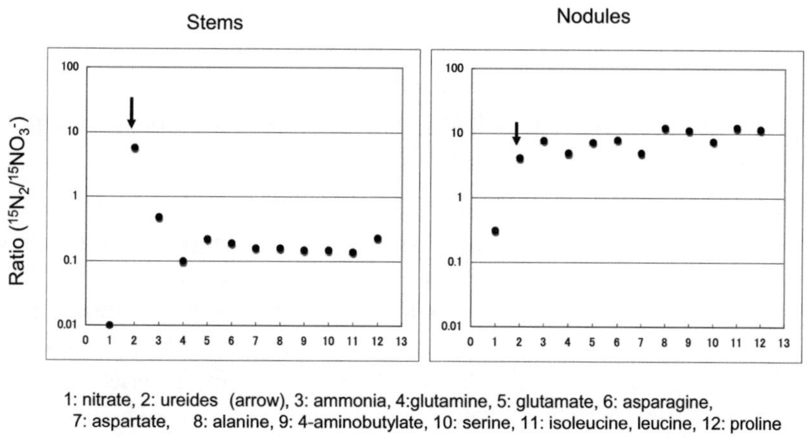

Figure 15. Comparison of the ratios of ^{15}N abundance from $^{15}N_2$ and $^{15}NO_3^-$ in stems and nodules. (A) Ratios in stems. (B) Ratios in nodules.

only ureides showed a high value nearly 10. This indicated that most of ureides in stems are derived from nitrogen fixation and amino acids in stem derived from nitrate absorption [19]. The ratios of ureides and amino acids in the nodules showed about 10.

6. Amino acid metabolism in soybean leaves

Soybean has trifoliolate leaves, in addition to a first pair of green cotyledons, the second pair of primary leaves, and the prophylls [30]. **Figure 16** shows the model of soybean leaflet of trifoliolate leaves (A) and the internal structure of soybean leaf tissue (B) [31]. Each leaflet is

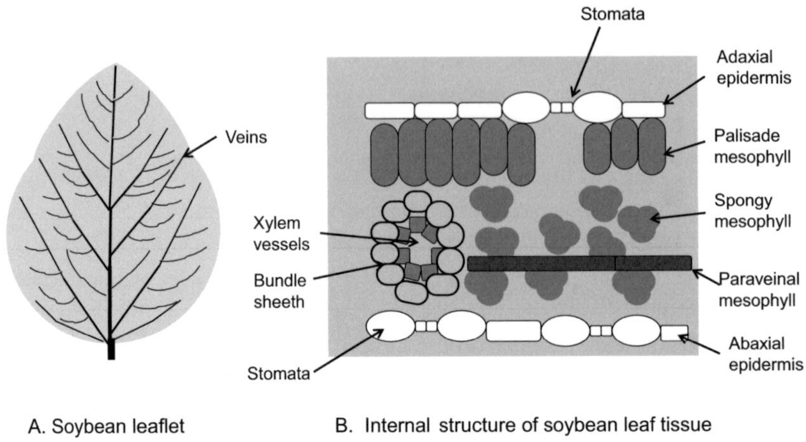

Figure 16. A model of soybean leaflet of trifoliolate leaves and the internal structure of soybean leaf. (A) Mature soybean leaflet of trifoliolate leaves with network of veins. (B) Model of the internal structure of soybean leaf tissue.

highly vasculated and as many as six orders of veins have been observed. All vascular bundles in a leaf are collateral, with adaxial xylem and abaxial phloem. The adaxial layer of mature leaflet is the upper epidermis, and the second and third layers are palisade tissue containing chloroplasts. A portion of fourth layer differentiates into veins and the paravenal mesophyll which is flanked by minor vain. The fifth and sixth layers become spongy mesophyll, and the abaxial layer becomes the lower epidermis [31].

Leaves are the organ of photosynthesis-producing sugar from carbon dioxide (CO_2) in the atmosphere and H_2O absorbed from roots. Also, leaves play an important role in N metabolism such as nitrate reduction and amino acid metabolism. The metabolic products in leaves are transported to the roots and apical buds to support their nutrition through phloem vessels. Evapotranspiration of water through stomata or leaf surface helps upward water flow and nutrient transport from root to shoots via xylem vessels. Xylem vessels are dead cell wall, but phloem vessels are living cells. Therefore, when petiole was treated by heat, phloem transport can be blocked (petiole-girdling treatment).

Petioles of upper or lower soybean leaves cultivated with solution culture with 10 mgN-NO_3^- at 69 days after planting were girdling treatment by hot steam. Then, after 10 hours of $^{15}N_2$ or $^{15}NO_3^-$ treatment, leaf blades are harvested for analysis [32].

Table 3 shows the concentration of total amino acid, ureides, nitrate, ammonium, and total soluble N in intact and girdled leaves. The ratios of ureides and nitrate were almost 1.0, indicating that the concentrations of these compounds did not change by the petiole girdling. This may be due that all the ureides and nitrate are metabolized in the leaf blades and not retransported via phloem. **Table 4** shows the sugar concentration in petiole girdled and intact leaves. The concentrations of fructose, glucose, and sucrose in the petiole-girdled leaves are 1.8–3.8 times higher than those in the intact leaves, irrespective of upper or lower leaves. These accumulations of sugars were due to a blockage of phloem transport from leaves to stems. On the other hand, the concentration of amino acids and ammonia (**Table 5**) increased about 1.5–2.6 times by petiole-girdling treatment, which is similar to the sugar concentration.

	Upper leaves			Lower leaves		
	Intact leaves	Stem girdling	Ratio	Intact leaves	Stem girdling	Ratio
Total amino acid-N	16	41	2.6	15	34	2.3
Ureides-N	36	36	1	45	50	1.1
Nitrate-N	36	44	1.2	37	32	0.9
Ammonium-N	60	88	1.5	56	92	1.6
Total soluble-N	769	1024	1.3	981	1065	1.1

Table 3. Concentration of amino acid, ureides, nitrate, ammonium, and 80% ethanol soluble fraction of intact and girdled leaves (μgN/gDW).

	Upper leaves			Lower leaves		
	Intact leaves	Stem girdling	Ratio	Intact leaves	Stem girdling	Ratio
Fructose	2.78	9.35	3.4	2.35	8.43	3.6
Glucose	2.84	7.31	2.6	1.86	7.1	3.8
Sucrose	4.74	8.35	1.8	6.03	11	1.8

Table 4. Concentration of 80% ethanol soluble sugar of soybean leaves (mgN/gDW).

	Upper leaves			Lower leaves		
	Intact leaves	Stem girdling	Ratio	Intact leaves	Stem girdling	Ratio
Asparagine	347	781	2.3	390	669	1.7
Threonine	287	772	2.7	222	677	3
Serine	782	1440	1.8	649	1283	2
Asparagine	928	8089	8.7	875	7760	8.9
Glutamate	1729	3290	1.9	1888	2597	1.4
Glutamine	0	914		0	774	
Proline	653	4.837	7.4	1008	2597	1.4
Alanine	492	1587	3.2	579	1106	1.9
Valine	775	1918	2.5	585	1509	2.6
Cysteine	270	294	1.1	221	282	1.3
Isoleucine	758	2674	3.5	730	2270	3.1
Leucine	733	2196	3	641	1824	2.8
Tyrosine	156	579	3.7	123	380	3.1
Phenylalanine	536	1426	2.7	404	896	2.2
GABA	3332	6267	1.9	3035	4683	1.5
Arginine	1582	1036	0.7	1043	1022	1

Data from SSPN [32].

Table 5. Composition of free amino acids in intact and girdled leaves of soybean plants.

Among amino acids, Asn was the highest ratio about 9.0 by the petiole-girdling treatment. Most of the other amino acids show the ratios 2–4, but only arginine showed the ratio 0.7 and 1.0 in upper leaves and lower leaves, respectively. Pate et al. [33] reported that asparagine and glutamine are predominant in phloem exudate obtained by phloem bleeding technique from legume fruits.

Figure 17 summarizes the flow of N in soybean leaves. Ureides, nitrate, and Asn transported to the leaf blades via xylem vessels are metabolized in leaves and assimilated into leaf protein. Then, the degradation products of leaf protein are retransported to the apical buds, roots, and pods via phloem.

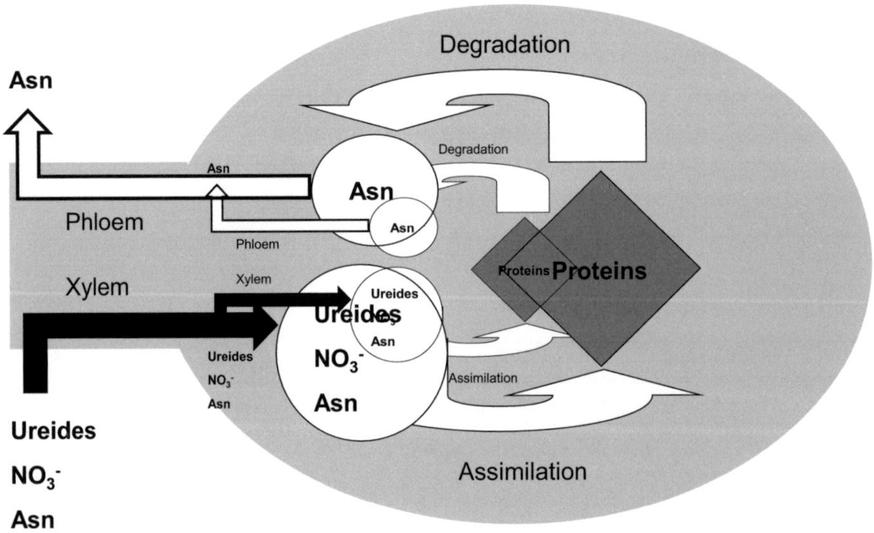

Figure 17. A model of nitrogen flow in soybean leaves.

7. Amino acid metabolism in soybean pods and seeds

Figure 18 shows the top (A) and side (B) views of a mature soybean seed and growing cotyledons in a pod (C, D) [31]. The mature soybean seed consists of a seed coat surrounding a large embryo. Seed coat has a hilum (seed scar), and there is a tiny hole (micropyle) at the end of the hilum. The tip of the hypocotyl radical axis is located just below the micropyle. There is a main vain at the dorsal part of a pod, and nutrients such as sugar and ureides and amino acids are transported through the vain. Seed coat has a funiculus connecting main vain and hilum. Nutrients are transported through vasucular bundles in seed coat; however, the vascular systems are not directly connected to the cotyledons. The cotyledons are cultured in a seed coat cavity. Therefore, nutrients are excreted to the cavity, and cotyledons absorb them by themselves.

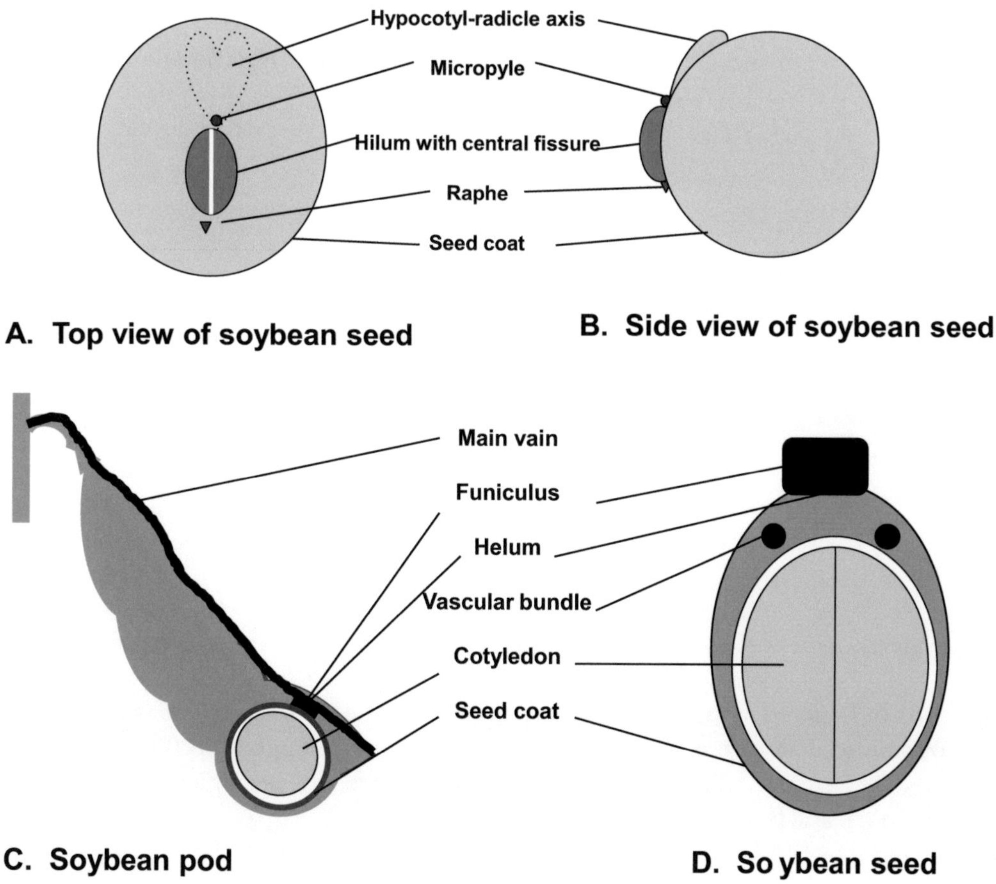

Figure 18. Structures of soybean seed and pod. (A) Top view of soybean seed. (B) Side view of soybean seed. (C) Soybean pod with seeds inside. (D) Soybean seed.

As shown in **Table 1**, young pods contained a high concentration of ureides both in the upper and the lower pods. The high accumulation of ureides in the pods may be due to that ureides are tentatively stored in the pods before transporting to the seeds. Seeds contain a high concentration of amino acids especially Asn and GABA (**Table 2**).

Figure 19 shows the changes in ureide-N and amino acid-N in seeds and pods of nodulated and nonnodulated soybean [34]. The concentrations of ureides in the pods of nodulated soybean were high at 1st September and decreased after 15th September. The ureide-N concentration kept low in the pods of nonnodulated soybean plants. Amino acid-N concentrations were similar between nodulated and nonnodulated soybeans and decreased linearly from 1st September to 10th October at maturing stage. On the other hand, the ureide-N concentrations in the seeds of nodulated and nonnodulated plants were constantly low. The amino acid-N concentrations were similar between nodulated and nonnodulated soybeans, decreased from 1st September to 22nd September, and then constant until maturity at 10th October.

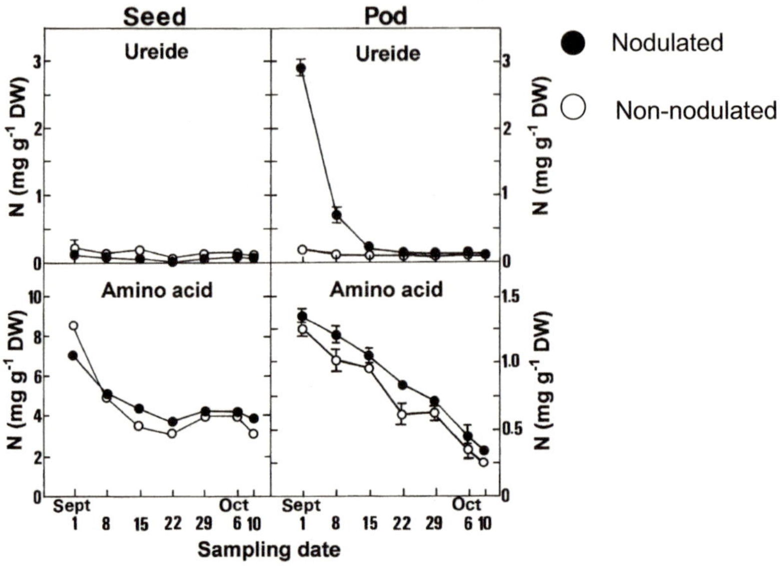

Figure 19. Changes in udeide-N and amino acid-N in pods and seeds of nodulated and nonnodulated soybean plants.

Most parts of N stored in matured seeds are storage proteins in the cotyledons (**Figure 20A**) [35]. Soybean seed storage protein consists of glycinin and β-conglycinin (**Figure 20B**). The glycinin is the hexamer, which has acidic and basic subunits. The β-conglycinin is the trimer, which has α′, α, and β subunits.

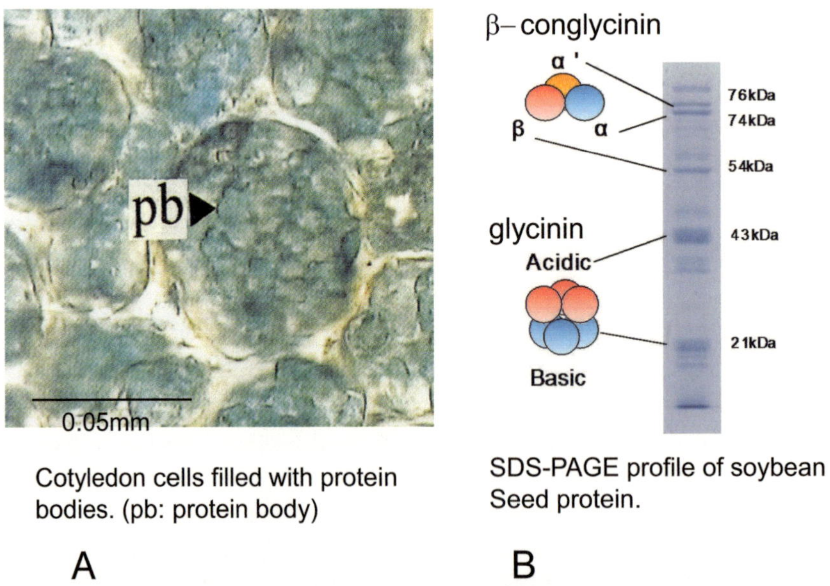

Figure 20. Microscopic picture of soybean cotyledon and soybean seed storage proteins. (A) Microscopic picture of thin slice of soybean cotyledon. Pb: protein body. (B) Soybean seed storage protein separated by SDS-PAGE and stained with CBB.

http://dx.doi.org/10.5772/intechopen.68992

Figure 21 summarizes the metabolism and transport of N from pod into seeds. The ureides, allantoin, and allantoic acid, transported from nodules, are tentatively accumulated in the young pods, and these are metabolized into amino acids such as Gln, Asn, and then, these amino acids are secreted into the cavity of seed coat. Then, cotyledons absorb amino acids and synthesize storage protein and sort to the protein body. Asn and amino acids from roots and leaves are also transported to the pods and secreted into seed coat, and then, cotyledons absorb them for storage protein synthesis.

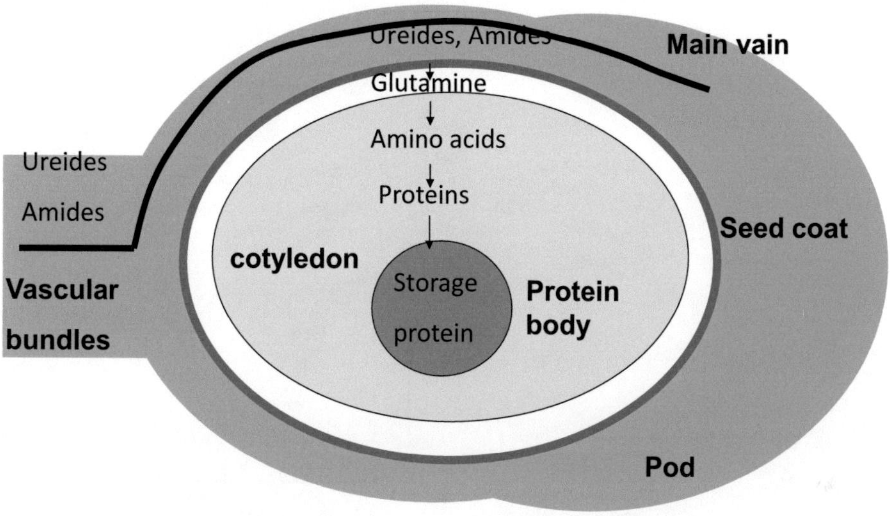

Figure 21. Model of the flow of ureides and amino acids in soybean pod to a seed.

8. Recycling of nitrogen from shoot to roots

Recycling of nitrogen from shoot to roots via phloem supports the initial growth of roots and nodules that need external nitrogen nutrients until nitrogen absorption and nitrogen fixation start to meet the N demand. The quantitative measurement of recycling of N in soybean cultivar Williams and hypernodulation mutant lines, NOD1-3, NOD2-4, and NOD3-7, was carried out by split root experiment in which a half root system was in the pot with ^{15}N-labeled solution and the other was in nonlabeled solution (**Figure 22**) [36]. The roots of soybean plants cultivated in culture solution were separated into two pots at 33 days after planting. At the next day, ^{15}N-labeled nitrate (10 mgN L^{-1}) was added in the one side of pot and non-labeled nitrate (10 mgN L^{-1}) in the other side of pot. After 2 days of treatment, plants were harvested and percentage of N from ^{15}N-labeled source (^{15}N%) was determined in each part. The ^{15}N was highest in the roots in ^{15}N-labeled pot (14.0%), followed by stems (6.0%) and leaves (3.9%). A small portion of recycling ^{15}N was detected both in roots (0.7%) and nodules (0.3%) in the nonlabeled pot after 2 days of split root treatment in Williams. The distribution of ^{15}N in sum of nodules and roots was the same in NOD1-3, NOD2-4, and NOD3-7 hypernodulation lines.

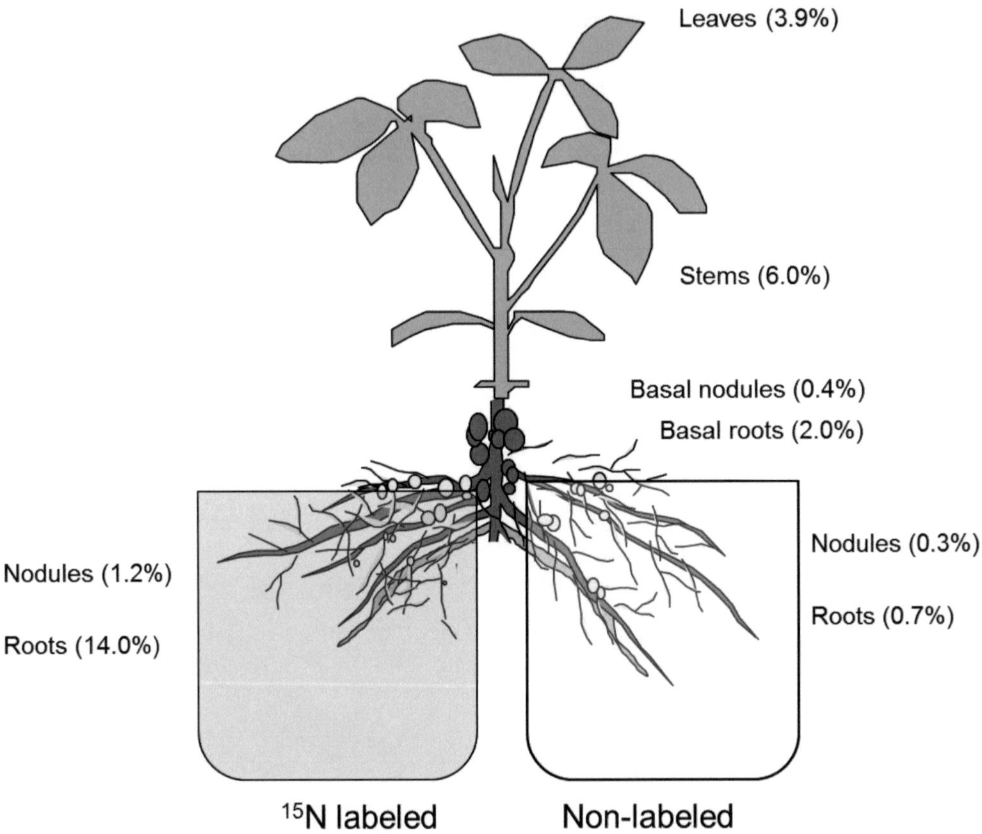

Figure 22. Recycling of N derived from a half root of soybean plants with ^{15}N-labeled nitrate. Percent in parenthesis is percentage of labeled N in total N in each part.

9. Overall nitrogen transport from fixed N and absorbed N

Figure 23 shows the model of initial flow of N in soybean plants originated from N_2 fixation (A) in nodules and NO_3^- absorption from roots (B).

The ureides produced by nitrogen fixation in nodules are transported to leaves and used for the leaf growth and metabolism. Mature leaves do not retransport ureides from phloem, but these are transported as amino acids, especially Asn. Some ureides are directly transported to the pods, and young pods accumulate a large amount of ureides, then it is used for seed growth via seed coat after decomposition into amino acids, such as Gln and Asn.

The absorbed nitrate in the roots is transported as NO_3^- or amino acids especially Asn after assimilated in the roots. The absorbed NO_3^- is transported to the leaves, then reduced by leaf nitrate reductase and nitrite reductase, then assimilated into amino acids. The amino acids are transported to the pods, and then seeds from the leaves.

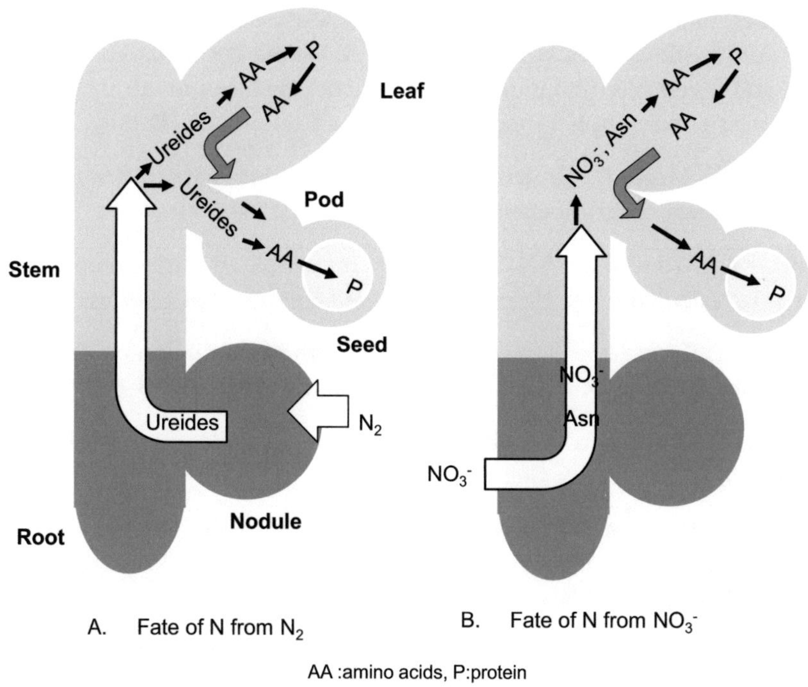

Figure 23. Model of the fate of N from N_2 fixation and NO_3^- absorption in soybean plants. (A) Fate of N from N_2. (B) Fate of N from NO_3^-.

Author details

Takuji Ohyama[1]*, Norikuni Ohtake[1], Kuni Sueyoshi[1], Yuki Ono[1], Kotaro Tsutsumi[1], Manabu Ueno[1], Sayuri Tanabata[2], Takashi Sato[3] and Yoshihiko Takahashi[1]

*Address all correspondence to: ohyama@agr.niigata-u.ac.jp

1 Faculty of Agriculture, Niigata University, Niigata, Japan

2 Agricultural Research Institute, Ibaraki University, Ibaraki, Japan

3 Faculty of Bioresource Sciences, Akita Prefectural University, Akita, Japan

References

[1] Sato T, Ohtake N, Ohyama T, Ishioka NS, Watanabe S, Osa A, Sekine T, Uchida H, Tsuji A, Matsuhashi S, Ito T, Kume T. Analysis of nitrate absorption and transport in non-nodulated and nodulated soybean plants with $^{13}NO_3^-$ and $^{15}NO_3^-$. Radioisotopes. 1999;**48**:450–458

[2] Fujikake H, Yamazaki A, Ohtake N, Sueyoshi K, Matsuhashi S, Ito T, Mizuniwa C, Kume T, Hashimoto S, Ishioka NS, Watanabe S, Osa A, Sekine T, Uchida H, Tsuji A, Ohyama T. Quick and reversible inhibition of soybean root nodule growth by nitrate involves a decrease in sucrose supply to nodules. Journal of Experimental Botany. 2003;**54**:1379–1388

[3] Tempest DW, Meers JL, Brown CM. Synthesis of glutamate in *Aerobacter aerogenes* by itherto unknown route. Biochemical Journal. 1970;**117**:405–407

[4] Kumazawa K, Yoneyama T, Arima Y, Muhammad SS. Assimilation of nitrogen by rice plant as revealed with ^{15}N. Proceedings of the Japan Academy Series B. 1987;**63**:219–222

[5] Yoneyama T, Kumazawa K. A kinetic study of the assimilation of ^{15}N-labelled ammonium in rice seedling roots. Plant & Cell Physiology. 1974;**15**:655–661

[6] Yoneyama T, Kumazawa K. A kinetic study of the assimilation of ^{15}N-labelled nitrate in rice seedling roots. Plant & Cell Physiology. 1975;**16**:21–26

[7] Arima Y, Kumazawa K. Evidence of ammonium assimilation via glutamine synthetase-glutamate synthase system in rice seedling roots. Plant & Cell Physiology. 1977;**18**:1121–1129

[8] Muhammad S, Kumazawa K. Assimilation and transport of nitrogen in rice. I. ^{15}N-labelled ammonium nitrogen. Plant & Cell Physiology. 1974;**15**:747–758

[9] Rosentahal GA. Plant Nonprotein Amino and Imino Acids: Biological, Biochemical, and Toxicological Properties. New York: Academic Press; 1982. pp. 1–273

[10] Done J, Fowden L. A new amino-acid amide in the groundnut plant (*Arachis hypogea*): Evidence of the occurrence of ɤ-methylene glutamine and ɤ-methylene glutamic acid. Biochemical Journal. 1952;**51**:451–458

[11] Zacharius RM, Polland JK, Steward FC. ɤ-methylene glutamine and ɤ-methylene glutamic acid in the tulip (*Tulipa gesneriana*). Journal of the American Chemical Society. 1954;**76**:1961–1962

[12] Ohyama T, Kera T, Ikarashi T. Occurrence of 4-methyleneglutamine and 2-oxo-4-methyl-3-pentene-1,5-dioic acid in leaves and stem of tulip plants. Soil Science and Plant Nutrition. 1988;**34**:613–620

[13] Ohyama T. Comparative studies on the distribution of nitrogen in soybean plants supplied with N_2 and NO_3^- at the pod filling stage. II. Assimilation and transport of nitrogenous constituents. Soil Science and Plant Nutrition. 1984;**30**:219–229

[14] Lea PJ, Miflin BJ. Transport and metabolism of asparagine and other nitrogen compounds within the plants. In: The Biochemistry of Plants. Vol. 5. Academic Press; San Diego, California, USA. 1980. pp. 569–607

[15] Ohyama T, Minagawa R, Ishikawa S, Yamamoto M, Hung NVP, Ohtake N, Sueyoshi K, Sato T, Nagumo Y, Takahashi Y. Soybean seed production and nitrogen nutrition. In: El-Shemy HA, editor. Soybean Physiology and Biochemistry. Rieka, Croatia: InTech; 2011

[16] Ohyama T, Kumazawa K. Incorporation of ^{15}N into various nitrogenous compounds in intact soybean nodules after exposure to $^{15}N_2$ gas. Soil Science and Plant Nutrition. 1978;**24**:525–533

[17] Ohyama T, Kumazawa K. Nitrogen assimilation in soybean nodules I. The role of GS/GOGAT system in the assimilation of ammonia produced by N_2 fixation. Soil Science and Plant Nutrition. 1980;**26**:109–115

[18] Ohyama T, Kumazawa K. Nitrogen assimilation in soybean nodules II. $^{15}N_2$ assimilation in bacteroid and cytosol fractions of soybean nodules. Soil Science and Plant Nutrition. 1980;**26**:205–213

[19] Ohyama T, Kumazawa K. Assimilation and transport of nitrogenous compounds originated from $^{15}N_2$ fixation and $^{15}NO_3^-$ absorption. Soil Science and Plant Nutrition. 1979;**25**:9–19

[20] Mizukoshi K, Nishiwaki T, Ohtake N, Minagawa R, Ikarashi T, Ohyama T. Nitrate transport pathway into soybean nodules traced by tungstate and $^{15}NO_3^-$. Soil Science and Plant Nutrition. 1995;**41**:75–88

[21] Sueyoshi K, Ishikawa S, Ishibashi H, Abdel-Latif S. Nitrate transport in barley. In Nitrogen Assimilation in Plants. Research Signpost; Kerala, India. 2010. pp. 67–81

[22] Ohyama T, Kato N, Saito K. Nitrogen transport in xylem of soybean plant supplied with $^{15}NO_3$. Soil Science and Plant Nutrition. 1989;**35**:131–137

[23] Ohyama T, Saito K, Kato N. Assimilation and transport of nitrate, nitrite, and ammonia absorbed by nodulated soybean plants. Soil Science and Plant Nutrition. 1989;**35**:9–20

[24] Ohya T, Tanoi K, Hamada Y, Okabe H, Rai H, Hojo J, Suzuki K, Nakanishi TM. An analysis of long-distance water transport in the soybean stem using $H_2{}^{15}O$. Plant & Cell Physiology. 2008;**49**(5):718–729

[25] Sakazume T, Tanaka K, Aida H, Ishikawa S, Nagumo Y, Takahashi Y, Ohtake N, Sueyoshi K, Ohyama T. Estimation of nitrogen fixation rate of soybean (*Glycine max* (L.) Merr.) by micro-scale relative ureide analysis using root bleeding xylem sap and apoplast fluid in stem. Bulletin of the Faculty of Agriculture, Niigata University. 2014;**67**:27–41

[26] Ohtake N, Nishiwaki T, Mizukoshi K, Minagawa R, Takahashi Y, Chinushi T, Ohyama T. Amino acid composition in xylem sap of soybean related to the evaluation of N_2 fixation by the relative ureide method. Soil Science and Plant Nutrition. 1995;**41**:95–102

[27] Herridge DF, Peoples MB. Ureide assay for measuring nitrogen fixation by nodulated soybean calibrated by ^{15}N methods. Plant Physiology. 1990;**93**:495–503

[28] Takahashi Y, Chinushi T, Nakano T, Ohyama T. Evaluation of N_2 fixation activity and N absorption activity by relative ureide method in field-grown soybean plants with deep placement of coated urea. Soil Science and Plant Nutrition. 1992;**38**:699–708

[29] Takahashi Y, Chinushi T, Ohyama T. Quantitative estimation of N_2 fixation activity and N absorption rate in field grown soybean plants by relative ureide method. Bulletin of the Faculty of Agriculture, Niigata University. 1993;**45**:91–105

[30] Lersten NR, Carlson JB. Vegetative morphology. In: Wilcox JR, editor. Soybeans: Improvement, Production, and Uses. 2nd ed. Madison, Wisconsin, USA: ASA, CSSA, SSSA publishers; 1987

[31] Carlson JB, Lersten NR. Reproductive morphology. In: Wilcox JR, editor. Soybeans: Improvement, Production, and Uses. 2nd ed. Madison, Wisconsin, USA: ASA, CSSA, SSSA publishers; 1987

[32] Ohyama T, Kawai S. Nitrogen assimilation and transport in soybean leaves: Investigation by petiole girdling treatment. Soil Science and Plant Nutrition. 1983;**29**:227–231

[33] Pate JS, Sharkey PJ, Lewis OAM. Phloem bleeding from legume fruits-A technique for study of fruit nutrition. Planta. 1974;**120**:229–243

[34] Ohtake N, Suzuki M, Takahashi Y, Fujiwara T, Chino M, Ikarashi T, Ohyama T. Differential expression of β-conglycinin genes in nodulated and non-nodulated isolines of soybean. Physiology Plant. 1996;**96**:101–110

[35] Ohtake N, Ikarashi Y, Ikarashi T, Ohyama T. Distribution of mineral elements and cell morphorogy in nodulated and non-nodulated soybean seeds. Bulletin of the Facululty of Agriculture, Niigata University. 1997;**49**:93–101

[36] Ohyama T, Ueno M, Ono Y, Ohtake N, Sueyoshi K, Sato T, Tanabata S. Recycling of nitrogen from shoots to underground parts in hypernodulation mutant lines of soybean by split-root experiment. Bulletin of the Faculty of Agricululture, Niigata University. (in press) 2017.

Chapter 9

Amino Acid Changes during Energy Storage Compounds Accumulation of Microalgae under the Nitrogen Depletion

Cao Xupeng, Xue Song and Fan Xuran

Additional information is available at the end of the chapter

http://dx.doi.org/10.5772/intechopen.68540

Abstract

The nitrogen depletion stress is widely used to promote energy storage compound (ESC) production of microalgae, such as starch and lipids. Previous reports show that the fast ESC's accumulation happens around the overall nitrogen content lowered to the half of normal cells. It indicates that the cells take an active nitrogen reassembly to rebalance the requirement of nitrogen, in which the amino acid conversion should play an important role. So here, using a marine strain, Isochrysis zhanjiangensis, as the model to give a detail view on metabolic, transcriptomic and proteomic levels during ESC's fast accumulation. The intracellular metabolite fluctuation within 32 h was profiled by GC-MS and LC-MS. These techniques identified and quantified the levels of 14 SMAs, 2 carbohydrates involved in the TCA cycle and glycolysis, and 28 free amino acids (AAs). The pulsed increase of pyruvate indicated a potential to produce more FAs. Although overall AAs showed a decreasing trend, Ala and Phe increased initially. Meanwhile, the transcriptomic and proteomic studies were utilized to show the nitrogen metabolic pathways changes. It is found that gamma-aminobutyric acid (GABA) and other nonprotein AAs play important roles in the regulation of energy metabolism.

Keywords: nitrogen reassembly, microalgae, energy storage compound, *Isochrysis*

1. Introduction

As the potential producer of the third-generation biofuels [1], and the origin of plenty of high-value products [2], microalgae have attracted more and more attentions from the last decade. An ideal algal biofuel production process should have high biofuel yield, with economic raw material utilization and proper biomass formulation. To meet these requirements, a two-stage culture method has been widely utilized, with nitrogen deficiency (N-deficiency) stress applied after the growth stage of the culture [3–7]. Nitrogen is the key element among the creatures, and the response to nitrogen deficiency varies among different plants [8]. For example, *Isochrysis zhangjiangensis* (synonym *Isochrysis zhanjiangensis*), a marine microalga with high carbon fixation capacity, will accumulate both polysaccharides and lipids as ESCs under N-deficiency conditions [3, 6, 9], while wild *Chlamydomonas reinharditii* stores starch and *Nannochloropsis oceanica* stores lipid preferentially.

However, the regulatory mechanism beneath the coordination of carbon and nitrogen metabolism is unclear, which is important for the design of high efficient process. For photoautotroph microalgae, carbohydrates are from the complex photosynthesis system. The synthesis of lipids is a more complex process than that of carbohydrates, which needs more ATP together with the reduction power from NADPH [10]. During the accumulation of ESCs, it can be reasonably assumed that plenty of enzymes are involved and a considerable portion of amino acids (AAs) are consumed or turned over to produce proteins (enzymes) for metabolism adjustment. Otherwise, some AAs involve the ESCs production directly. For instance, branched AAs (leu, Ile and Val) take part in the Ac-CoA production [11, 12], which is the raw material for the fatty acid synthesis; Glu is the precursor of chlorophyll synthesis. The clear understanding of carbohydrate and lipid accumulation coordinating process will contribute to oriental enrichment of bio-products according to the interest of industry.

The fast development of various "-omics" analyses provides versatile tools for probing complex biological problems. Different levels of "-omics" are combined to show a multidimensional information and widen our views on the essence of biological processes. Therefore, it gives us the opportunity to investigate the facts involving in this "golden period" of ESC's production, especially here, focused on the AA-related processes.

2. The methods used for "-omics" studies of *I. zhangjiangensis*

2.1. Strain and cultivation

I. zhangjiangensis FACHB-1750 was maintained by the Freshwater Algae Culture Collection of the Institute of Hydrobiology, Chinese Academy of Sciences. The microalga had been previously cultivated in F/2 medium without silica under 14:10 light/dark cycle [3]. Algal cells were harvested in the exponential phase and re-suspended in nitrate-free medium to a final concentration of 3.0–4.0 × 10^6 cells/mL. For flask cultures, 200 mL seeds were inoculated in 500-mL glass shaking flasks with hand shake after inoculation and sampling. For bioreactor cultures, 500 mL seeds were inoculated in 600-mL glass bubbling reactors and aired by CO_2 enriched air (2%, v/v) at the speed of 100 mL/min. The cultivation was under the control of self-made Algal Station Platform for reproducible growth (http://www.mbpe.dicp.ac.cn/yjcg/kyjz/kyjzpage.html).

2.2. "-omics" analysis procedures

I. zhangjiangensis cells from standard 7 day's cultivation were used as sample pools for the transcriptomic database construction. Sequencing was performed on Illumina HiSeq™ 2000 by BGI Tech (Shenzhen, China). Reads were assembled to unigenes (7511 clusters and 16,712 singleton transcripts) by Trinity [13] and further annotated against NCBI NR database (non-redundant protein database) and using Nt (non-redundant nucleotide database), SwissProt, KEGG and COG database by blastx (e-value $<10^{-5}$). The calculation of unigene expression uses FPKM method (Fragments Per kb per Million fragments) [14]. The COG cluster enrichment analysis was performed basing on the expression. The RNA-seq data are available with accession No. PRJNA217946 on NCBI.

The proteomics analysis was performed according to previous report [15] against above RNA-seq database. Totally 1862 proteins were identified and quantified.

The AAs and other small molecular acids (SMAs) were inspected by SIM LC-MS and GC-MS, respectively. The experimental details can be found from Zhang et al. [16]. Furthermore, by the pulse-isotope label of ^{15}N on nitrogen containing compounds (NCCs) during the "golden period," the turnover of nitrogen between AAs was investigated.

2.3. Growth, photosystem II (PS II) activity and other biochemical composition analyses

Other biochemical and physiological parameters were detected as previous description [3, 6, 9]. In brief, dry weight was determined gravimetrically after filtration by Whatman GF/C filters (47 mm diameter) and air dried in the air until constant weight achieved. Nitrate analysis was conducted with SEAL Analytical AutoAnalyzer 3 following the manufacturer's instructions. Lipid analysis was performed using Nile Red or GC according to Wang's method [6, 9]. The chlorophyll fluorescence measurements were performed using a chlorophyll fluorometer (Water-PAM, Heinz Walz GmbH).

3. Overall of cell growth and turnover of AAs

3.1. The newly synthesized AAs during the fast accumulation of ESCs by pulsing label

The assimilation of nitrogen starts from the reduction of nitrate and the formation of glutamate. Without nitrate supply from the medium, free AAs were turn over between different NCCs, including proteins, nucleic acids and AAs themselves. By pulsing adding 12.3 mg/L ^{15}N (in form of $Na^{15}NO_3$) at day 2, when initial 24.6 mg/L N was nearly used out, the labeled ratio of free AAs was detected within 32 h.

It has been reported, while the external nitrogen supply depleted, the protein, especially ribulose-1,5-bisphosphate carboxylase/oxygenase (Rubisco), was partially degrade as a nitrogen pool to reassemble AAs and other NCCs [17]. So, the final labeled ratio of AAs was expected near 33% (according to total nitrogen supplied). Most of AAs (15/23) consisted with the theoretical value with about 33% labeled ratio. However, some AAs showed interesting changing patterns (**Figure 1**).

The labeled AAs were newly synthesized, and therefore, most of free AAs show a high labeled ratio at the beginning (day 3-1 of **Figure 1**), except Arg, which increased stepwise. The Arg is mostly from the conversion of ornithine. The low level of labeled Arg indicates a slower than average AAs' synthesis rate of it in the early of exponential growth phase.

Among 23 AAs detected, 6 of them (His, taurine, Asn, Gln, Arg, and Trp) show a the final label ratio higher than 35%, indicating more than average newly synthesized of them were left inform of free AAs. Together with the low label level of Arg, the nitrogen assimilation may mostly form Gln branch.

The first undetectable labeled AA was methionine sulfoxide, which disappeared within 24 h. It is the oxidation form of methionine and formed post-translationally. The content of methionine sulfoxide is about one-fifth of that of Met in *I. zhangjiangensis* [16].

The second disappeared labeled AA was ornithine. It may be the result of the increasing labeled Arg and function involving in the urea cycle [11].

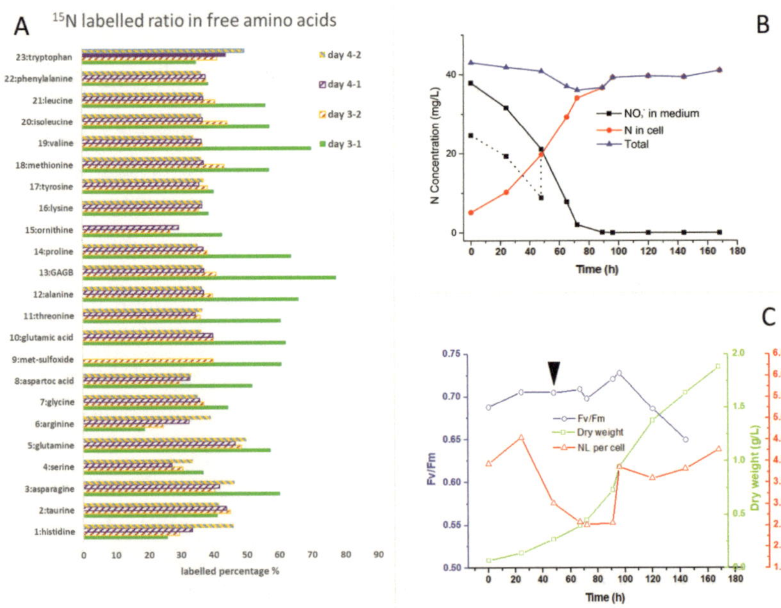

Figure 1. ^{15}N labeling percentage of AAs during the ESC fast accumulation period. (A) The labeling ratio of AAs changed within 32 h. The theoretical labeled ratio is about 33%. (B) The nitrogen changes during a standard bioreactor cultivation. The dotted line is nitrate concentration in medium of pulsing label experiment. (C) The dry weight, F_v/F_m and neutral lipids content changes in above cultivation. The black arrow indicates the start of pulsing label of ^{15}N.

3.2. The change of free AAs together with SMAs

The AAs metabolism relates to other SMAs closely. Zhang et al. have reported a detailed metabolic network change, with 18 protein AAs, 11 non-protein AAs, 14 SMAs and 2 carbohydrates [16]. As shown in **Figure 1B**, the total nitrogen was steady during the fast ESC accumulation stage, so the per cell-based qualification was used to tracing the changes. The cell number doubled after nitrogen depletion, and it is reasonable that the AA content level decreased to about half level. However, Phe had an evident 3-time increase during the early stage of nitrogen depletion. Other AAs maintained a steady state or increased in quantity slightly at the beginning and then decreased below their initial amounts.

The relative constant of protein AAs may contribute to the release of AAs from Rubisco. **Table 1** is the comparison of AA content in Rubisco large subunit and whole cell of *I. zhangjiangensis*. The data for Rubisco are from the putative AA sequence translated based on RNA-seq and that for whole cell is measured by MS from nitrogen-rich cultured cells. They share a similar composition. Only Glu shows a great difference.

In non-protein AAs, only Gamma-aminobutyric acid (GABA) shows a profoundly increase after nitrogen depleted, indicating an important role in the response of nitrogen stress.

Together with the changes of AAs, including the metabolites of glycolysis (D-glucose-1p,dihydroxyacetone-p, glyceraldehyde-3p, D-fructose-6p) and the glyoxylate or TCA cycles (citrate, cis-aconitate, isocitrate, α-ketoglutarate) showed a pattern of initial increase in quantity followed by subsequent decrease. Malate, fumarate and succinate, all from the TCA cycle, shared the same tendency of steady decline. Glucose and lactate showed steady increase in quantity [16].

In general, nitrogen has positive effects on the growth of microalgae, while having a negative impact on the accumulation of lipids, for aims of biofuel production. When carbon is not a limiting factor, intracellular energy substrates will accumulate under nitrogen starvation. Evidence shows that glutamate dehydrogenase as a metabolism shunt plays an important role in ensuring that the nitrogen metabolism does not affect the function of mitochondria and nitrogen recycling [18]. As a consequence, Gln, Glu, Asp, and Asn form the basis of several other organic nitrogen compounds, especially AAs [18].

3.3. AA metabolism–related transcriptomics and proteomics change

The overall transcriptomics and proteomics annotation results by KEGG pathways show opposite changes, up-regulated in transcriptomics while down-regulated in proteomics (**Figure 2**). The transcription and translation are controlled complex mechanisms. Their trends difference is a result of different level responses to the nitrogen deficiency.

To have a more global view of cell response, basing on the Clusters of Orthologous Groups (COGs) enrichment analysis of differentially expressed genes, amino acid transport and metabolism cluster (E) is the most significantly down-regulated cluster (**Figure 3**).

AA	Mole % in		Mass % in	
	Rubisco large subunit	Whole cell	Rubisco large subunit	Whole cell
His	1.5%	0.3%	1.8%	0.4%
Cys	1.8%	0.4%	1.7%	0.4%
Trp	1.8%	0.0%	2.9%	0.0%
Glu	3.4%	9.0%	3.8%	10.9%
Met	3.7%	2.1%	4.2%	2.6%
Phe	3.8%	4.8%	4.9%	6.5%
Asn	4.0%	/	4.1%	/
Ile	4.1%	5.3%	4.2%	5.7%
Tyr	4.3%	2.1%	6.0%	3.1%
Gln	4.4%	0.7%	5.0%	0.8%
Pro	4.7%	4.5%	4.2%	4.2%
Lys	4.9%	6.2%	5.5%	7.4%
Ser	5.0%	7.9%	4.1%	6.9%
Arg	6.0%	4.8%	8.1%	6.9%
Thr	6.4%	4.8%	5.9%	4.7%
Asp	6.6%	4.5%	6.8%	4.9%
Gly	7.3%	13.0%	4.3%	8.0%
Val	7.6%	7.4%	6.9%	7.2%
Leu	8.0%	9.0%	8.1%	9.7%
Ala	10.6%	13.2%	7.3%	9.7%

*The bold italic underlined AAs are first 7 abundant AA in each group.

Table 1. The AA profiling in Rubisco large subunit and whole cell*.

3.4. Glu vs. α-ketoglutarate

Glu is the key AA for photosynthesis cell as the precursor of chlorophyll. Also, it is the second AA products after the series reduction of nitrate to ammonia. It is the most abundant AA (in weight ratio) in *I. zhangjiangensis* (**Table 1**). α-Ketoglutarate plays two vital roles in the biochemical pathways of microalgae. One role is its participation in the TCA cycle, and the other is its involvement in the synthesis of glutamate. α-Ketoglutarate accumulates in the early stage of nitrogen deficiency in *Phaeodactylum tricornutum* and *I. zhangjiangensis* [10, 16]. α-Ketoglutarate accumulated incipiently instead of participating in the synthesis of glutamate, as indicated by the smooth decline in glutamine levels during the same time period. Then, both the level of glutamate and glutamine decreased significantly and the high level of α-ketoglutarate then resulted in feedback inhibition for its precursors, leading to their accumulation and apparent

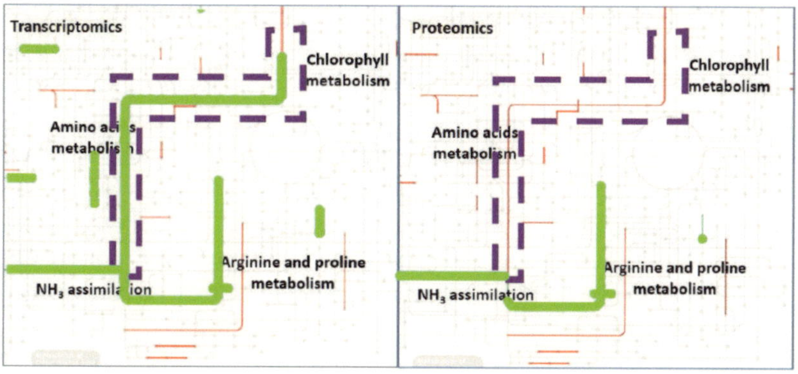

Figure 2. The mapping of significant changed AA metabolism related genes and proteins on KEGG metabolic pathways (map01100). The thin line indicates the significantly down-regulated elements, while the thick line indicates the significantly up-regulated elements.

increase (isocitrate, cis-aconitate, citrate) consequently [10, 16]. Nitrogen supply status showed directly influence on nitrogen assimilation, and by way of α-ketoglutarate intrigued the rebalance the carbon metabolism on a certain degree. AKG-Gln-Glu is the linker between nitrogen assimilation and central carbon metabolism in *I. zhangjiangensis*' response to nitrogen starvation.

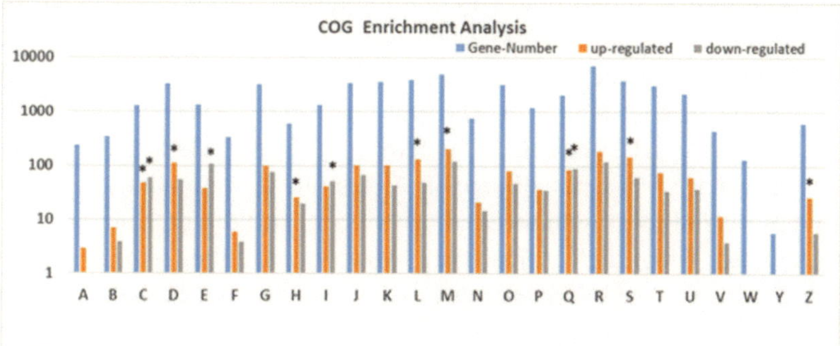

Figure 3. Enrichment of COG clusters analysis. Bars indicated the number of identified genes in different COG clusters and up or down regulated gene number in each cluster. The significant enrichment classes were mark with asterisk (*) on the top of the bar. The meaning of letters (A–Z) is listed below and significantly up or down regulated clusters were mark with up* or down* after their names correspondingly and the number after * was the *Q*-value. RNA processing and modification (A); chromatin structure and dynamics (B); energy production and conversion (up*, 2.95E–02) (down*, 1.00E–05) (C); cell cycle control, cell division, chromosome partitioning (up*, 1.05E–02) (D); amino acid transport and metabolism (down*, 0.0) (E); nucleotide transport and metabolism (F); carbohydrate transport and metabolism (G); coenzyme transport and metabolism (up*, 2.95E–02) (H); lipid transport and metabolism (down*, 1.88E–03) (I); translation, ribosomal structure and biogenesis (J); transcription (K); replication, recombination and repair (up*, 1.93E–03) (L); cell wall/membrane/envelope biogenesis (up*, 2.42E–11) (M); cell motility (N); posttranslational modification, protein turnover, chaperones (O); inorganic ion transport and metabolism (P); secondary metabolites biosynthesis, transport and catabolism (up*, 6.57E–05) (down*, 1.54E–07) (Q); general function prediction only (R); function unknown (up*, 3.60E–07) (S); signal transduction mechanisms (T); intracellular trafficking, secretion, and vesicular transport (U); defense mechanisms (V); extracellular structures (W); nuclear structure (Y); Cytoskeleton (up*, 2.54E–02) (Z).

It's reported that Gln and Glu subject influence on the intermediates of the TCA cycle. The transcription repressor of aconitase, CcpC, can be suppressed by glutamate, when nitrate is assimilated in phototrophic eukaryotes, glutamate or its precursors can arrest the glutamate synthase operon, which is closely affected by both carbon and nitrogen sources, and induced in the presence of ammonium and glycolytic carbon sources to form a feedback inhibition cycle [19]. The nitrogen source for the increased glutamine was considered as the recycling of internal nitrogen, indicating by the higher level of a key intermediate in the urea cycle, citrulline. Proteolysis and AA metabolism consume plenty of energy and release CO_2, resulting in the low level of intermediates in central carbon metabolism in the early stage of nitrogen deficiency [16].

Furthermore, in *I. zhangjiangensis*, the photosynthesis brought enough glucose to cell after the growth was inhibited by nitrogen depletion, and further stored in form of β-1,3 glucan [9, 16]. The wholly increased intermediates in glycolysis process inhibit glyoxylate cycle from the very beginning. By sharing the intermediates with TCA, the influence also causes the transitory increase of α-ketoglutarate, and further influence Gln and Glu metabolism.

3.5. Succinate and Asp, pyruvate and Ala: two secondary intersections of carbon and nitrogen metabolism

In most microalgae species, aspartate and glutamate usually constitute a large proportion of the total amino acid content, while GABA, Trp, Met, His, ornithine and others only represent a small proportion [20]. Asp makes up 6.9% or more of the total AA content and approximately 1.5–1.8% of total detected free AAs in *I. zhangjiangensis*. Asp serves as the source of nitrogen for transaminations and as the early product of the carbon fixation pathway for carbon storage in blue-green algae [21]. Succinate and Asp are intermediates between the glyoxylate cycle and the TCA cycle in cells. The glyoxylate cycle is involved in the formation of glucose and further AAs and nucleotides, whereas the TCA cycle targets the generation of ATP as an energy carrier. For *I. zhangjiangensis,* when nitrogen was limited, some cell physiological activities related to growth receded or stopped, and the product of carbon fixation, glucose, accumulated, while the surplus energy was stored in the form of polysaccharose or lipids [3, 6, 9]. The increase of intermediates of glycolysis and the TCA cycle, as well as AMP and ADP [10], blocked the glyoxylate and TCA cycles. Most of metabolites of the TCA cycle in the inhibitor-treated sample showed a stable concentration at the beginning, except succinate and fumarate. Subsequently, with a further decrease in photosynthetic activity, the TCA cycle activity increased to produce more energy, and the levels of all TCA cycle metabolites decreased. While lipid accumulation increased lately, malate, α-ketoglutarate and other metabolites in TCA decreased. More energy consumed in lipid synthesis than that of polysaccharose may be the reason for this.

Due to the share of intermediates of above cycles, citrate, cis-aconitate and isocitrate accumulated rapidly. In one glyoxylate cycle, two molecules of Ac-CoA are consumed, while one molecular Ac-CoA is consumed during one TCA cycle. Otherwise, from the stable level of pyruvate, it is postulated the initiative synthesis of Ac-CoA was intact. Therefore, the block of above two cycles induced a transitory net accumulation of Ac-CoA, which redirected to FAs, which also consists with our previous work [6, 9].

http://dx.doi.org/10.5772/intechopen.68540

Pyruvate and Ala are another pair of metabolites at the carbon-nitrogen metabolism intersection. Pyruvate is the key linker between different sub cell component (**Figure 4**), and the precursor of acetyl-CoA, the precursor for FAs synthesis and TCA cycle. As the most abundant AA in Rubisco, which contributes about 3% dry weight of normal *I. zhangjiangensis* cells, Ala was considered to release ammonium after nitrogen depleted by consuming α-ketoglutarate and producing Glu and pyruvate. The free Ala kept decrease in this period.

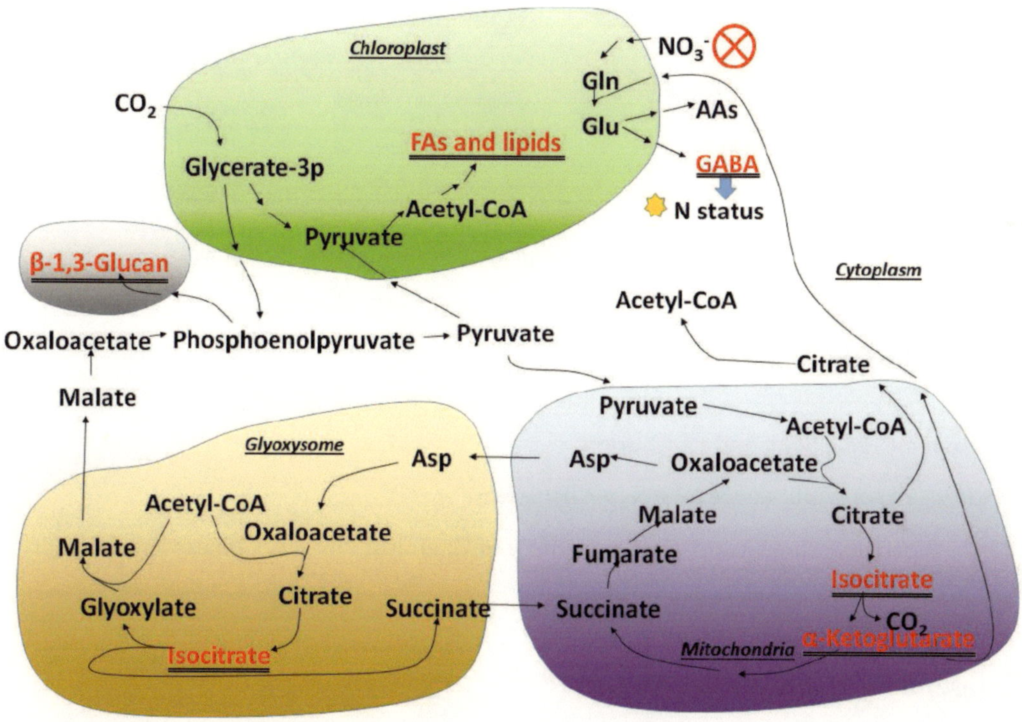

Figure 4. Postulated carbon partition of *I. zhangjiangensis* under during the "golden period." While exogenous nitrogen depleted, the nitrogen assimilation stopped and α-ketoglutarate as the reactant of Gln-Glu cycle had a transitory increase. The synthesis of Gln was reduced, and proteolysis was enhanced, while GABA acted as a transient N buffer as well as a signal to indicating the nitrogen level in cells to stimulate subsequent response. The slowdown of growth without exogenous nitrogen also caused the accumulation of carbohydrates, such as glucose or glucan, which and whose derivants in glycolysis together with α-ketoglutarate inhibited the activities in TCA and glyoxylate cycle. Partial of carbon flowed to pyruvate for further FAs or lipids. Words with doubled underline: increased, others: no significant change. Drafted based on Zhang et al.

4. Conclusion

The AA metabolism changes during the fast ESCs accumulation process of *I. zhangjiangensis* was detailed illustrated by "-omics" analysis. An overall changing model is raised from above data, which will help us to understand the function of different AAs and promote the new regulation methodology for the bioenergy development. A draft global change of AA-related metabolism is shown in **Figure 5**.

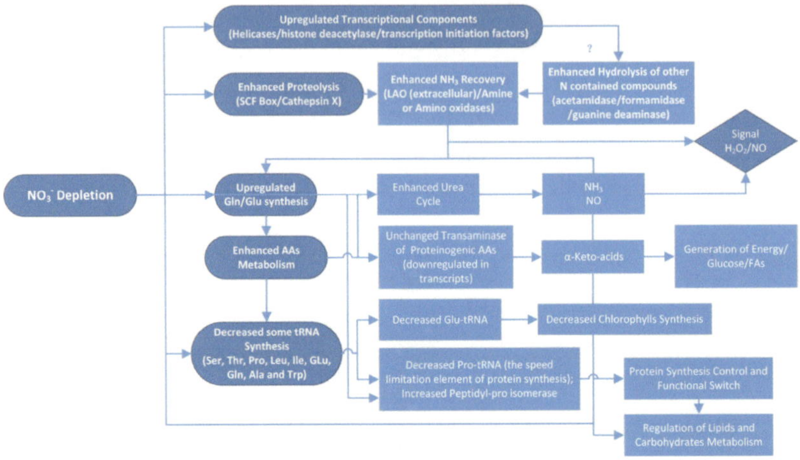

Figure 5. The possible response of AA metabolism of *I. zhangjiangensis* based on omics analysis in the fast ESCs accumulation period.

Author details

Cao Xupeng*, Xue Song and Fan Xuran

*Address all correspondence to: c_x_p@dicp.ac.cn

Marine Bioengineering Group, Dalian Institute of Chemical Physics, Chinese Academy of Sciences, Dalian, China

References

[1] Gonzalez-Fernandez C, Ballesteros M. Linking microalgae and cyanobacteria culture conditions and key-enzymes for carbohydrate accumulation. Biotechnology Advances. 2012;**30**(6):1655-1661. DOI: 10.1016/j.biotechadv.2012.07.003

[2] Vidoudez C, Pohnert G. Comparative metabolomics of the diatom *Skeletonema marinoi* in different growth phases. Metabolomics. 2011;**8**(4):654-669. DOI: 10.1007/s11306-011-0356-6

[3] Feng D, Chen Z, Xue S, Zhang W. Increased lipid production of the marine oleaginous microalgae *Isochrysis zhangjiangensis* (Chrysophyta) by nitrogen supplement. Bioresource Technology. 2011;**102**(12):6710-6716. DOI: 10.1016/j.biortech.2011.04.006

[4] Chisti Y. Constraints to commercialization of algal fuels. Journal of Biotechnology. 2013;**167**:201-214. DOI: 10.1016/j.jbiotec.2013.07.020

[5] Hu Q, Sommerfeld M, Jarvis E, Ghirardi M, Posewitz M, Seibert M, et al. Microalgal triacylglycerols as feedstocks for biofuel production: Perspectives and advances. The Plant Journal. 2008;**54**:621-639. DOI: 10.1111/j.1365-313X.2008.03492.x

[6] Wang H, Yao C, Ai J, Cao X, Xue S, Wang W. Identification of carbohydrates as the major carbon sink of the marine microalga *Isochrysis zhangjiangensis* (Haptophyta) and optimization of its productivity by nitrogen manipulation. Bioresource Technology. 2014;**171**:298-304. DOI: 10.1016/j.biortech.2014.08.090

[7] Yao C, Ai J, Cao X, Xue S, Zhang W. Enhancing starch production of a marine green microalga Tetraselmis subcordiformis through nutrient limitation. Bioresource Technology. 2012;**118**:438-444. DOI: 10.1016/j.biortech.2012.05.030

[8] Huppe HC, Turpin DH. Integration of carbon and nitrogen metabolism in plant and algal cells. Annual Review of Plant Physiology and Plant Molecular Biology. 1994;**45**:577-607. DOI: 10.1146/annurev.pp.45.060194.003045

[9] Wang H, Meng Y, Cao X, Ai J, Zhou J, Xue S, et al. Coordinated response of photosynthesis, carbon assimilation, and triacylglycerol accumulation to nitrogen starvation in the marine microalgae *Isochrysis zhangjiangensis* (Haptophyta). Bioresource Technology. 2015;**177**:282-288. DOI: 10.1016/j.biortech.2014.11.028

[10] Guerra LT, Levitan O, Frada MJ, Sun JS, Falkowski PG, Dismukes GC. Regulatory branch points affecting protein and lipid biosynthesis in the diatom *Phaeodactylum tricornutum*. Biomass and Bioenergy. 2013;**59**:306-315. DOI: 10.1016/j.biombioe.2013.10.007

[11] Allen AE, Dupont CL, Obornik M, Horak A, Nunes-Nesi A, McCrow JP, et al. Evolution and metabolic significance of the urea cycle in photosynthetic diatoms. Nature. 2011;**473**:203-207. DOI: 10.1038/nature10074

[12] Ge F, Huang W, Chen Z, Zhang C, Xiong Q, Bowler C, et al. Methylcrotonyl-CoA carboxylase regulates triacylglycerol accumulation in the model diatom *Phaeodactylum tricornutum*. Plant Cell. 2014;**26**(4):1681-1697. DOI: 10.1105/tpc.114.124982

[13] Grabherr MG, Haas BJ, Yassour M, Levin JZ, Thompson DA, Amit I, et al. Full-length transcriptome assembly from RNA-Seq data without a reference genome. Nature Biotechnology. 2011;**29**:644-652. DOI: 10.1038/nbt.1883

[14] Mortazavi A, Williams BA, McCue K, Schaeffer L, Wold B. Lorian Schaeffer and Barbara Wold Mapping and quantifying mammalian transcriptomes by RNA-Seq. Nature Methods. 2008;**5**:621-628. DOI: 10.1038/nmeth.1226

[15] Wang FJ, Dong J, Jiang XG, Ye ML, Zou HF. Capillary trap column with strong cation-exchange monolith for automated shotgun proteome analysis. Analytical Chemistry. 2007;**79**(17):6599-6606. DOI: 10.1021/ac070736f

[16] Zhang YS, Liu Y, Cao XP, Gao P, Liu X, Wang X, et al. Free amino acids and small molecular acids profiling of marine microalga *Isochrysis zhangjiangensis* under nitrogen deficiency. Algal Research. 2016;**13**:207-217. DOI: 10.1016/j.algal.2015.12.001

[17] Losh JL, Young JN, Morel FMM. Rubisco is a small fraction of total protein in marine phytoplankton. New Phytologist. 2013;**198**:52-58. DOI: 10.1111/nph.12143

[18] Perez-Garcia O, Escalante FM, de-Bashan LE, Bashan Y. Heterotrophic cultures of microalgae: Metabolism and potential products. Water Research. 2011;**45**(1):11-36. DOI: 10.1016/j.watres.2010.08.037

[19] Commichau FM, Forchhammer K, Stulke J. Regulatory links between carbon and nitrogen metabolism. Current Opinion on Microbiology. 2006;**9**(2):167-172. DOI: 10.1016/j.mib.2006.01.001

[20] Brown MR. The amino-acid and sugar composition of 16 species of microalgae used in mariculture. Journal of Experimental Marine Biology and Ecology. 1991;**145**(1):79-99. DOI: 10.1016/0022-0981(91)90007-J

[21] Macko SA, Fogel ML, Hare PE, Hoering TC. Isotopic fractionation of nitrogen and carbon in the synthesis of amino acids by microorganisms. Chemical Geology. 1987;**65**(1):79-92. DOI: 10.1016/0168-9622(87)90064-9

Chapter 10

Bioactive Molecules Profile from Natural Compounds

Adina-Elena Segneanu, Silvia Maria Velciov,
Sorin Olariu, Florentina Cziple, Daniel Damian and
Ioan Grozescu

Additional information is available at the end of the chapter

http://dx.doi.org/10.5772/intechopen.68643

Abstract

Currently, wide world research is focused on sustainable development and the demand for innovative clean technologies, nevertheless natural potential reconsideration could represent a viable solution for the identification and design of new pharmacological agents from renewable resources. The main reason consists of special properties of these natural derivates: immunomodulating activity with continuously perfectible selectivity and efficiency. Plants and herb extracts have been used for centuries as traditional medicines, throughout the entire world. Romanian phytotherapy represents practically a very important part of our traditional knowledge and heritage. Therapeutic properties of plant active principles still continue to be the subject of many researches. In this chapter, an overview of plant bioactive molecules from the perspective of modern phytochemistry is presented. A special part is devoted to a very special medicinal plant, Viscum *album*, in particular identification of amino acids and thionins from mistletoe.

Keywords: phytochemicals, secondary metabolites, analytic methods

1. Introduction

Since ancient times, people have searched and found in nature remedies for various diseases [1, 2]. Romanian tradition pays a special attention to plants which attributes them the properties of living beings (soul, feeling, hearing and sight). Also there is an extraordinary relation between human beings and nature, an almost mystical interdependence. Most often the healing herbs were considered sacred. Phytotherapy origins are lost in the mists of time. In Romania, the traditional medicine has a very long history. Platon, Herodot and Pedanos Dioscoride have mentioned about the herbal medical system from Dacia and medicinal plants used by our ancestors [1]. In Romanian tradition, there is a ritual harvesting these herbs which requires strict compliance with the optimal schedule at a specified date and time. Such is the case of belladonna (Atropa *belladonna)* that is harvested on full moon only from April–May period, before Pentecost. Medicago *falcate* known as earth vortex must be collected only on harvest time. Melilotus *officinali* is plucked only on Sanziene holiday and on Cross day, two important Romanian holidays. It is believed that after this period the plant loses its properties. Romanian traditional medicine involves a very large number of heal plants: twigs, buds, bark and leaves of trees (alder, sambucus), flowers, seeds, stems or roots from plants. Some of the healing herbs were specific to Romanian herbal medicine: Salicornia *herbacea,* Anchusa *officinalis,* Actaea *spicata,* Symphytum *officinale,* Verbascum *thapsus,* Urtica *dioica,* Cicuta *virosa,* Typha *angustifolia,* Chelidonium *majus,* Bryonia *alba L.,* Thymus *vulgaris L., Alisma plantago-aquatica L.,* Hyoscyamus *niger L.,* Verbascum *phlomoides L.,* Achillea *millefolium L.,* Veratrum *album,* Clemantis *vitalba L.,* Potentilla *reptans L.,* Lappa *maior Gartn.,* Datura *stramonium L.,* Dipsacus *pilosus L.,* Erythraea *centaurium Pers.,* Mentha *piperita L.,* Cynoglossum *officinale L.,* Lithospermum *arvense L.* and Galim *verum* [3]. But then their use was spread throughout Balkan areal and Europe. Currently, it is widely used for *Symphytum officinale for its* anti-inflammatory and wound healing activity. Withal, this plant has a high content of allantoin, one of the active principles of the plant it became more important as an ingredient in cosmetics [4–7].

Recent studies on medicinal plants assigned the therapeutic capacity of medicinal plants to their complex structure composed mainly from highly bioactive compounds, minerals, vitamins, etc. [2].

Generally, medicines contain just one active substance, synthetically, whereas medicinal plants are practically a mixture of over dozens or even hundreds of chemicals that act synergistically [2–3]. Moreover, medicinal plants contain a large amount of vitamins and minerals, easily assimilated by human body. Many recent studies demonstrate that vitamins and minerals obtained through chemical synthesis have not the same beneficial effect as similar natural products. It may be due to the fact that in natural products there is a synergistic and complementary action between vitamins, minerals and enzymes, while synthetic compounds (vitamins or minerals) are isolated and even obtained as a different enantiomeric form [8–10]. On the other hand, drugs present other major disadvantages compared with medicinal plants: (i) various side effects; (ii) contraindications; (iii) interactions with other substances; (iv) drug resistance (drug dependence); (v) expensive and (vi) long time consuming research [8]. In comparison, natural compounds present a superior structural diversity, complex structure and multiple stereocenters [10–12]. These are just few arguments that may tilt the scales in favor of herbal medicines. Moreover, World Health Organization (WHO) aims to increase the integration of traditional medicine in order to improve health care system [13].

2. Plant metabolite

Paramount importance of botanic products for humanity is due mainly to their phytocompounds, active principles with therapeutic properties. Several studies have investigated these plant-derived compounds [14–19]. Depending on the role they hold in living organisms, natural substances are divided in the next major categories: (i) *primary metabolites*, molecules common to all biological systems (proteins, fats, sugars) and (ii) *secondary metabolites*, compounds that could be specific for different species as a direct result of the evolution process of a particular phylogenetic group [16, 18–20]. **Figure 1** shows a schematic representation of plant metabolites [16–20].

Bioactive molecules are basically those secondary metabolites exhibiting therapeutic, preventing, toxicological and immunostimulating activity [16–20]. The most known plant-derived bioactive compounds are presented in **Figure 2**.

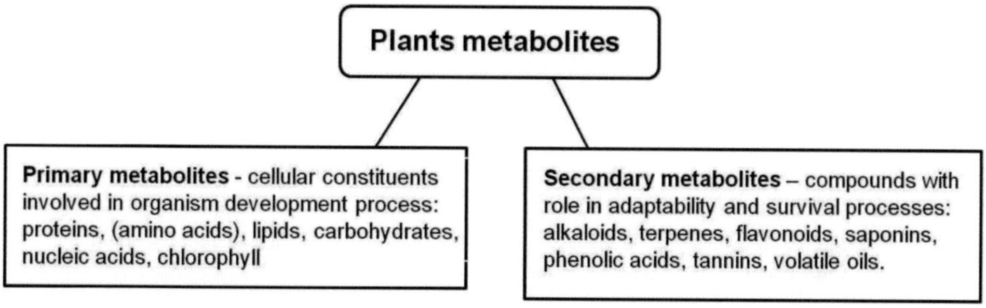

Figure 1. Plant metabolites.

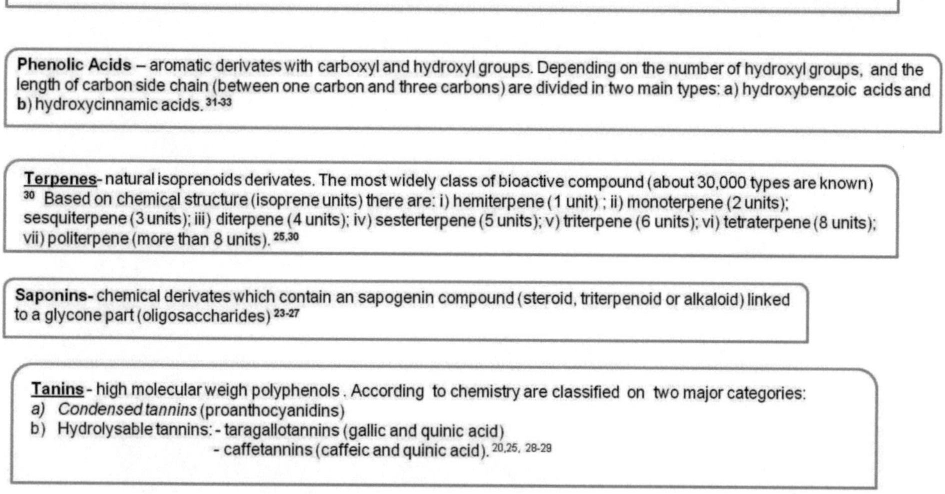

Figure 2. Schematic representation of plant bioactive compounds.

Biological activity of these compounds has been extensively investigated in particular in the last decades [4–32]. Thus, it demonstrated that there is a close connection between the chemical structure of the natural active principles (functional group types, number and position related to carbon skeleton, substitution in aromatic ring, stereochemistry, side chain length, saturation, etc.) [17, 20, 22, 25, 27, 34]. The role of metabolites in human organism is briefly presented in **Table 1**. And some examples of these compounds are shown in **Table 2**.

Compound type	Pharmacological properties
Terpenoid	Antimicrobial, antiviral, antiviral, anthelmintic, antibacterial, anticancer, antimalarial, anti-inflammatory [15, 34]
Phenolics acids	Anticarcinogenic and antimutagenic, anti-inflammation and anti-allergic [16, 20, 25, 31–35]
Alkaloids	Antispasmodic, antimalarial, analgesic, diuretic activities, local anesthetic, antihypertensive, antiasthma, antimalarials, diuretic, bactericidal [14–16, 20, 21]
Flavonoids	Antioxidant activity, cardiovascular protective, anti-inflammatory, hepatoprotective, antiviral, antibacterial [20, 22–24, 34]
Saponins	Antitumor, antiviral, antifungal, anti-inflammatory, immunostimulant, antihypoglycemic, antihepatotoxic and hepatoprotective, anticoagulant, neuroprotective, antioxidant [16, 20, 24–27, 34]
Tannins	Antioxidant, anti-carcinogenic, diuretics, hemostatic, anti-mutagenic, metal ion-chelators, antiseptic, [14, 16, 20, 25, 28–32]

Table 1. Biologic activity of main groups of natural compounds.

Secondary metabolites	Important molecules	References
Alkaloids	Caffeine, piperine, atropine, berberine, morphine, quinine, cocaine, nicotine, strychnine, codeine, ephedrine, dopamine, serotonine, vinblastine, vincristine, brucine, capsaicin, solanine, tomatine, choline, etc.	[15, 21, 34]
Terpenes	*Hemiterpene*: isoprene, isovaleric acid	[15, 34]
	Monoterpene: limonele, eucalyptol, menthol, nerol, citral	
	Sesquiterpene: zinziberene, farnesol	
	Diterpene: cafestol, retinal, retinol	
	Sesterterpenes: bulgarene, farnesol, lindarene	
	Triterpene: provitamin A, betulin, cymarin	
	Tetraterpene: lycopen, α si β carotenoids	
	Polyterpene: vitamin E, gutta-percha	
Flavonoids	*Flavones:* luteolin, diosmetin, apigenin	[15, 22, 23]
	Flavonols: quercetin, myricetin, rutin, kaempferol	
	Flavanones: hesperetin, naringenin	
	Flavanonol: silymarin, taxifolin	
	Isoflavones: daidzin, genistin	
	Anthocyanidin: cyanidin, delphinidin, peonidin, petunidin	

Secondary metabolites	Important molecules	References
Phenolic acids	Cinnamic acid, benzoic acid, ferulic acid, coumaric acid, caffeic acid, salicylic acid, gallic acid	[15, 33]
Saponins	Panaxadiol, diosgenin	[15]

Table 2. Some well-known examples of plant metabolites.

3. Profiling of plant bioactive molecule

Achievement of the natural plant bioactive molecules profile involves more consecutive stages (**Figure 3**) [14, 17, 18].

3.1. Selection of plant species

First and foremost stage is required to evaluate the existing ethnomedicinal studies, chemotaxonomical data regarding a particular medicinal plant, information collected from different historic documents, traditional knowledge from even local quacks and specialists [14, 37].

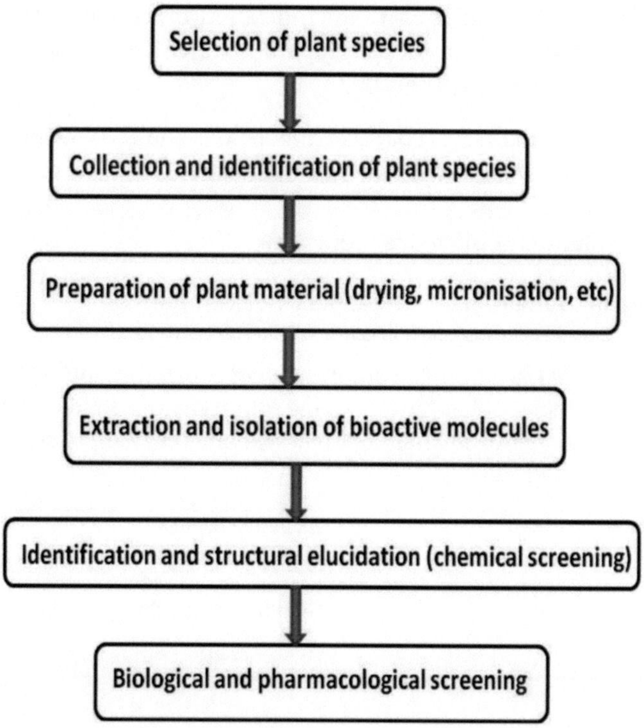

Figure 3. Flowchart of plant bioactive molecules profiling.

3.2. Collection and identification of plant species

This represents a key stage required to afford a reliable profile of natural active principles. And involve the next steps:

(a) Procurement of botanic component only from sources with guaranteed good agriculture and collection practices. An essential step demand to investigate a possible microbial, pesticide or heavy metals contaminations to avoid adversely affect the results of the chemical screening of bioactive metabolites, increased the time and cost of studies [18, 36, 37]. **Table 3** presents the main analytical techniques used to detect a possible plant contamination.

(b) Plant taxonomic or genetic identification [18, 36, 37]. A modern method for authentification the botanic precursor use genomic analysis (DNA barcoding method) [38]. Research has been shown that biodiversity and plant growth environmental conditions (temperature, humidity, soil physic and chemical properties) could influence the bioactive molecules profile [39].

3.3. Preparation of plant material (drying, micronisation, etc.)

The botanical material processing is needed to avoid the degradation of plant bioactive compounds [14]. The drying is recommended to be performed in areas-controlled atmosphere (absence of humidity, well-ventilated, constant temperature).

The dried botanic material is subjected to micronization process through mechanical techniques. The other methods of plant sample preparation involve: (i) botanic material homogenization or (ii) plant maceration [14, 39, 42].

This step aims to minimize the sample particle dimensions and thus to enhance the extraction yield [14].

3.4. Extraction and isolation of bioactive molecules

This is the key stage in evaluation of natural bioactive compounds.

(a) *Extraction and separation techniques*: In literature, there are many studies on extraction of certain groups of plant metabolites. However, the selectivity of conventional extraction methods (soxhlet extraction, hydrodistillation, maceration, percolation, steam distillation, etc.)

Plant contamination assay	Analytical method
Heavy metals	Atomic absorption spectroscopy, ICP-MS, etc.
Pesticide or/and herbicide residues	GC-MS, mass spectrometry, HPLC-MS, etc.
Microbial content	HPLC-MS, etc

Table 3. Plant contamination: chemical assays.

are at least moderate and economically inefficient (energy, hazardous reagents consumption, time and temperature) [18, 39–42]. The other main disadvantages of these techniques are (i) not environment friendly; (ii) high possibility of degradation of thermostable active principles and (iii) additional steps (extract concentration, cleanse) [39–42]. Advanced extraction processes (solid-phase extraction, ultra-sound-assisted extraction, microwave-assisted extraction, supercritical fluid extraction, pulsed electric field extraction, pressurized liquid extraction, enzyme-assisted extraction, surfactant-mediated extraction) have minimized many of these shortcomings. Usually, the separation of a particular group of bioactive compounds from a complex natural product required a selective separation strategy based on phytochemicals partition in several different polarity solvents [43]. Nevertheless, natural product chemistry research concerns the development of new and highly efficient extraction techniques. Recent studies have reported that calixarenes could represent an attractive opportunity in this regard [44].

(b) *Isolation methods*: The physical properties (solubility, molecular weight, stability, dipole moment, etc.) of targeted bioactive compounds are essential for an efficient isolation method [39, 41, 42]. Another important factor is the nature of extraction solvent [39]. Generally, based on existing databases, the plant metabolites isolation are carried out through chromatographic methods: thin chromatography (TLC), flash chromatography, high performance liquid chromatography (HPLC), high-performance thin-layer chromatography (HPTLC), gas chromatography (GC) or Fourier transform infrared spectroscopy (FT-IR) [14, 39, 41, 42]. A biological material previously uninvestigated and is in demand to develop an appropriate isolation procedure that require following additional steps: (i) phytochemical evaluation; (ii) bioassay (immunoassay (monoclonal antibodies) [14, 39].

3.5. Identification and structural elucidation (chemical screening)

This is the forefront but also the most difficult step in natural product chemistry. Achievement of the bioactive molecules complete profile requires the cutting-edge technology and advanced knowledge specialists. Investigation on new natural compounds entails a larger work volume determined mainly by the absence of plant scientific data [14, 39, 45–48]. Plant bioactive molecules profiling is based on various spectroscopic techniques, advanced chromatographic (hyphenated techniques) methods and a complete morphostructural characterization procedure using X-ray crystallographic techniques, polarimetry and electronic microscopy (**Table 4**) [14, 39, 45–48]. An optimal strategy based on high-tech technology provides fast and highly efficient complete structural information about the targeted compounds [39, 42, 47, 48]. **Table 5** shows the main analytical techniques applied in natural bioactive compounds chemical screening [14, 39, 45–48].

3.6. Biological and pharmacological screening

There are various methods designed to investigate the biological activity of a targeted natural compounds. An optimal procedure must fulfill several criteria: fast, simple, reliable, high sensibility and selectivity, availability and low cost. Bioactivity evaluation for a

Spectroscopic methods	**UV-Vis spectroscopy**
	Fourier transform infrared spectroscopy
	Mass spectroscopy:
	(a) Electron impact mass spectrometry (EIMS)
	(b) Chemical ionization mass spectrometry (CIMS)
	(c) Electrospray ionization mass spectrometry (ESIMS)
	(d) Electrospray ionization mass spectrometry (ESIMS)
	(e) Fast atom bombardment mass spectrometry (FABMS)
	Nuclear Magnetic Resonance **(NMR) spectroscopy:**
	(a) *One-dimensional techniques:* ^{1}HNMR, ^{13}CNMR, ^{13}CDEPT, ^{13}CPENDANT,^{13}C J mod.
	(b) *Two-dimensional techniques:* ^{1}H-^{1}H COSY, ^{1}H-^{1}H DQF-COSY, 1H-1H COSY-lr, 1H-1H NOESY, ^{1}H-^{1}H ROESY, ^{1}H-^{1}H TOCSY, ^{1}H-^{13}C HMBC, ^{1}H-^{13}C HMQC, ^{1}H-^{13}C HSQC,HSQCTOCSY
Chromatography methods	**Gas-chromatography**: GC, GC-MS, GC-TOF-MS; GC-MS/MS, two-dimensional GC coupled with mass spectrometry (GC×GC-MS), GC-FTIR, GC-NMR
	Liquid chromatography: LC/UV; LC/MS; LC/UV/MS; LC/MS-MS; LC/NMR, LC-UV-DAD, HPLC-NMR
Other analytic techniques	XRD; TEM; polarimetry

Table 4. A brief overview of bioactive molecules profiling tools [39, 42, 47, 48].

Plant sample	Propose structure	Abbreviation	SIM (selected-ion monitoring)
V_1 (hexane)	Cystine	C-C	41, 42
	Glutamic acid	Glu	38, 40
	Phenylalanine	Phe	56, 57
	Ornithine	Orn	59,60,61
	Histidine	His	84, 89
	Tyrosine	Tyr	61, 63, 94
	Glycine	Gly	116, 74
	Homoserine	HSER	102,128, 143
	Asparagine	Asn	155, 69
	Isoleucine	Ile	171, 129
	Valine	Val	158, 116
	Threonine	Thr	160, 101
	β-Alanine	β Ala	158, 98
	Valine	Val	158,72
	β-Alanine	β Ala	129, 158, 98
	Homoserine	HSER	102, 128, 143
	Asparagine	Asn	155, 69

Plant sample	Propose structure	Abbreviation	SIM (selected-ion monitoring)
V_2 (CCl_4)	Asparagine	Asn	155, 69
	Cystine	C-C	41,42
	Alanine	Ala	130, 70
	Glutamic acid	Glu	38, 40
	Ornithine	Orn	59,60,61
	Tryptophan	Trp	130
	β-Alanine	β Ala	129, 158, 98
	Phenylalanine	Phe	56, 57
	Tyrosine	Tyr	61, 63, 94
	Homoserine	HSER	102,128, 143
	Valine	Val	158,72
	Lysine	Lys	170, 129
	Glycine	Gly	116, 74
	Isoleucine	Ile	170, 130
	Hystidine	Hys	84, 87
V_3 (petroleum ether)	Glutamic acid	Glu	38, 40
	Cystine	C-C	41,42
	Phenylalanine	Phe	56, 57
	Glycine	Gly	116, 74
	Leucine	Leu	172, 86
	β-Alanine	β Ala	129, 158, 98
	Isoleucine	Ile	170, 130
	Cysteine	Cys	248, 162, 206
	Tyrosine	Tyr	61, 63, 94
	Hystidine	Hys	84, 87
	Glutamine	Gln	84, 187
	Lysine	Lys	170, 129
	Tryptophan	Trp	130
	Valine	Val	158,72
	Aspartic acid	Asp	216, 130
	Methionine sulfoxide		229,182,138
	S-Carboxymethyl-cysteine		144,203,262
	Proline-hydroxyproline (dipeptide)	PHP	156, 186
	Lysine-alanine (dipeptide)	LYS-ALA	170, 224, 153
	3-Methyl-cysteine	1MHIS	172,259,130
	Arginino succinic acid	ARG-SUC	441, 326
	Methionine	Met	203, 277
	Cystathionine	CTH	203, 272

Plant sample	Propose structure	Abbreviation	SIM (selected-ion monitoring)
V_4 (acetone)	Cystine	C-C	41,42
	Glutamic acid	Glu	38, 40
	Phenylalanine	Phe	56, 57
	β-Alanine	β Ala	129, 158, 98
	Ornithine	Orn	59,60,61
	Glycine	Gly	116, 74
	Isoleucine	Ile	170, 130
	Histidine	Hys	84, 87
	Glutamine	Gln	84, 187
	Valine	Val	158,72
	Tyrosine	Tyr	61, 63, 94
	Lysine	Lys	170, 129
	Homoserine	HSER	102,128, 143
	Proline-hydroxyproline (dipeptide)	PHP	156, 186
	3-Methyl-cysteine	1MHIS	172,259,130
	Homocysteine	HCYS	142, 203
	Glycyl-glycine (dipeptide)	Gly-Gly	117, 144, 201

Table 5. Compounds identified through GC-MS analysis.

plant extraction (plant fraction) is usually performed through *in vitro* or/and *in vivo* studies [14, 49, 50]. Most often, *in vitro* studies are focused on the evaluation of specific cell biology (cell count, growth rate, metabolic rate, cell function and protein expression). *In vitro* tests are conducted on various animal or human cell cultures, enzymes, depending on targeted natural compound biological activity [14, 49, 50]. For instance, the bioassays for antitumor activity are conducted on tumor experimental models. Complementary, the immunological activity on normal cell culture should be monitored. The cells will be analyzed by fluorescence microscopy and will be quantified to establish the degree of apoptosis and implicitly the cell viability. Also, the time-lapse video microscopy can be used to evaluate the bioactive phytochemicals [43]. The *in vivo* biotests are applied on animals (mice, rats, pigs, etc.).

Natural compounds bioassay can be demonstrated also using computational chemical methods: quantitative structure-activity relationship (2D or 3D QSAR) and structure-activity relationship (SAR) [75, 76].

Regarding the antioxidant activity of natural compounds, literature demonstrates the existence of a considerable number of studies using two analytical techniques: electron spin resonance (ESR) and chemiluminescence. But the obtained results depend on the type of reactant (specific free radical) used [51]. Electrochemistry, especially by the instrumentality of voltammetry has been shown to be a useful method for the investigation of the antioxidant activity of different targeted compounds [52].

4. Natural compounds in Viscum *album* as an example of medicinal plant

One of the most renowned medicinal plants is *Viscum album* L., which has very different applications: tonic, cardiotonic, antiviral, cancer, etc. In different European countries, mistletoe extracts are prepared and commercially available (*Iscador, Isorel, Eurixor, Plenesol, Vysorel, Lektinol, Helixor,* etc.) as alternative treatment for cancer therapy [53–58].

First information on the use of this plant for its benefits on the human body dates back to ancient times. The druids and Celts considered as sacred mistletoe that grows on oak. Over time, peoples were attributed a special symbolism to this evergreen plant: immortality, knowledge, wisdom, universal panacea, love, fortune, fertility, etc. [54, 57]. There are considered that magical properties of mistletoe are kept only if the complied both the collection ceremony: a golden knife in a special moment of day before full moon, on right period (summer or winter solstice) [54].

In traditional medicine, Viscum are used for various health benefits: poison antidote, anti-age, anti-inflammatory, fertility, antitumor, headaches, preventing epilepsy, cure for plague, erysipelas, etc. [53–55].

Many studies have been carried out for determination of the outstanding biological effects: antiproliferative activity, antitumor activity, antiviral activity, cardiovascular, immunostimulant and antidiabetic [56, 58–65]. But the extremely complex chemical composition of this plant has not been precisely determined yet. Nevertheless, several secondary metabolites such as flavonoids, alkaloids, steroids, terpenoids were detected [66]. However, research has demonstrated that viscum chemical composition varies depending on (i) the type of host tree on which it grows (oak, maples, acacia, robinia, poplar, etc.), (ii) time of harvesting, (iii) environmental conditions and (iv) extraction method [56, 67].

The attempts to establish the compounds responsible for biological, immunomodulating and cytotoxic activity had targeted especially the lectins and viscotoxins as active components [56, 67]. Nevertheless, these compounds represent only a small content of percent from the entire plant peptide content which is not fully understood in terms of chemical structure and biological activity. Relatively recent research had emphasized on the presence of other peptide derivate, viscumamide with antitumor activity [68]. However, there are still many compounds pharmacologically active that can be found. Continuous development of analysis techniques can provide important information about new highly bioactive compounds isolated from plant extracts.

4.1. Importance of natural small peptide

From the multitude of classes of biomolecules isolated from natural compounds, a special attention has been given to amino acids and small peptides due to their remarkable properties (high solubility, strong antioxidant, reduce high blood pressure, analgesic, anti-tumor, immunomodulatory, etc.). In addition, these biologically active compounds have various applications in pharmacology, cosmetics, sports and food.

In plants, these biomolecules are involved also in defense mechanisms against various classes of pathogens (bacteria, fungi, parasites, etc.) [69, 70].

Given that cancer is the second leading cause of death in European countries, and one of the most imminent health problems in the developed world [71–73], there is an overwhelming interest for new efficient antitumor agents with high bioavailability and minimal side effects. In this context, research on plant bioactive molecules with putative antitumor activity is even more justified.

Thionins represent a special class of small peptide with multiple disulfide bonds [43, 68, 69]. They have shown cytotoxicity and antitumor activity [69, 70]. Research has reported that mistletoe contains several types of thionins: viscothionin A1, viscothionin A2, viscothionin A3, viscothionin B, viscothionin C1, viscothionin D, viscothionin E, viscothionin P1 [69, 70].

4.2. Determination of amino acids and thionins from Viscum *album*

In an effort to detect the amino acids and thionins from Viscum *album* a selective partition strategy based on solvents with different polarities (methanol, hexane and carbon tetrachloride) was developed [43]. The plant material (Viscum *album* leaves and young leaves from Quercus *robur*) was obtained from a collection taken in December 2015 in Timis, Romania. Plant sample was identified at Victor Babes University of Medicine and Pharmacy Timisoara. The botanical material was dried and then finely ground in a ball mill. A plant sample (3 g) was placed in a 100 mL volumetric flask containing 50 mL of methanol. The result mixture was sonicated for 60 min at 40°C, with a frequency of 50 kHz. Then the solution was filtered through a 0.30 μm pore size *filter* and subsequently extracted with the following organic solvents: *n*-hexane (V_1) and carbon tetrachloride and (V_2). The separation of thionins was carried on the next experiment: 2 g of sample was extracted successively with petroleum ether (30 mL) and acetone (30 mL) [43]. Identity of the compounds from the obtained viscum fractions: V_1 (hexane), V_2 (CCl_4), V_3 (petroleum ether) and respectively, fraction V_4(acetone) was performed using GC-MS and TOF MS methods.

4.3. GC-MS analysis

The GC-MS chromatograms for mistletoe extract fraction V_1–V_5 are presented in **Figure 4(a)–(d)**.

The results of design isolation strategy based on different solvent polarity were analyzed through GC-MS [43]. The identified compounds are presented in **Table 5**; after a careful comparison with spectral database, NIST/NBS was used to compare the results of analysis [43].

4.4. TOF-MS analysis

The mass spectra of mistletoe fractions V_1–V_4 (acquired in positive ion mode, in a mass range of 100–3000 m/z) are presented in **Figure 5(a)–(d)**.

http://dx.doi.org/10.5772/intechopen.68643

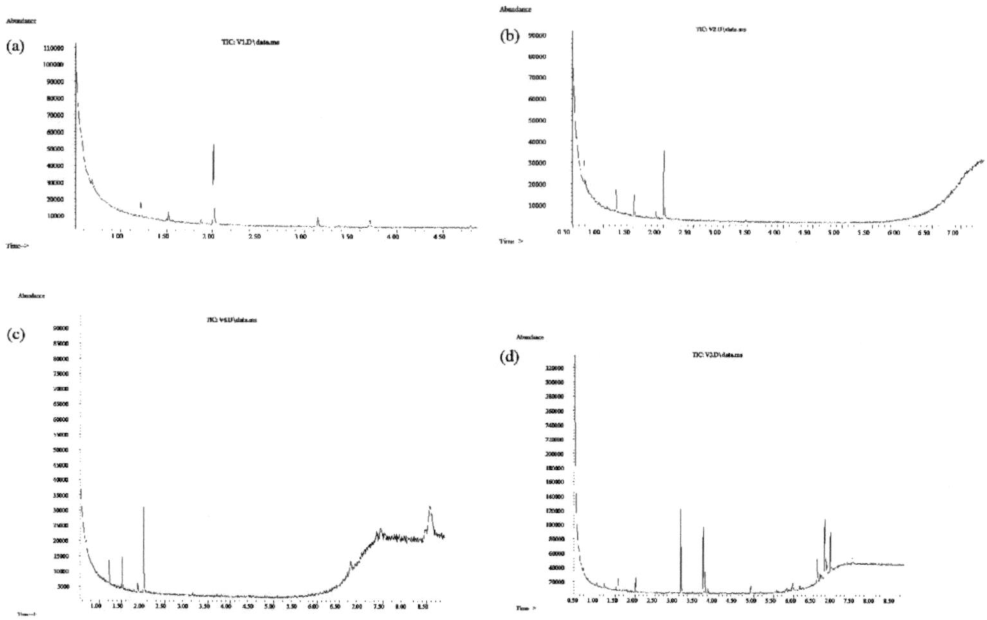

Figure 4. TIC of (a) V_1 extract, (b) V_2 extract, (c) V_3 extract and (d) V_4 extract.

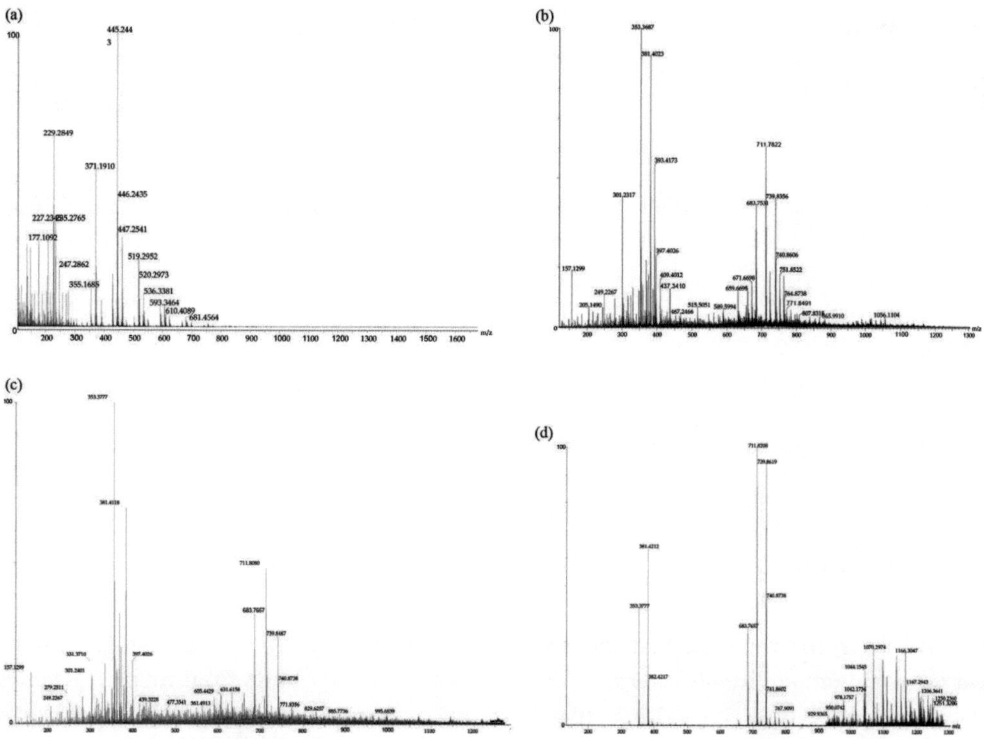

Figure 5. Positive ion mode TOF-MS of (a) V_1extract, (b) V_2extract, (c) V_3extract and (d) V_4extract.

4.5. FT-IR spectroscopy

The solid (fine grounded) sample of mistletoe was analyzed also through FT-IR spectroscopy (**Figure 6**). It has been aimed to identify the absorptions bands specific to amino acids and peptides from: (i) 3400 cm^{-1} (O-H and N-H bonds); (ii) 3330–3130 cm^{-1} (NH_3+ groups); (iii) symmetric absorption at 2080–2140 cm^{-1}or 2530–2760 cm^{-1}; (iv) 1500–1600 cm^{-1} (ammonium group deformation vibrations); (v) 1610–1660 cm^{-1} (carboxylate group); (vi) 1724–1754 cm^{-1} (carbonyl vibrations) and (vii) vibrations bands characteristic for thionins (1687, 1675, 1663, 1654, 1644, 1632, 1621, 1611) [45, 74].

The FT-IR spectra were recorded using a Universal ATR accessory (UATR) and mistletoe samples 20 mg and 30 mg, respectively, mixed with KBr.

From the spectra analysis, the presence of bands specific to amino acids, thionins and peptides can be noticed.

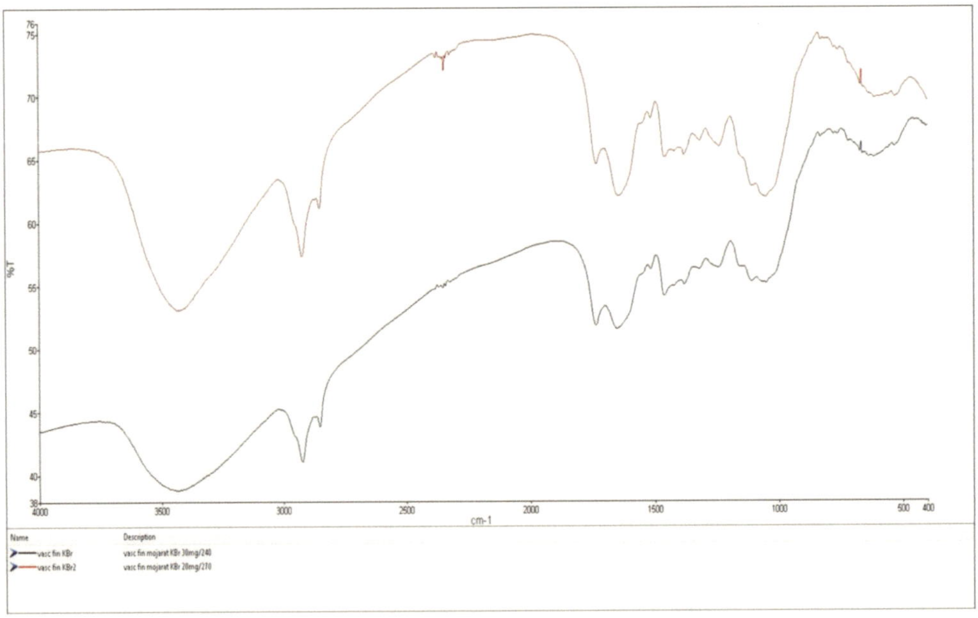

Figure 6. The FT-IR spectra for the mistletoe sample.

5. Conclusions

The collective results suggest that chosen separation solvent and analytic strategies are efficient for isolation and identification of targeted natural compounds from mistletoe sample. Further studies on mistletoe extract are necessary to gain insight into the complete bioactive molecules profile with high antitumor activity.

Continuous development of analysis techniques can provide important information about highly bioactive molecules isolated from natural compounds. Particular importance must be paid to the choice of optimal separation methods which must be simple but highly selective and efficient for separation of a certain class of natural metabolites. A special emphasis has been given to identify the peptides because it was considered that nature of amino acids, their quantity in plant and the ratio to known peptides for their high bioactivity may be relevant to their anticancer action. Research on small peptide with pharmacological activity continues to be a topic of great interest to the current science due to their special high biological activity, chemical stability, bioavailability, etc. From this perspective, further research will allow to predict the formulation of the peptide profile from natural extract with a specific biological effect with application in cancer prevention or therapy.

Acknowledgements

We like to thank Dr. Andrei Bunaciu for his contribution to this chapter.

Author details

Adina-Elena Segneanu[1]*, Silvia Maria Velciov[2], Sorin Olariu[2], Florentina Cziple[3], Daniel Damian[4] and Ioan Grozescu[4]

*Address all correspondence to: s_adinaelena@yahoo.com

1 Scient Analytics, SCIENT, Research Center for Instrumental Analysis, Cromatec, Plus, Ilfov, Romania

2 Victor Babes University of Medicine and Pharmacy, Timisoara, Romania

3 Eftimie Murgu University, Resita, Romania

4 University Politehnica Timisoara, Timişoara, Romania

References

[1] Danciu ET, Dusca AI. The Spirituality of the Geo-Dacian People—between Truth and Legend. Revista de Stiinte Juridice. 2008;141

[2] Sandberg F, Corrigan D. Natural Remedies. Their Origins and Uses. New York: Taylor & Francis; 2001

[3] Crisan IH. Medicina in Dacia. Ed. Dacica; 2013. ISBN 978-973-88076-2-4

[4] Bojor O. Ghidul plantelor medicinale si aromatice de la A la Z, Ed. Fiat Lux; 2003. ISBN 973-9250-68-8

[5] Neagu E, Roman GP, Radu GL. Antioxidant capacity of some *Symphytum officinalis* extracts processed by ultrafiltration. Romanian Biotechnological Letters. 2010;**15**(4):5505–5511

[6] Alkan FU, Anlas C, Ustuner O, Bakirel T, Sari AB. Antioxidant and proliferative effects of aqueous and ethanolic extracts of *Symphytum officinale* on 3T3 Swiss albino mouse fibroblast cell line. Asian Journal of Plant Science and Research. 2014;**4**(4):62–68

[7] Thornfeldt C. Cosmeceuticals containing herbs: Fact, fiction, and future. Dermatologic Surgery. 2005;**31**:873–880

[8] Taylor L. The Healing Power of Rainforest Herbs. Square One Publishers Inc, New York, USA. 2005. ISBN: 0-7570-0144-0

[9] Clement B. Nutri-Con: The Truth About Vitamins & Supplements. The Vitamin Myth Exposed. Hippocrates Health Institute & OCA; 2005. https://www.organicconsumers.org/news/nutri-con-truth-about-vitamins-supplements

[10] Nguyen LA, He H, Pham-Huy C. Chiral drugs: An overview. International Journal of Biomedical science. 2006;**2**:85–100

[11] Lahlou M. The success of natural products in drug discovery. Pharmacology & Pharmacy. 2014;**4**:17–31

[12] Phillipson JD. Phytochemistry and medicinal plants. Phytochemistry. 2001;**56**:237–243

[13] World Health Organization (WHO). Traditional Medicine Strategy 2014–2023. Hong-Kong, China: World Health Organization; 2013. pp.1-78 1-78. ISBN 9789241506090

[14] Chikezie PC, Ibegbulem CO, Mbagwu FN. Bioactive principles from medicinal plants. Research Journal of Phytochemistry. 2015;**9**(3):88–115

[15] Shakya AK. Medicinal plants: Future source of new drugs. International Journal of Herbal Medicine. 2016;**4**(4):59-64

[16] Capasso F, Gaginella TS, Grandolini G. Phytotherapy—A Quick Reference to Herbal Medicine. Berlin Heidelberg: Springer-Verlag; 2003. ISBN 978-3-540-00052-5

[17] Soetan KO, Aiyelaagbe OO. The need for bioactivity-safety evaluation and conservation of medicinal plants—A review. Journal of Medicinal Plants Research. 2009;**3**(5):324-328

[18] Azmir J, Zaidul ISM, Rahman MM, Sharif KM, Mohamed A, Sahena F, Jahurul MHA, Ghafoor K, Norulaini NAN, Omar AKM. Techniques for extraction of bioactive compounds from plant materials: A review. Journal of Food Engineering. 2013;**117**:426-436

[19] Dias DA, Urban S, Roessne U. A historical overview of natural products in drug discovery. Metabolites. 2012;**2**(2):303–336

[20] Anulika NP, Ignatius EO, Raymond ES, Osasere OI, Abiola AH. The chemistry of natural product: Plant secondary metabolites. International Journal of Technology Enhancements and Emerging Engineering Research. 2016;**4**(8):1–8. ISSN 2347-4289

[21] Woolley JG. Plant alkaloids. In: Encyclopedia of Life Sciences. Nature Publishing Group, John Wiley & Sons. 2001. pp. 1–11

[22] Kumar S, Pandey AK. Chemistry and biological activities of flavonoids: An overview. The Scientific World Journal. 2013:Article ID 162750

[23] Falcone Ferreyra ML, Rius SP, Casati P. Flavonoids: Biosynthesis, biological functions, and biotechnological applications. Frontiers in Plant Science-Plant Physiology. 2012;**3**:Article 222

[24] Negi JS, Negi PS, Pant GJ, Rawat SM, Negi SK. Naturally occurring saponins: Chemistry and biology. Journal of Poisonous and Medicinal Plant Research. 2013;**1**(1):001–006

[25] Saxena M, Saxena J, Nema R, Singh D, Gupta A. Phytochemistry of medicinal plants. Journal of Pharmacognosy and Phytochemistry. 2013;**1**(6):168–182

[26] Kareru PG, Keriko JM, Gachanja AN, Kenji GM. Direct detection of triterpenoid saponins in medicinal plants. African Journal of Traditional, Complementary and Alternative Medicines. 2008;**5**(1):56–60

[27] Man S, Gao W, Zhang Y, Huang L, Liu C. Chemical study and medical application of saponins as anti-cancer agents. Fitoterapia. 2010;**81**:703–714

[28] Hassanpour S, Maheri-Sis N, Eshratkhah B, Baghbani Mehmandar F. Plants and secondary metabolites (Tannins): A review. International Journal of Forest, Soil and Erosion. 2011;**1**(1):47–53

[29] Khanbabaee K, van Ree T. Tannins: Classification and definition. Natural Product Reports. 2001;**18**:641–649

[30] Breitmaier E. Terpenes. Weinheim: Wiley-VCH Verlag GmbH & Co. KGaA; 2006. ISBN: 3-527-31786-4

[31] Lattanzio V. Phenolic compounds: Introduction. In:Ramawat KG, Merillon JM, editors. Natural Products. Berlin Heidelberg: Springer-Verlag; 2013. pp.1543–1580

[32] Balasundram N, Sundram K, Samman S. Analytical, nutritional and clinical methods—Phenolic compounds in plants and agri-industrial by-products: Antioxidant activity, occurrence, and potential uses. Food Chemistry. 2006;**99**:191–203

[33] Robbins RJ. Phenolic acids in foods: An overview of analytical methodology. Journal of Agricultural and Food Chemistry. 2003;**51**:2866–2887

[34] Kabera JN, Semana E, Mussa AR, He X. Plant secondary metabolites: Biosynthesis, classification, function and pharmacological properties. Journal of Pharmacy and Pharmacology. 2014;**2**:377–392

[35] Ozcan T, Akpinar-Bayizit A, Yilmaz-Ersan L, Delikanli B. Phenolics in human health. International Journal of Chemical Engineering and Applications. 2014;**5**(5):393–396

[36] van Breemen Richard B, Fong Harry HS, Farnsworth NR. The role of quality assurance and standardization in the safety of botanical dietary supplements. Chemical Research in Toxicology. 2007;**20**(4):577–582

[37] Schwikkard SL, Mulholland DA. Useful methods for targeted plant selection in the discovery of potential new drug candidates. Planta Medica. 2014;**80**(14):1154–1160

[38] Balachandran KRS. Mohanasundaram S, Sathishkumar R. DNA barcoding: A genomic-based tool for authentication of phytomedicinals and its products. Botanics: Targets and Therapy. 2015;**5**:77–84

[39] Sarker SD, Latif Z, Gray AI. Natural products isolation. 2nd ed. New Jersey, USA: Humana Press Inc; 2006. ISBN 1-58829-447-1

[40] Segneanu AE, Macarie CA, Pop RO, Balcu I. Combined microwave–acid pretreatment of the biomass. In: Shaukat SS, editor. Progress in Biomass and Bioenergy Production. Croatia: In Tech; 2011. pp.223–238. ISBN 978-953-307-491-7

[41] Sasidharan S, Chen Y, Saravanan D, Sundram KM, Yoga Latha L. Extraction, isolation and characterization of bioactive compounds from plants' extracts. African Journal of Traditional, Complementary and Alternative Medicines. 2011;**8**(1):1–10

[42] Kaufmann B, Christen P. Recent extraction techniques for natural products: Microwave-assisted extraction and pressurized solvent extraction. Phytochemical Analysis. 2002;**13**(2):105–113

[43] Segneanu AE, Grozescu I, Cziple F, Berki D, Damian D, Niculite CM, Florea A, Leabu M. *Helleborus purpurascens*—Amino acid and peptide analysis linked to the chemical and antiproliferative properties of the extracted compounds. Molecules. 2015;**20**:22170–22187

[44] Segneanu AE, Damian D, Hulka I, Grozescu I, Salifoglou A. A simple and rapid method for calixarene-based selective extraction of bioactive molecules from natural products. Amino Acids, Springer. 2016;**48**:849–858

[45] Neda I, Vlazan P, Pop RO, Sfarloaga P, Grozescu I, Segneanu A-E. Peptide and amino acids separation and identification from natural products. In: Krull IS, editors. Analytical Chemistry. Croatia: Intech; 2012. pp.135–146. ISBN 978-953-51-0837-5

[46] Segneanu AE, Gozescu I, Dabici A, Sfirloaga P, Szabadai Z. Organic compounds FT-IR spectroscopy. In: Uddin J, editor. Macro to Nano Spectroscopy. Croatia: InTech; 2012. pp.145–164. ISBN 978-953-51-0664-7

[47] Guo X, Lankmayr E. Hyphenated techniques in gas chromatography. In: Mohd MA, editor. Advanced Gas Chromatography—Progress in Agricultural, Biomedical and Industrial Applications. InTech; Croatia, 2012. pp.5–26. ISBN: 978-953-51-0298-4

[48] Ibekwe NN, Ameh SJ. Hyphenated techniques in liquid chromatography as current trends in natural products analysis. International Research Journal of Pure &Applied Chemistry. 2015;**7**(3):132-149

[49] Wijesekera ROB. The Medicinal Plant Industry. CRC Press; Taylor & Francis Inc, Bosa Roca, USA. 1991. ISBN 9780849366697

[50] Vlietinck AJ, Apers S. Biological screening methods in the search for pharmacologically active natural products. In: Tringali C, editor. Bioactive Compounds from Natural

Sources Isolation, Characterisation and Biological Properties. London: Taylor & Francis; 2001. ISBN 0-203-26972-1;

[51] Rior RLP, Wu X, Schaich K. Standardized methods for the determination of antioxidant capacity and phenolics in foods and dietary supplements. Journal of Agricultural and Food Chemistry. 2005;**53**(10):4290–4302

[52] Sochor J, Dobes J, Krystofova O, Ruttkay-Nedecky B, Babula P, Pohanka M, Jurikova T, Zitka O, Adam V, Klejdus B, Kizek R. Electrochemistry as a tool for studying antioxidant properties. International Journal of Electrochemical Science. 2013;**8**:8464–8489

[53] Taiga A. Quantitative phytochemical properties of mistletoe (*Viscum album*) from five different plants. Research Journal of Agricultural and Environmental Management. 2013;**2**(6):150–153

[54] Frazer Sir JG. The Golden Bough. 3rd ed. London: Macmillan and Co, Limited St. Martin's Street; 1920

[55] Chernyshov VP, Heusser P, Omelchenko LI, Chernyshova LI, Vodyanik MA, Vykhovanets EV, Galazyuk LV, Pochinok TV, Gaiday NV, Gumenyuk ME, Zelinsky GM, Schaefermeyer H, Schaefermeyer G. Immunomodulatory and clinical effects of *Viscum album* (Iscador M and Iscador P) in children with recurrent respiratory infections as a result of the Chernobyl nuclear accident. American Journal of Therapeutics. 2000;**7**(3):195–203

[56] Twardziok M. Mechanism of action of *Viscum album* L. extracts in Ewing sarcoma [PhD thesis]. Freie Universität Berlin; 2015

[57] Vicas SI, Rugina D, Socaciu C. Antioxidant activity of European mistletoe (*Viscum album*). In: Rao V, ed. Phytochemicals as Nutraceuticals-Global Approaches to Their Role in Nutrition and Health. InTech, Croatia; 2012. pp.115–134. ISBN 978-953-51-02038

[58] Hajto T, Fodor K, Perjesi P, Nemeth Peter. Difficulties and perspectives of immunomodulatory therapy with mistletoe lectins and standardized mistletoe extracts in evidence-based medicine. Evidence-Based Complementary and Alternative Medicine. 2011;**6**:Article ID 298972. Hindawi Publishing Corporation

[59] Gray AM, Flatt PR. Insulin-secreting activity of the traditional antidiabetic plant *Viscum album* (mistletoe). The Journal of Endocrinology. 1999;**160**(3):409–414

[60] Adeeyo AO, Adefule AK, Ofusori DA, Aderinola AA, Caxton-Martins EA. Antihyperglycemic effects of aqueous leaf extracts of mistletoe and *Moringa oleifera* in streptozotocin-induced diabetes Wistar rats. Diabetologia Croatica. 2013;**42**(3):81–88

[61] Marvibaigi M, Supriyanto E, Amini N, Adibah F, Majid A, Jaganathan SK. Preclinical and clinical effects of mistletoe against breast cancer. BioMed Research International. 2014;**15**:Article ID 785479

[62] Kuttan G, Vasudevan DM, Kuttan R. Effect of a preparation from *Viscum album* on tumor development in vitro and in mice. Journal of Ethnopharmacology. 1990;**29**(1):35–41

[63] Valentiner U, Pfuller U, Baum C, Schumacher U. The cytotoxic effect of mistletoe lectins I, II and III on sensitive and multidrug resistant human colon cancer cell lines in vitro. Toxicology. 2002;**171**(2-3):187–199

[64] Antony S, Kuttan R, Kuttan G. Role of natural killer cells in iscador mediated inhibition of metastasis by adoptive immunotherapy. Immunological Investigations. 2000;**29**(3):219-231

[65] Eno AE, Ibokette UE, Ofem OE, Unoh FB, Nkanu E, Azah N, Ibu JO. The effects of a Nigerian specie of *Viscum album* (Mistletoe) leaf extract on the blood pressure of normotensive and Doca-induced hypertensive rats. Nigerian Journal of Physiological Sciences. 2004;**19**(1-2):33–38

[66] Li Y, Zhao YL, Yang YP, Li XL. Chemical constituents of *Viscum album* var. Meridianum. Biochemical Systematics and Ecology. 2011;**39**:849–852

[67] Luczkiewicz M, Cisowski W, Kaiser P, Ochocka R, Piotrowski A. Comparative analysis of phenolic acids in mistletoe plants from various hosts. Acta Poloniae Pharmaceutica-Drug Research. 2001;**58**(5):373–379. ISSN 0001-6837

[68] Poojary B, Belagali SL. Synthesis, characterisation and biological evaluation of cyclic peptides: Viscumamide, yunnanin A and evolidine. Zeitschrift fur naturforschung section b-a journal of chemical sciences, 2005;**60**(12):1313–1320

[69] Larsson S. Mistletoes and thionins as selection models in natural products drug discovery.. Digital Comprehensive Summaries of Uppsala Dissertations from the Faculty of Pharmacy 49. Uppsala: Acta Universitatis Upsaliensis; 2004. 65pp. ISBN 978-91-554-6824-8;

[70] Guzmán-Rodríguez JJ, Ochoa-Zarzosa A, López-Gómez R, López-Meza JE. Plant antimicrobial peptides as potential anticancer agents. BioMed Research International. 2015:**11**:Article ID 735087

[71] Colegate SM, Molyneux RJ. Bioactive Natural Products—Detection, Isolation, and Structural Determination. 2nd ed. Taylor & Francis Group, LLC, USA; 2008

[72] Simard EP, Engels EA. Cancer as a cause of death among people with AIDS in the United States. Clinical Infectious Diseases. 2010;**51**(8):957–962

[73] Tantry MA. Plant natural products and drugs: a comprehensive study. Asian Journal of Traditional Medicines. 2009;**4**(6):241–249

[74] Giudici M, Pascual R, de la Canal L, Pfüller K, Pfuller U, Villalain J. Interaction of viscotoxins A and B with membrane model systems: Implications to their mechanism of action. Biophysical Journal. 2003;**85**:971–981

[75] Hemmateenejad B, Javidnia K, Nematollahi M, Elyasi M. QSAR studies on the antiviral compounds of natural origin. Journal of the Iranian Chemical Society. 2009;**6**(2):420–435

[76] Kuramochi K. Synthetic and structure-activity relationship studies on bioactive natural products. Bioscience, Biotechnology, and Biochemistry. 2013;**77**(3):446–454

Section 3

Role of Amino Acids in Animal

Chapter 11

Amino Acid for Japanese Quails: Methodologies and Nutritional Requirement

Danilo V.G. Vieira, Fernando G.P. Costa,
Matheus R. Lima, José G.V. Júnior,
Talita P. Bonaparte, Danilo T. Cavalcante,
Sarah G Pinheiro, Marilu S Sousa,
Ana C M Conti and Érika M Figueireido

Additional information is available at the end of the chapter

http://dx.doi.org/10.5772/intechopen.68547

Abstract

The methodologies applied to chickens and laying hens, to determine the digestibility and requirement of protein and amino acids are used with quails, however, they need a more careful evaluation due to peculiarities inherent to the Coturnix genus, in order to provide consistent results. The nutritional requirements of the birds are determinate using the dose-response and the factorial method. Several mathematical models and techniques of diet formulation are allied to the dose-response method in determining nutritional requirements. The curvilinear (hyperbolic) models better portray population behaviour in response to increasing nutrient doses in diets. The reading model, allow a better estimation of the requirement, in relation to the mathematical models used in the dose-response method. The techniques of comparative slaughter and nitrogen balance are effective in determining the nutritional requirements of quails, however, the latter need to be corrected by the loss of nitrogen in the feathers in determining the requirements of crude protein and amino acids for maintenance. The protein-free diet, coupled with the industrial amino acid supplementation, provides more robust digestibility values, since it more effectively predicts the endogenous excretion pattern.

Keywords: amino acids, physiology, Japanese quails, methodologies, requirements

1. Introduction

Created for various purposes (hunting, meat, ornamentation, eggs) the production of quail is a reality worldwide. Countries such as Spain, France, China and the United States stand out for the production of meat, however, when the production is intended to egg production, countries, as China, Japan and Brazil are highlights.

Quail farming in Brazil in the year 2015 reached a total of 21.99 million head, either for meat or for eggs and 447.47 million dozens of eggs [1], which means an increase of 8.1 and 13.9%, respectively, in relation to 2014.

The success of the activity in Brazil is due to the large producing companies that have settled in the territory and to the creation of research groups inserted in Academic Center, with studies directed to the genetic improvement, management and production, and nutrition of quail. The Centers that stand out are: Group of Studies and Poultry Technologies—Federal University of Paraiba, Areia, PB; Nucleus of Fish and Bird Studies—Federal University of Paraiba, Bananeiras, PB; Nucleus of Studies in Poultry Science and Technology—Federal University of Lavras, Lavras, MG. As well as research groups located at the University of Espirito Santo, Alegre—ES, Federal University of Minas Gerais, Belo Horizonte, MG, Maringa State University, Maringa, PR, and Federal University of Viçosa, Viçosa, MG, and the last three groups differ from the firsts, because they also present a breeding program. Worldwide countries such as India, France, Spain and Egypt also stand out with quail research.

[2] World studies on quails date back to 1992, and since 2002 the number of studies in the various research Centers has increased, both in the world and in Brazil. This advance in the Brazilian researches is concomitant with advances in methodologies for food evaluation and nutritional requirements [3, 4], in the knowledge of cellular biochemistry, physiology and animal nutrition, in the development of laboratories in the Research Centers, and in industrial manufacturing of amino acids, premix, etc.

Although of the same family (Phasianidae), commercial poultry, broilers, chickens and quail are of different genres. The latter belong to the genus Coturnix, while the former are of the Gallus genus. Faced with this taxonomic difference, quails have peculiar digestive physiology, and in addition, growth and early reproductive activity, and among others, have low feed intake, which gives them a higher rate of passage in the gastric tract. These differences denote a specific nutritional requirement, mainly protein and amino acid.

Several methodologies applied to chickens and laying hens [4] are effective in quail use; however, they need a more careful evaluation, due to peculiarities inherent to the Coturnix genus, in order to provide consistent results. Another aggravating factor is the lack of quail standard lineages that makes nutrition dynamism even more peculiar with quail. There are few reputable and reputable companies in Brazil that work on quail breeding and provide genetic material for sale. There are few reputable companies in Brazil that work on quail breeding and provide genetic material for sale.

There are two basic methods (dose response and factorial method) for determining the nutritional requirements of birds. However, several mathematical models and techniques for formulating diets that are allied, to the dose-response method, and techniques such as comparative slaughter (CS) and nitrogen balance (NB), used in the factorial model to predict the nutritional requirement values of crude protein and amino acids for birds.

In this chapter, we will discuss the peculiarities of Japanese quails in relation to broilers, laying hens and heavy matrices, and the need to use with criteria, the methodologies to estimate digestibility and requirement, and also review the use of mathematical models, diet formulation and the methodologies of CS and NB.

2. Peculiarities and methodologies

2.1. Peculiarities

Part of the requirements for amino acids and protein for maintenance (laying hens and quail) is directly related to precocity, intestinal gastric tract size (IGT), feather production, development of the reproductive tract, and part of the gain requirement is related to egg weight (inside the same species), and also, rate of muscle deposition (maturity).

Quails, whether intended for laying or cutting, have early maturity and are related to growth rate, and also to size of animals [5, 6], and thus, smaller animals have higher growth rates and lower age to maturity.

Precocity in growth is related to the time the animal takes to achieve sexual maturity, is a guiding parameter in breeding programs, and also denotes different requirements for animals. In this sense, the models that describe growth curves [7–10] validate the premise that each species/lineages and animal category have different nutritional requirements.

Comparing the Gompertz growth curves for Japanese quails [9], meat quails [7, 9], light and semi-heavy laying hens [11] and broilers [8, 10], it is worth mentioning, that between maturity rates (0.720, 0.0594 and 0.0694, 0.0245 and 0.0230, 0.0373 and 0.0411), respectively, and Japanese quails have the highest maturity rate, which refers to higher nutritional needs, protein and amino acids.

Japanese quails have a lower weight of IGT than chickens, laying hens and heavy matrices, but, a higher relative weight in relation to body weight and this factor predisposes a higher rate of passage of the digest by IGT [12–14].

Japanese quail [15] presented weight absolute and relative oviduct of 10.18 g and 3.05%, and ovary of 6.36 g and 2.16%, that are lower in relative to laying hens [16] that presented absolute and relative oviduct weight of 76.98 g and 6.58%, and absolute and relative ovary weight of 36.04 g and 3.08%. However, the relative weight of quail eggs is higher, and may reach 10% of body weight. The weight of eggs of quails has a mean value of 12 g [17–20] and eggs of laying hens around 65 g [21–24].

2.2. Crude protein: methodologies and requirement

The protein and amino acids requirements for quails can be defined by the method of dose and factorial method. The most common is dose-response method and has generated a lot of information's. However, considering the more accurate method, in predicting the requirement of amino acids and crude protein, in this topic of proteins, we will approach a subject only on the factorial model. The approach of the dose-response method will be in the topic about amino acids.

Some studies have been carried out to estimate crude protein (CP) requirements for commercial bird keeping, gain and production using CS and NB techniques.

There studying the requirements of CP for maintenance and gain with Japanese quails in production using the CS technique, [18] obtained the following equation: CP (g/bird/day) = 6.71 × body weight$^{0.75}$ + 0.615 × weight gain + 0.258 × egg mass.

The requirement of CP for maintenance and gain for growing Japanese quails was estimated in the period from 01 to 32 days of age, through the CS technique. The predicted equations were: CP (g/bird/day) = 2.845 × body weight$^{0.75}$ + 0.461 × weight gain for quails aged 01–12 days of age and CP (g/bird/day) = 4.752 × body weight$^{0.75}$ + 0.843 × weight gain for quails in the period from 15 to 32 days of age [25, 26].

The CP requirement for maintenance and gain using the BN technique was determined by the following equation for the 52 week old Lohmann LSL® laying hens: CP (g/bird/day) = 1.94 × body weight$^{0.75}$ + 0.480 × weight gain + 0.301 × egg mass [27].

Using the NB technique to determine the maintenance and gaining needs of Ross® broilers, at 7 days of age of 56, [28] the following equation was obtained: CP (g/bird/day) = 1.323 × body weight$^{0.75}$ + 0.272 × weight gain for males and the following equation for females: CP (g/bird/day) = 1.748 × body weight$^{0.75}$ + 0.277 × weight gain.

Working with 5-week-old Hubbard® matrices, [29] determined the CP (g/bird/day) values for maintenance using the CS and NB techniques, and values their obtained, respectively, were 3.77 and 2.02 × body weight$^{0.75}$, and the mean value for CP requirement for gain was 0.406 × weight gain, for techniques CS.

Working with light replacement pullets, Lohmann LSL®, from the age of 42–63 days, using the CS technique, [30] found CP (g/bird/day) values for maintenance and gain of: 4.7625 × body weight$^{0.75}$ and 0.313 × weight gain.

When evaluating laying hens Hubbard® at age 36–46 weeks of age, [31] estimated the following equation to predict protein requirements: CP (g/bird/day) = 2.282 × body weight$^{0.75}$ + 0.356 × weight gain + 0.262 × egg mass.

It is known that nutritional needs are changed according to species, animal category, room temperature, diet composition and animal density. However, another important factor that changes the nutritional needs is the methodologies used [18, 25–31], such as the CS and NB techniques used in the elaboration of prediction equations.

In an attempt to elucidate the effects of the two techniques in determining PB requirements for maintenance and gain, the values predicted by these two techniques will be compared.

The CP (g/bird/day) requirements for maintenance were predicted by the CS technique, with growing animals, in the studies [26, 29, 30], which, respectively, used: pullets (42–63 days of age), heavy matrices (3–20 weeks of age) and Japanese quails (15–32 days of age). The values are similar between the species, 4.765 and 4.752 × body weight$^{0.75}$ for pullets and quails; however, they are discrepant when compared to heavy matrices 3.77 × body weight$^{0.75}$.

Using the NB technique to determine CP (g/bird/day) requirements for maintenance, the values predicted by the authors, [27, 28, 31], who, respectively, worked with broiler chickens (7–56 days of age), laying hens, and heavy matrices, were: 1.323; 1.94 and 2.28 × body weight$^{0.75}$. It can be observed that there is no similarity between the all determined values. However, the values are consistent when analyzing animals in the same category [27, 31] which were: 1.94 and 2.28 × body weight$^{0.75}$.

It can be observed that the net requirement of CP (g/bird/day) for gain, determined by the two techniques (CS and NB) and reported in the works of [18, 25–31], is, respectively: 0.615; 0.461; 0.843; 0.480; 0.272; 0.406; 0.313 and 0.356 × weight gain. Comparing the requirements of CP (g/bird/day) to gain, with the CS technique, the values are: 0.406; 0.461; 0.615 and 0.843 × weight gain. Those predicted in the NB technique are: 0.272; 0.356 and 0.480 × weight gain.

A relevant comparison is to analyze the same technique and animal's age, growth and posture. Within the CS technique, with growing animals, the values were: 0.313; 0.461 and 0.843 × weight gain, respectively, pullets (42–63 days of age), quails (01–12 days) and quails (15–32 days of age). In NB technique for growing animals, the values were, respectively: 0.272 × weight gain, for broilers; and for the animals in posture were: 0.356 and 0.480 × weight gain, respectively for, laying hens and heavy matrices.

The values of requirement of CP (g/bird/day) estimated for egg mass production in laying hens and Japanese quails were, respectively, 0.301 × egg mass [26], and 0.258 × egg mass [18]. However, the first one presents a lower requirement of amino acids and CP, evidencing that the greater requirement of quails is related to the higher maturity rate, that is, higher precocity [9–11]. Corroborating the findings of [18, 25–27, 29], where these authors found a requirement of CP for greater maintenance and gain for Japanese quails in relation to laying hens and heavy matrices.

It is clear from the aforementioned works that the CS and NB techniques used to determine PB requirements for maintenance and gain provide conflicting, inter and intraspecific values, which makes comparison difficult. In an attempt to elucidate this difference between the methodologies, [32] described the potential of nitrogen retention in laying pullets by analyzing the two techniques: CS and NB. The authors describe that excreted nitrogen measured in the BN technique does not seem to contain all possible physiological effects, except for amino acid oxidation, in relation to the CS technique. This factor suggested by the authors seems to be the accounting for the nitrogen lost in feathers (NF), which is not measured when used in the NB technique, and with that, the CP requirement values for maintenance between the two techniques are more discrepant.

For [33], the use of the BN technique is even more aggravating, because in this technique, diets are formulated with different levels of protein to generate deficiency and excess CP in the animals' diet. In this sense, the relationship between the protein level and the loss of NF was established, described by the equation $NF = 0.3007 + 0.0086\,N$, where, for each gram of increase in the nitrogen concentration of the diet, there was a loss of 8.6 mg of nitrogen in feathers, that is, the deficiency in proteins leads to less deposition of amino acids in the feathers, thus, changes the requirement of maintenance the animals. [34] Also verified the influence of nitrogen losses on feathers on the need for maintenance, using the BN technique.

Using the correction value of nitrogen losses in the feathers found by [33], [32] in their work using this correction could conclude that differences between CP needs for maintenance between the two techniques, CS and NB, decreased fell from of 1.56: 1 for 1.28: 1, comparing CS: NB.

2.3. Amino acids: methodologies, digestibility and requirement

The requirements of amino acids have been described by two methodologies: empirical and factorial method [4]. To evaluate the nutritional requirements in the dose-response or empirical method, the diets are formulated with increasing levels amino acid, gradually, and observed the response of the animals through polynomials (linear and quadratic), broken line and hyperbolic and analyzed in order to estimate the requirements of birds. In the factorial method, the requirements are described in function of the maintenance, growth and production, and relate to the metabolic weight, weight gain and eggs mass production. This method was described in the topic of proteins.

In addition to the methodologies, there are also techniques for formulating diets that also change the requirements. One of the techniques consists of gradual increases of the nutrient tests [35], the other prioritizes the dilution of the diets, which consists of formulating a diet free of the test nutrient and another diet with the same nutrient in excess, and the nutrient levels studied will be obtained by the dilution of the two diets [36].

The success in determining the requirements is a thin line, that is, the robustness of the proposed models and determined requirement are allied to the knowledge and interpretation of each physiological factor of the animals, and mathematical model, in order to promote satisfactory performance to birds.

2.3.1. Amino acids methodologies: techniques for preparing diets

In the technique proposed by [35], the supplementation of a single amino acid generates imbalances in the relations between amino acids and amino acids/lysine. This point is crucial, since it refers to the ideal protein concept proposed by [37], where the diet needs to have an optimal balance of amino acids to provide maximum performance to the animals with lower nitrogen excretion.

The supplementation technique is widely used [38–45] and has generated a large number of discrepant nutritional information. The main factor is the imbalance between amino acids. The main antagonist relationships between amino acids are: arginine and lysine and the relationship between leucine, isoleucine and valine.

The excess of lysine in the diets, when using the technique proposed by [35], to assessing the lysine requirements, promotes an increase in serum lysine levels, and consequently, a greater loss of arginine by renal catabolism due to the increase in enzyme arginase [46], and generating confusion in the determination of the optimal levels of lysine.

Ref. [45] evaluated different levels of digestible arginine in the diet of Japanese quails, with diets formulated by the supplementation technique, estimated an ideal dietary arginine level of 1.148% in diets with 1.083% digestible lysine and relation arginine/lysine of 1.06.

Refs. [47, 48] evaluated the requirement of digestible lysine with Japanese laying quails, using the supplementation technique, but these authors maintained the relationships between the amino acids of the diets. The authors found digestible lysine levels of 1.117 and 1.120%, respectively, in diets with arginine/lysine ratios of 1.26 and 1.16, respectively.

Ref. [17] found levels of digestible lysine for Japanese quails in production of 1.030% in diets formulated by the supplementation technique and without correction of the arginine/lysine ratio.

The Brazilian Poultry and Swine Table [49] and the Table for Japanese and European Quails [50] present, respectively, digestible lysine values of 1.083 and 1.030% and digestible arginine of 1.256 and 1.260%, respectively, with arginine/lysine ratios of 1.16 and 1.22.

Looking at the data, mentioned above, it is evident that the imbalance of the diets promotes different results among the authors [17, 47, 48]. However, [47, 48] found values equal, but maintained the relationship between the major amino acids; however, this practice of supplementing all amino acids to maintain relationships raises the cost of formulating diets.

Another known, but poorly studied amino acid relationship is branched-chain amino acids (isoleucine, leucine and valine). These three amino acids compete for the same intestinal transporter and for the same enzymes in cell metabolism [46].

When applying the concept of protein reduction and ideal protein, the basal diets composed of corn and soybean meal have increased maize levels, with this there is an increase in dietary leucine levels in relation to isoleucine and valine. Excess leucine [46] in the diet depresses the use of valine and isoleucine by animals, decreasing their performance.

Ref. [51] observed that high concentrations of isoleucine and low levels of valine and leucine affected the performance of laying hens in the laying phase. This same effect was verified by [52] when evaluating valine/lysine and isoleucine/lysine relations for Japanese quails in production. The author recommends relations, respectively, of 0.75 and 0.82, for isoleucine and valine.

Analyzing the diets [52] of experiment I, where valine/lysine relations were evaluated, and the level of isoleucine in the diets was 1.0%. In experiment II, where the ideal isoleucine/lysine relations was verified, the level of valine in the diets was 0.75%, the latter was determined in experiment I. The higher levels of isoleucine (1.0%) used in the diets of experiment I, may have promoted lower performance in the animals, even in diets with higher levels of valine (0.75, 0.80, 0.85, 0.90, 0.95 and 1.05%).

This assumption is found in Experiment II, where the ideal level of isoleucine was 0.82%, when the diet contained 0.75% valine, indicating that excess isoleucine (1.0%—in experiment I) affected performance of the birds, not allowing to verify improvement, even with higher levels of valine, or even the level of 0.82% of isoleucine that promoted the best performance the birds was limited to the value of 0.75% valine in the diet, since in experiment II levels of isoleucine were of 0.65, 0.70, 0.75, 0.80, 0.85 and 0.90%, corroborating the findings of [51]. In addition, in both experiments (I and II), the diets contained near levels of leucine, respectively, 1.597 and 1.537%.

Aiming to understand the relationship between valine/isoleucine and recommend the best level of valine and isoleucine in the diet of Japanese laying quails [53], proposed the following methodology. In experiment I, were studied valine levels of 0.74, 0.81, 0.88, 0.95 and 1.02%, with fixed level of isoleucine (0.70%). In experiment II, the same levels of valine were evaluated, now, with different levels of isoleucine (0.64, 0.70, 0.76, 0.82 and 0.88%). The author recommends valine levels of 0.74 and 0.64% of isoleucine in the diet of Japanese quails in production. In addition, the leucine level used in both experiments was 1.47%.

It is noteworthy that the interpretations of the results of [53] do not repeat with those of [52], and these findings show that there are other factors involved in the study of the relationship between branched chain amino acids, intestinal transporters and metabolic enzymes, and which have not yet been described.

Comparing the two techniques of diet formulation [54], in his work proposed to study the technique of supplementation and dilution of diets, and to evaluate the levels of digestible lysine for broilers from 01 to 42 days of age. The author suggests the most appropriate dilution technique to formulate the diets, since it promotes better performance to the animals, and this technique reduces the use of supplemental amino acids to maintain the relationship between amino acids, since many of them have high cost of supplementation.

2.3.2. Amino acids methodologies: requirement

The two methodologies used to evaluate the amino acid requirements for poultry are the empirical method and the factorial method, and have as diet formulation techniques, supplementation and dilution, discussed above. In this topic, we will address the methodologies, specifically the dose-response method, since the factorial method has already been described in the topic on proteins.

In the empirical method, the requirement is determined through the addition of the nutrient test in the diets. The levels studied should promote a response curve where they can observe deficiency, gain, stability and toxicity [55]. The response curve can be interpreted by several mathematical models [4], and the choice of them can change the value of the animal requirements.

The models used are: first and second degree polynomials, the broken line model and exponential.

The first polynomial models and the interrupted line model describe the linear performance of the animal due to the addition of nutrients. In addition, the interrupted line model predicts that, from a given level of nutrient supplementation, there is no effect, establishing whether a plateau, where the requirement is determined by the intercept of the line with the plateau. In the first model (first-degree polynomial), there is no predict an optimal level, but only data behavior, increasing or decreasing, and it is not possible to infer whether the behaviors of the line will be kept at lower levels or higher doses high.

The description of the behavior of the data in a linear way is the premise of the response of a single animal however, the population response pattern tends to be curvilinear, since the animals have different responses, even those of the same genetics and age [56], and thus, linear models do not accurately predict the requirements of animals.

The quadratic model presents an advantage in relation to the two models already mentioned, since the answer is curvilinear, describing the population pattern; however, in this model, the optimum point tends to be in the middle of the points studied, since there is a tendency of symmetry between the points to generate the response curve, so the authors work with the estimated value of 95% as the requirement of the animals.

For [57, 58], the models used to predict the requirements must have biological and mathematical meaning.

Nonlinear models predict that the animal's response tends to decrease as it reaches maximum performance or asymptotic point. However, in this type of model, the exponential, the maximum performance would never reach, that is, it never reaches the asymptotic point, so, the authors suggest assigning a percentage ranging from 95 to 99% of the asymptotic response [4] as being the requirement of the animals.

Several are the works that use the empirical method to determine the requirements of amino acids with Japanese quail, using the most diverse mathematical models. The choice of model should be judicious, and the model should most accurately describe the animal's response.

For [59] the linear, polynomial and exponential models, within their limitations, present good adjustments; however, the answers are varied, with this, there is indecision about the best to be recommended level. In this way, [59–61] propose the use of the reading model in an attempt to overcome the indecision generated in the choice of the mathematical model to estimate the nutritional requirements of amino acids, since this model allows a better interpretation of the behavior of the population in function of the levels studied.

Ref. [62] reviewed the reading model and noted that it would allow better estimation, in relation to the mathematical models used in the dose-response method. However, other factors that affect nutritional requirements such as temperature and type of lodging are not possible to include in the model.

As previously reported, the factorial method, described in the topic on proteins, is considered the most appropriate, since in this methodology, it is possible to fractionate the requirements in maintenance, gain and production, and it is possible to add other factors such as temperature, etc.

The mathematical models of prediction with amino acids resemble their construction, with the models already described in the topic on protein, through the factorial method. All peculiarities inherent to quails in relation to broilers and laying hens need to be weighed in the construction of the prediction equations for amino acids.

Due to the scarcity of work in these molds for Japanese quails, and especially with amino acids, no research data will be presented for comparison and elucidation of the techniques, since the premises discussed in the models of protein requirements are the same.

The Brazilian Poultry and Swine Tables [49] indicate the lysine requirements for Japanese quails in posture by the factorial method, but are approximate data of other species.

2.3.3. *Amino acids Methodologies: digestibility*

The digestibility of the amino acids can be influenced by the physiology of the animal and the technique/methodology used. The digestibility is measured by comparing the amount of amino acids present in the test feed, and the intake of the same by the animals and the difference of what are recovered in the excreta.

Quails have a higher relative weight of large intestine in relation to broilers and laying hens. In the large intestine of the animals, there is microorganism that ferments the cecal content and with this can contribute to cecal production of amino acids and or nitrogen, underestimating the digestibility of the amino acids and altering the nitrogen balance. In this sense, quails would present values of amino acid digestibility, less than roosters, laying hens and broilers [63, 64] and allied to this factor, the greater passage rate would contribute to greater escape of protein/amino acids to the large intestine, greater amino acid excretion and fecal nitrogen, further underestimating the results.

To avoid increased cecal amino acid production, ileal content collection, cecectomy, and accurate feeding techniques are suggested to predict amino acid digestibility [49, 50, 65–67]. In addition, fasting [65], used in the precise feeding technique, is criticized by several authors, since fasting animals have patterns of endogenous loss of amino acids different from fed animals. Values of digestible amino acids determined with quails and using the above techniques are scarce.

Using the precise feeding technique, with intact and cecectomized roosters and intact Japanese quail, [68] studied the amino acid digestibility of different foods (maize, low tannin sorghum) and verified that the digestibility of amino acids with cecectomized roosters is greater in relation to intact roosters for most of the amino acids present in maize, with the exception of the amino acids: cystine, threonine, arginine and histidine. The digestibility of the amino acids present in the sorghum did not change due to the cecectomy, except for methionine, where the cecectomized roosters had a higher value. When comparing quails with intact roosters, the authors concluded that the digestible amino acid values with quails are larger, analyzing the corn, but with sorghum, there was no difference except for the amino acid histidine.

Although the authors [68] did not present statistical data comparing the amino acid digestibility values of cecectomized roosters and quails, in absolute values, for maize, the data presented similarities, but when comparing sorghum, values with Japanese quails showed the lower digestibility values. For the amino acid proline (2% points) and histidine (36% points), the other amino acids on average the difference were around seven percentage points less in the digestibility values for quails. These data for maize suggest that although quails have proportionately larger ceca, this factor did not interfere with digestible amino acid values. Another important factor is that using digestible amino acid values of intact roosters for quails is not recommended.

Some authors [69, 70] have suggested, respectively, some methodologies to stabilize the endogenous loss of amino acids by the animals, such as protein-free diet (PFD) and enzymatically hydrolyzed casein (EHC) techniques.

All these methodologies were worked with broilers and laying hens, and several authors criticized their use [71, 72]. However [73] suggest the PFD technique associated with amino acid supplementation, as being the one that best estimates the endogenous loss of amino acids by birds.

Studies evaluating these techniques with quails are scarce, especially those that evaluate the EHC and PFD and PFD techniques associated with industrial amino acid supplementation, as well as the technique of cecectomy and collection of ileal content.

The above-mentioned propositions suggest that formulating diets based on recommendations of digestible amino acids determined with intact and cecectomized roosters is not recommended for Japanese quails and should make considerations about the digestive physiology of quails, as well as the methodologies described. In quail nutrition, there are rare papers describing mathematical models to predict the requirement for amino acids.

3. Conclusions

Quails present physiological and behavioral peculiarities in relation to laying hens, heavy matrices and broilers. The differences between species of industrial poultry are premised to develop specific feeding programs for each species, lineage and animal category.

The nitrogen balance and comparative slaughter technique provide discrepant data, but are more consistent and close when the correction factor for nitrogen deposited in feathers is used.

The mathematical models used to describe nutritional requirements, in dose-response method, must be used with discretion, since the ideal model must have not only mathematical meaning but also biological meaning.

Prediction equations developed with broilers and laying hens should not be used to predict the protein and amino acids requirement for quails, should developing models appropriate for the each species, and animal category.

Author details

Danilo V.G. Vieira[1,8,9]*, Fernando G.P. Costa[2,9], Matheus R. Lima[3,9], José G.V. Júnior[4], Talita P. Bonaparte[5], Danilo T. Cavalcante[6,9], Sarah G Pinheiro[7,9], Marilu S Sousa[1,8], Ana C M Conti[1,8] and Érika M Figueireido[1]

*Address all correspondence to: danilovargaszoo@hotmail.com

1 Federal University of Tocantins, Araguaina, Brazil

2 Federal University of Paraiba, Areia, Brazil

3 Federal University of South Bahia, Teixeira de Freitas, Brazil

4 Federal University of Espirito Santo, Alegre, Brazil

5 Federal University of Bahia, Salvador, Brazil

6 Federal University of Paraiba, Areia, Brazil

7 Federal University of Ceara, Fortaleza, Brazil

8 Group of Studies and Research in Poultry, Araguaina, Tocantins, Brazil

9 Poultry Technology Studies Group, Areia, Paraiba, Brazil

References

[1] Brazilian Institute of Geography and Statistics (BIGE). Municipal livestock production. Rio de Janeiro, 2016;**43**:1–49. Available in: http://biblioteca.ibge.gov.br/visualizacao/periodicos/84/ppm_2015_v43_br.pdf.

[2] Minvielle F. The future of Japanse quail for research and production. World's Poultry Science Journal. 2004;**60**:500–507. DOI: 10.1079/WPS200433

[3] Rostagno HS, Bünzen S, Sakomura NK, Albino LFT. Methodological improvements in feedstuffs evaluation and nutritional requirements for poultry and swine. Brazilian Journal of Animal Science. 2007;**36**:295–304. DOI: 10.1590/S1516-35982007001000027

[4] Sakomura NK, Rostagno HS, editors. Research Methods in Monogastric Nutrition. 3nd ed. Jaboticabal: FUNEP; 2016. p. 262. ISBN 979-85-7805-154-9

[5] Arango JA, Van Vleck LD. Size of beef cows: Early ideas, new developments. Genetics and Molecular Research. 2002;**1**:51–63

[6] Tholon P, Paiva RDM, Mendes ARA, Barrozo D. Use of linear and nonlinear functions to fit the growth of Santa Gertrude cattle raised under grazing. Ars Vetirnary. 2012;**28**:234–239. DOI: 10.15361/2175-0106.2012v28n4p234-239

[7] Drumond ESC, Gonçalves FM, Veloso RC. Curvas de crescimento para codornas de corte. Rural Science. 2013;**43**:1872–1877. DOI: 10.1590/S0103-84782013001000023

[8] Winkelstroter LK. Growth of three commercial genotypes of broilers [thesis]. Diamantina: Federal University of Vales do Jequitinhonha e Mucuri; 2013.

[9] Mota LFM, Alcântara DC, Abreu LRA, Costa LS, Pires AV, Bonafé CM, Silva MA, Pinheiro SRF. Growth comparison of different genetic groups using nonlinear models. Brazilian Journal of Veterinary and Animal Science. 2015;**67**:1372–1380. DOI: 10.1590/1678-4162-7534

[10] Demuner LF. Growth comparison of different genetic groups using nonlinear models [doctoral dissertation]. Pirassununga: University of São Paulo; 2016

[11] Neme R, Sakomura NK, Fukayama EH, Freitas ET, Fialho FB, Resende KT, Fernandes JBK. Growth curves and deposition of body components in pullets of different strains. Brazilian Journal of Animal Science. 2006;**35**:1091–1100. DOI: 10.1590/S1516-359820060 00400021

[12] Flauzina LP. Perfomance anda organ measure of Japanese quail fed with different crude protein levels [thesis]. Brasília: University of Brasília; 2007.

[13] Marcato ISM, SakomuraI NK, Fernandes JBK, Siqueira JC, Dourado LRB, Freitas ER. Growth and nutrients deposition on organs of two commercial strains of broiler chickens. Brazilian Journal of Animal Science. 2010;**39**:1082–1091. DOI: 10.1590/S1516-35982010000500019

[14] Grieser DO. Growth and composition study body of codornas lines cutting and posting [thesis]. Maringá: State University of Maringá; 2012.

[15] Santos ICL, Maciel WC, Gomes VS, Sampaio FP, Machado DN, Lima SVG, Lopes ES, Silva RCR, Bezerra WGA, Teixeira RSC. Regression of the reproductive tract of European quails (*Coturnix coturnix*) submitted to forced molting by wheat bran diet. Acta Veterinaria Brasilica. 2014;**8**:101–106. DOI: 10.21708/avb.2014.8.2.3256

[16] Jardim Filho RM, Stringhini JH, Andrade MA, Nunes AB, Leandro NSM, Barcellos B. Egg quality, blood biochemical parameters and reproductive tract development for Lohmann LSL hens fed increasing levels of digestible lysine. Acta Scientiarum Animal Sciences. 2008;**30**:25–31. DOI: 10.4025/actascianimsci.v30i1.3596

[17] Costa FGP, Rodrigues VP, Goulart CC, Neto RCL, Souza JG, Silva JHV. Digestible lysine requirements for laying Japanese quails. Brazilian Journal of Animal Science. 2008;**37**:2136–2140. DOI: 10.1590/S1516-35982008001200009

[18] Jordão Filho J, Silva JHV, Costa FGP, Sakomura NK, Silva CT, Chagas NA. Prediction equations to estimate the demand of energy and crude protein for maintenance, gain and egg production for laying Japanese quails. Brazilian Journal of Animal Science. 2011;**40**:2423–2430. DOI: 10.1590/S1516-35982011001100020

[19] Vieira DVG, Barreto SLT, Valeriano MH, Jesus LFD, Silva LFF, Mencalha R, Barbosa KS, Mendes RKV, Cassuce MR, Melo T. Requirements for calcium and phosphorus available for Japanese quail of the 26 to 38 weeks of age. Brazilian Journal of Animal Health and Production. 2012;**13**:204–213. DOI: 10.1590/S1519-99402012000100018

[20] Lima RC, Costa FGP, Goulart CC, Cavalcante LE, Freitas ER, Silva JHV, Dantas LS, Rodrigues VP. Nutritional requirement of crude protein for Japanese quail (*Coturnix coturnix japonica*) in the production phase. Brazilian Archive of Veterinary Medicine and Zootechnics. 2014;**66**:1234–1242. DOI: 10.1590/1678-6414

[21] Cardoso AS, Costa FGP, Lima MR, Nogueira ET, Santos CS, Sousa RB, Lima RC, Vieira DVG. Nutritional requirement of digestible threonine for white egg layers of 60 to 76 weeks of age. Journal Applied Poultry Research. 2014;**23**:724–728. DOI: 10.3382/japr.2014-01011

[22] Cardoso AS, Costa FGP, Silva JHV, Sariava EP, Nogueira ET, Santos CS, Lima MR, Vieira DVG. Nutritional requirement of digestible tryptophan for white-egg layers of 60 to 76 weeks of age. Journal Applied Poultry Research. 2014;**23**:729–734. DOI: 10.3382/japr.2014-01012

[23] Bezerral RM, Costa FGP, Givisiez PEN, Goulart CC, Santos RA, Lima MR. Glutamic acid supplementation on low protein diets for laying hens. Acta Scientiarum Animal Sciences. 2015;**37**:129–134. DOI: 10.4025/actascianimsci.v37i2.25911

[24] Pastore SM, Gomes PC, Barreto SLT, Viana GS, Silva EA, Almeida RL, Barbosa LVB, Oliveira WP. Nutritional requirement of digestible lysine for white-egg laying hens in production. Rural Science. 2015;**45**:1496–1502. DOI: 10.1590/0103-8478cr20140661

[25] Silva JHV, Silva MB, Jordão Filho J, Silva EL, Andrade IS, Melo DA, Ribeiro MLG, Rocha MRF, Costa FGP, Júnior WMD. Maintenance and weight gain of crude protein and metabolizable energy requirements of Japanese quails (*Coturnix coturnix japonica*) from 1 to 12 days of age. Brazilian Journal of Animal Science. 2004;**33**;1209–1219. DOI: 10.1590/S1516-35982004000500013

[26] Silva JHV, Silva MB, Jordão Filho J, Silva EL, Andrade IS, Melo DA, Ribeiro MLG, Rocha MRF, Costa FGP, Júnior WMD. Maintenance and weight gain in crude protein and metabolizable energy requirements of Japanese quails (*Coturnix coturnix japonica*) from 15 to 32 days of age. Brazilian Journal Animal Science. 2004;**33**:1120–1230. DOI: 10.1590/S1516-35982004000500014

[27] Sakomura NK, Basaglia R, Resende KT. Modelling protein utilization in laying hens. Brazilina Journal Animal Science. 2002;**31:**2247–2254. DOI: dx.doi.org/10.1590/S1516-35982002000900013

[28] Longo FA, Sakomura NK, Figueiredo AN, Rabello CBV, Ferraudo AS. Prediction Equations for Protein Requirements on Broilers. Brazilina Journal Animal Science. 2001;**30**:1521–1530. DOI: 10.1590/S1516-35982001000600020

[29] Filardi RS, Sakomura NK, Basaglia R. Prediction equation of crude protein requirements for growing heavy matrices. Brazilina Journal Animal Science. 2000;**29**:2308–2315. DOI: 10.1590/S1516-35982004000400010

[30] Albino LFT, Fialho FB, Bellaver C, Hara C, Paiva GJ. Estimates of energy and protein requirements for laying pullets in rearing. Brazilian Agricultural Research. 1994; **29**:1625–1629

[31] Rabello CBV, Sakomura NK, Longo FA. Equação de predição da exigência de proteína bruta para aves reprodutoras pesadas na fase de produção. Brazilian Journal Animal Science. 2002;**31**:1204–1213. DOI: 10.1590/S1516-35982002000500016

[32] Silva EP, Sakomura NK, Bonato MA, Donato DCZ, Peruzzi NJ, Fernandes JBKF. Description of the potential for nitrogen retention in pullets by different methodologies: Minimum retention. Rural Science. 2014;**44**:333–340. DOI: 10.1590/S0103-84782014000200022

[33] Silva EP. Modeling of responses of broiler to different lysine Intakes [doctoral thesis]. Jaboticabal: State University of Paulista; 2012.

[34] Leveille GA, Fisher H. The amino acid requirements for maintenance in the adult rooster: I. Nitrogen and energy requirements in normal and protein-depleted animals receiving whole egg protein and amino acid diets. Journal Nutrition. 1958;**66**:441–453

[35] D'Mello JPF. A comparison of two empirical methods of determining amino acid requirements. World's Poultry Science Journal. 1982;**38**:114–119. DOI: 10.1017/WPS19820010

[36] Fisher C, Morris TR. The determination of the methionine requirement of laying pullets by a diet dilution technique. British Poultry Science, 1970;**11**:67–82. DOI: 10.1080/00071667008415793

[37] Mitchell HH. Comparative Nutrition of Man and Domestic Animals. New York: Academic Press; 1964. p. 701. ISBN: 978-0-12-395551-7

[38] Ribeiro MLG, Silva JHV, Dantas MO, Costa FGP, Oliveira SF, Jordão Filho J, Silva EL. Nutritional requirement of lysine for laying quails in function of the level of diets crude protein. Brazilian Journal of Animal Science. 2003;**32**:156–161. DOI: 10.1590/S1516-35982003000100020

[39] Rodrigues VP. Basis of different feed formulation of industrialand amino acid supplementation for Japanese quails [thesis]. Areia: Federal University of Paraíba; 2011.

[40] Lima MR. Nutritional recommendations of threonine and tryptophan for Japanese quail and light layers in production [thesis]. Areia: Federal University of Paraíba; 2012.

[41] Cavalcante DT. Determination of the second limitant amino acid for Japanese Quails [thesis]. Areia: Federal University of Paraíba; 2013.

[42] Reis RS. Relationship Isoleucine: Lysine and lysine levels in diets for laying Japanese quails [doctoral thesis]. Viçosa: Federal University of Viçosa; 2013.

[43] Santos GC. Valine, isoleucine and arginine levels in diets with low protein level for Japanese quail in production [doctoral thesis]. Botucatu: State University of Paulista; 2013.

[44] Lima HJD, Bareto SLT, Donzele JL, Tinoco IFF, Ribas NS. Ideal ratio of digestible methionine plus cystine to digestible lysine for growing Japanese quails. Colombian Journal of Livestock Sciences. 2015;**28**:313–322. DOI: 10.17533/udea.rccp.v28n4a03

[45] Maurício TV, Vargas Junior JG, Souza MF, Barboza WA, Nunes LC, Soares RTN, Nascimento HS. Digestible arginine levels in diets for Japanese quails. Semina: Agrarian Sciences. 2016;**37**:2453–2462. DOI: 10.5433/1679-0359.2016v37n4Supl1p2453

[46] D'Mello JPF, editor. Amino Acids in Animal Nutrition. 2nd ed. Wallingford: CABI Publishing; 2003. p. 544. DOI: 10.1079/9780851996547.0000

[47] Pinto R, Ferreira AS, Donzele JL, Silva MA, Soares RTRN, Custódio GS, Pena KS. Lysine requirement for laying Japanese quails. Brazilian Journal of Animal Science. 2003;**32**:1182–1189. DOI: 10.1590/S1516-35982003000500019

[48] Ribeiro CLN. Digestible lysine level in feed for laying Japanese quails during the production [thesis]. Visoça: Ferderal University of Viçosa; 2011.

[49] Rostagno HS, editor. Brazilian Tables for Poultry and Swuine: Food composition and Nutritional Requirements. 3rd ed. Viçosa: Federal University of Viçosa. 2011. p. 252

[50] Silva JHV, Costa FGP, editors. Table for Japanese and European Quails. 2nd ed. Jaboticabal: FUNEP; 2009. p. 110

[51] Peganova S, Eder K. Interactions of various supplies of isoleucine, valine, leucine e tryptophan on the performance of laying hens. Poultry Science. 2003;**82:**100–105. DOI: 10.1093/ps/82.1.100

[52] Paula E. Relationships of valine and isoleucine to lysine in diets for laying Japanese Quails [thesis]. Viçosa: Federal University of Viçosa; 2011.

[53] Petrucci FB. Nutritional levels of valina and isoleucine digestible for Japanese quails in Production [thesis]. Alegre: Federal University of Espirito Santo; 2013.

[54] Siqueira JC. Estimates of lysine requirements of broilers by dose response and factorial methods [doctoral thesis]. Jaboticabal: State University of Paulista; 2009.

[55] Mercer LP. The determination of nutritional requirements: Mathematical modeling of nutrient-response curves. The Journal of Nutrition. 1992;**122:**706–708

[56] Fisher C, Morris TR, Jennings RC. A model for the description and prediction of the response of laying hens to amino acid intake. British Poultry Science. 1973;**14**:469–484. DOI: 10.1080/00071667308416054

[57] Mercer LP, May HE, Dodds SJ. The Determination of nutritional requirements in rats: Mathematical modeling of sigmoidal, inhibited nutrient-response curves. The Journal of Nutrition. 1989;**119**:1465–1471

[58] Mercer LP, Yi T, Dodds SJ. Determination of nutritional requirements in rats: Variation with time of weight gain responses to indispensable amino acids. The Journal of Nutrition. 1993;**123**:964–971

[59] Fisher C, Morris TR, Jennings RC. A model for the description and prediction of the response of laying hens to amino acid intake. British Poultry Science. 1973;**14**:469–484. DOI: 10.1080/00071667308416054

[60] Curnow RN. A smooth population response curve based on an abrupt threshold and plateau model for individuals. Biometrics. 1973;**29**:1–10. DOI: 10.2307/2529671

[61] Pilbrow PJ, Morris TR. Comparison of lysine requirements amongst eight stocks of laying fowl. British Poultry Science. 1974;**15**:51–73. DOI: 10.1080/00071667408416081

[62] Silva EP, Sakomura NK, Hauschild L, Gous RM. Reading model to estimate optimum economic intakes of amino acids for poultry. Rural Science. 2015;**45**:450–457. DOI: 10.1590/0103-8478cr20120799

[63] Sakamoto MI, Murakami AE, Souza LMG, Franco JRG, Bruno LDG, Furlan AC. Energy value of some alternative feedstuffs for Japanese quails. Revista Brasileria de Zootecnia. 2006;**35**:818–821. DOI: 10.1590/S1516-35982006000300026

[64] Fachinello MR, Pozza PC, Furlan AC, Paula VRC, Bonagurio LP, Marcato SM, Leal IF, Huepa LMD. Nutritional evaluation of passion fruit seed meal for meat quails. Brazilian Journal of Health and Animal Production. 2016;**17**:202–213. DOI: 10.1590/S1519-99402016000200008

[65] Sibbald IR. A bioassay for true metabolizable energy in feedingstuffs. Poultry Science. 1976;**55**:303–308. DOI: 10.3382/ps.0550303

[66] Green S, Bertrand SL, Duron MJC. Digestibilities of amino acids in soybean, sunflower and groundnut meals determined with intact and cecectomized cockerels. British Poultry Science. 1987;**28**:643–652. DOI: 10.1080/00071668708416999

[67] Parsons CM. Influence of caecectomy and source of dietary fibre or starch on excretion of endogenous amino acids by laying hens. British Journal Nutrition. 1984;**51**:541–548. DOI: 10.1079/BJN19840059

[68] Vasan P, Dutta N, Mandal AB, Sharma K, Kadam MM, Elangovan AV. Comparative digestibility of amino acids of cereals in caecectomized cockerels and Japanese quails. British Poultry Science. 2008;**49**:176–180. DOI: 10.1080/00071660801969499

[69] Rostagno HS, Rogler JC, Featherstn WR. Studies on the nutritional values of sorghum grains with varying tannin content for chickens. Amino acids digestibility studies. Poultry Science. 1973;**52**:772–776. DOI: 10.3382/ps.0520772

[70] Moughan PJ, Darragh AJ, Smith WC, Butts CA. Perchloric and trichloraacetic acids as precipitants of rotein in endogenous ileal digesta from the rat. Journal Science Food Agriculture. 1990;**52**:13–21. DOI: 10.1002/jsfa.2740520103

[71] Moughan PJ, Butts CA, Rowan AM, Deglaire A. Dietary peptides increase endogenous amino acid losses from the gut in adults. The American Journal Clinical Nutrition. 2005;**81**:1359–1365

[72] Adedokun SA, Parsons CM, Lilburn MS. Comparison of ileal endogenous amino acid flows in broiler chicks and turkey poults. Poultry Science. 2007a;**86**:1682–1689. DOI: 10.1080/00071660902951313

[73] Brito CO, Albino LFT, Rostagno HS, Gomes PC, Nery LR, Silva EA. Aminoacidic diet in determining Ileal endogenous losses in broiler chickens: A methodological proposal. Revista Brasileira de Zootecnia. 2009;**38**:2161–2166. DOI: 10.1590/S1516-35982009001100013

Chapter 12

Effects of Excess Dietary Tryptophan on Laying Performance, Antioxidant Capacity and Immune Function of Laying Hens

Xinyang Dong and Xiaoting Zou

Additional information is available at the end of the chapter

http://dx.doi.org/10.5772/intechopen.68546

Abstract

Present study was conducted to establish Tryptophan (Trp) needs of Xinyang green-shell laying hens by evaluating its effect on laying performance, egg quality, antioxidant capacity, and the immune functions. A total of 525 laying hens, 28 weeks of age, were randomly allocated to five treatment groups, each of which included 5 replicates of 21 hens. Hens were fed the basal diet based on corn and soybean meal for 12 weeks. L-Trp was added to the control diet at 0.0 0.02, 0.04, 0.06, and 0.08%, respectively, to achieve 0.15, 0.17, 0.19, 0.21 or 0.23% Trp. Laying rate, average egg weight, and feed conversion ratio (FCR) were significantly increased by Trp levels from 0.19 to 0.23%. Dietary Trp from 0.17 to 0.19% increased egg internal quality (albumen height and haugh unit) rather than external quality. Supplementing with Trp increased glutathione peroxidase and total superoxide dismutase activity, total antioxidative capacity concentration and decreased malondialdehyde concentration. Serum IgA concentration increased at 0.21–0.23% dietary Trp, while serum IgM increased linearly in response to dietary Trp levels. We suggest that the optimum level of dietary Trp was ranged from 0.19 to 0.21% for Xinyang green-shell laying hens under the current study conditions.

Keywords: laying hens, Tryptophan, laying performance, egg quality, antioxidant capacity, immunoglobulins

1. Introduction

Tryptophan (Trp) is a nutritionally essential amino acid in animals with a wide range of physiological roles. It is considered to be a substrate for protein synthesis [1], a feed intake enhancer in livestock and poultry [2], a contributor to improved growth performance [3], and a factor in the generation of hormone-like substances [4]. In addition, it has been reported that a deficiency of Trp decreased antibody production in rats [5], indicating that Trp may have a role in immune function. Apart from being a structural component of all proteins, Trp is a precursor of serotonin [5-hydroxytryptamine (5-HT)]. Serotonin (a neurotransmitter) has many functions in the central nervous system to inhibit aggression and modulates stress response, including social and environmental adaptability [6]. Recent findings suggested that dietary Trp may have beneficial effects on the enzymatic and non-enzymatic antioxidant capacity in laying hens [7], rats [8], and fish [9].

Tryptophan concentration in animals is the lowest of all the amino acids, and thus, it can easily become rate-limiting for protein anabolism [10]. Practical diets composed of vegetable protein sources typically result in the essential amino acids such as Trp being limiting to a similar extent after that of total sulphur amino acids, lysine, threonine and isoleucine in poultry [11]. Although many researchers conducted studies to evaluate the requirements of Trp in poultry, results of dose-response studies addressing the Trp need are variable [7, 11]. This might due to the effect of a variety of factors, such as genotype, age, and diet. Present study was conducted to establish Trp needs of chicks using Xinyang green-shell laying hens, a local strain hybridized by varieties of White Leghorns (female) and domestic green-shell (male), by evaluating the effects of different levels of Trp supplementation on their laying performance, egg quality, antioxidant capacity, and the immune functions.

2. Material and methods

The experiment was conducted in accordance with the Chinese guidelines for animal welfare and approved by the Animal Welfare Committee of Animal Science College, Zhejiang University.

2.1. Birds and housing

Xinyang green-shell laying hens (n = 525), 28 weeks of age, were randomly allocated to five treatment groups, each of which included 5 replicates of 21 hens. Hens were fed the basal diet based on corn and soybean meal. L-Trp (Ajinomoto, Japan) was added to the control diet at 0.0 0.02, 0.04, 0.06, and 0.08%, respectively, to achieve 0.15, 0.17, 0.19, 0.21 or 0.23% Trp. Ingredient composition and calculated nutrients are presented in **Table 1**. Hens were kept in three-layer complete ladder cages (3 birds per cage) under the same managerial conditions in a ventilated room. The temperature inside the barn was 21–27°C, and relative humidity was 60–70%. The photoperiod was 16L: 8D throughout the experiment. Each cage was equipped with two nipple drinkers and one feeder. Diets were offered twice daily for ad libitum intake and laying hens had free access to water. The experiment lasted 12 week, including a one-week acclimation period and an 11-week experimental period.

2.2. Laying performance parameters and egg quality

During the experimental period, feed residues were collected and weighted weekly to enable estimation of average daily feed intake (ADFI). Eggs from each replicate were counted and

Ingredients	%	Nutrient levels[2]	%
Corn	62.00	ME/(MJ/Kg)	10.86
Soybean meal	6.60	CP	16.36
Peanut meal	14.50	Ca	3.30
Limestone	9.00	TP	0.32
Wheat bran	5.60	Lys	0.72
Met	0.10	Met	0.30
Lys·HCl	0.20	Thr	0.52
Premix[1]	2.00	Trp	0.15
Total	100.00		

[1]Premix provided the following per kilogram: Vitamin A, 9900 IU; vitamin D3, 2625 IU; vitamin E, 49.5 mg; vitamin K3, 6 mg; vitamin B1, 3 mg; vitamin B2, 10.5 mg; vitamin B6, 6 mg; vitamin B12, 0.03 mg; niacin, 60 mg; folic acid, 3 mg; pantothenic acid, 18 mg; biotin, 0.3 mg; Cu, 9 mg; Fe, 120 mg; Mn, 140 mg; Zn, 120 mg; I, 1.1 mg; Se, 0.4 mg.
[2]Values were calculated from data provided by Feed Database in China (2013).

Table 1. Ingredients and nutrient composition of basal diet.

weighted daily to calculate laying rate and average egg weight. Egg mass was calculated by multiplying egg weight by egg production. Feed conversion ratio (FCR) was calculated as grams of feed intake per gram of egg mass produced. Health status and mortalities were visually observed and recorded daily during the entire experimental period. The magnitude of performance parameters such as laying rate was adjusted for hen mortalities.

At the end of the experiment, 30 eggs from each treatment were randomly collected to assess egg quality parameters. The eggs were weighed and cracked, and albumen height, haugh units, yolk colour, eggshell thickness, and eggshell strength were measured with a digital egg tester (DET-6000, NABEL, Kyoto, Japan). Eggshell thickness (without the shell membrane) was measured at the middle part of the egg.

2.3. Blood sampling and laboratory analyses

At the end of the experiment, 12 h after feed withdrawal, two birds were randomly selected from each replicate, and blood samples were collected from the axillary vein. Blood samples were drawn into Eppendorf tubes (10 ml) and centrifuged at 3000 × g for 10 min to separate out serum. The obtained serum was stored in 1.5-mL Eppendorf tubes at −70°C until analyses and thawed at 4°C before analysis. Serum concentrations or activities of total superoxide dismutase (T-SOD), catalase (CAT), glutathione peroxidase (GSH-Px), total antioxidative capacity (T-AOC), and malondialdehyde (MDA) were measured spectrophotometrically (UV-2000, Unico Instruments Co. Ltd., Shanghai, China) using commercial diagnostic kits (Nanjing Jiancheng Bioengineering Institute, Nanjing, China). Immunoglobulins in the serum were analysed by a microplate reader (SpectraMax M5, Molecular Devices, Sunnyvale, CA) using a sandwich enzyme linked immunosorbent assay (ELISA) using chicken specific IgA, IgG, IgM ELISA quantitation kits (R&D company, System, Inc., McKinley Place NE Minneapolis, MN), respectively, according to the instructions of the manufacturer, and absorbance was measured at 450 nm.

2.4. Statistical analysis

The data were expressed as means ± SE and analysed statistically by one-way ANOVA, using SPSS 18.0 for Windows (SPSS Inc., Chicago, IL). When significant differences were found ($P < 0.05$), Tukey post hoc tests were performed.

3. Results

The laying performance parameters (**Table 2**) were significantly ($P < 0.05$) affected by dietary Trp levels. Laying rate, average egg weight, and FCR in hens fed 0.21–0.23, 0.19–0.21, and 0.21% dietary Trp were significantly improved ($P < 0.05$), respectively, compared with those hens fed on the control diet. No significant difference of ADFI was observed among all the groups ($P > 0.05$).

Results of egg quality characteristics are provided in **Table 3**. The highest value of albumen height ($P < 0.05$) came from 0.19% Trp group. Haugh unit in 0.17 to 0.19% Trp group was significantly higher ($P < 0.05$) than that of other groups. No significant difference of yolk colour, eggshell strength, and eggshell thickness was observed among all the groups ($P > 0.05$).

Results of antioxidant parameters in serum (**Table 4**) showed that supplementing with Trp increased ($P < 0.05$) GSH-Px and T-SOD activity, T-AOC concentration, and decreased ($P < 0.05$) MDA concentration ($P > 0.05$) but had no effect on CAT activity ($P > 0.05$). Briefly, serum GSH-Px activity in 0.17 and 0.19% Trp group was significantly increased by 27.62% ($P < 0.05$) and 27.42% ($P < 0.05$), respectively, compared with the control group. With the increase of supplemental Trp levels, serum T-SOD activity and T-AOC contents were gradually increased, whereas serum MDA content was gradually decreased.

Regarding serum antibodies (**Table 5**), serum IgA concentration increased ($P < 0.05$) at 0.21 to 0.23% dietary Trp, compared with those receiving 0.15% Trp. Serum IgM concentration increased linearly ($P < 0.05$) in response to dietary Trp levels. No significant effect was observed for serum IgG concentration due to dietary Trp levels ($P > 0.05$).

Items[1]	**Dietary L-tryptophan levels, %**				
	0.15	**0.17**	**0.19**	**0.21**	**0.23**
Laying rate, %	63.76 ± 0.65^{b}	64.15 ± 0.67^{b}	65.17 ± 1.86^{b}	66.70 ± 0.38^{a}	66.91 ± 1.34^{a}
ADFI, g/hen	85.78 ± 1.14	87.68 ± 2.62	$86.81.43 \pm 1.18$	84.49 ± 2.55	85.52 ± 0.44
Egg weight, g	45.57 ± 0.14^{b}	46.02 ± 0.06^{ab}	46.30 ± 0.15^{a}	46.60 ± 0.04^{a}	46.09 ± 0.36^{ab}
FCR	2.88 ± 0.03^{a}	2.88 ± 0.02^{a}	2.85 ± 0.01^{a}	2.79 ± 0.01^{b}	2.81 ± 0.02^{a}

[1]ADFI = average daily feed intake; FCR = feed conversion ratio.
Means sharing different letters (a, b) in the same row are significantly different ($P < 0.05$).

Table 2. Effect of dietary L-tryptophan on laying performance.

Items	Dietary L-tryptophan levels, %				
	0.15	0.17	0.19	0.21	0.23
Albumen height, mm	4.47 ± 0.17[b]	4.50 ± 0.05[b]	4.96 ± 0.22[a]	4.66 ± 0.13[b]	4.41 ± 0.17[b]
Haugh units	68.94 ± 1.26[b]	70.96 ± 0.60[a]	73.41 ± 1.88[a]	69.10 ± 0.48[b]	68.22 ± 1.54[b]
Yolk colour, points	7.56 ± 0.17	7.22 ± 0.22	7.34 ± 0.19	7.26 ± 0.14	7.46 ± 0.16
Eggshell strength, Kgf	3.67 ± 0.20	3.35 ± 0.35	3.60 ± 0.35	3.58 ± 0.29	3.69 ± 0.32
Eggshell thickness, mm	0.31 ± 0.00	0.32 ± 0.01	0.32 ± 0.01	0.32 ± 0.01	0.32 ± 0.00

Means sharing different letters (a, b) in the same row are significantly different ($P < 0.05$).

Table 3. Effect of dietary L-tryptophan on egg quality.

Items[1]	Dietary L-tryptophan levels, %				
	0.15	0.17	0.19	0.21	0.23
GSH-Px, U/ml	3236.74 ± 164.11[b]	4130.87 ± 198.83[a]	4124.35 ± 322.76[a]	3503.48 ± 56.12[b]	3388.70 ± 369.18[b]
T-SOD, U/L	394.85 ± 14.48[b]	396.65 ± 7.25[b]	440.11 ± 8.38[a]	415.92 ± 11.62[a]	418.37 ± 8.8[a]
CAT, U/ml	2.81 ± 0.21	2.67 ± 0.16	2.69 ± 0.13	2.87 ± 0.19	2.83 ± 0.24
MDA, nmol/ml	3.45 ± 0.17[a]	3.12 ± 0.37[a]	2.60 ± 0.34[b]	2.47 ± 0.23[b]	2.44 ± 0.20[b]
T-AOC, U/L	3.85 ± 0.84[b]	4.67 ± 0.42[a]	4.71 ± 0.40[a]	4.49 ± 0.53[a]	5.15 ± 0.46[a]

[1] T-SOD = total superoxide dismutase; CAT = catalase; GSH-Px = glutathione peroxidase; T-AOC = total antioxidative capacity; MDA = malondialdehyde.
Means sharing different letters (a, b) in the same row are significantly different ($P < 0.05$).

Table 4. Effect of dietary L-tryptophan on antioxidant parameters in serum.

Items	Dietary L-tryptophan levels, %				
	0.15	0.17	0.19	0.21	0.23
IgA, ng/ml	1779.98 ± 100.03[c]	1830.65 ± 158.94[bc]	2034.70 ± 121.37[abc]	2226.91 ± 94.59[ab]	2329.05 ± 118.36[a]
IgG, ng/ml	413.24 ± 36.50	448.67 ± 36.83	451.85 ± 34.55	475.49 ± 36.83	499.89 ± 34.09
IgM, μg/mL	1743.24 ± 24.63[b]	1999.39 ± 61.72[a]	2022.86 ± 155.02[a]	2057.07 ± 169.52[a]	2244.93 ± 107.29[a]

Means sharing different letters (a, b) in the same row are significantly different ($P < 0.05$).

Table 5. Effect of dietary L-tryptophan on serum immunoglobulins concentrations.

4. Discussion

Tryptophan plays a significant role in laying hen nutrition because it is considered to be the third-limiting amino acid, after the sulphur-containing amino acids and lysine. Most commercial diets are calculated on an amino acid basis, rather than a protein basis [12]. Therefore, it is important to have an accurate value of amino acids to use as a requirement when formulating diets. Results of dose-response studies addressing the Trp need are variable, although the [13] suggests a requirement of 160 mg per hen per d for the commercial layer. In a study using Hy-Line hens (53 week), egg production was similar among groups receiving 0.17, 0.21 or 0.23 g/kg Trp, but at 0.19 g/kg Trp, egg production had maximized significantly compared to those given 0.17 g/kg Trp [12]. By using Rhode Island Red × White Leghorn layers (25 week), researchers found that 500 mg Trp/kg diet improved the egg production rate in laying hens [14]. However, in Babcock Brown layers (40 week), the authors found that supplementing Trp had no effect on laying performance [7]. In the current study, we found that the laying performance parameters (i.e. laying rate, egg weight, and FCR) were significantly improved by dietary Trp levels (from 0.19 to 0.23%) in Xinyang green-shell laying hens (28 week). Combining to all the aforementioned, it is clearly that the variable need for Trp in laying hens is partially due to the genotype and age.

Egg quality is important for consumer appeal and encompasses several aspects related to the shell (external quality) and to the albumen and yolk (internal quality). Egg quality has a genetic basis, and the parameters of egg quality vary between strains of hens [15, 16]. However, egg quality is also influenced by diet nutrition such as dietary protein and amino acid content [17]. Limited research has been conducted on the effects of supplemental Trp on egg quality of laying hens. In Ref. [7], the authors found that adding 0.2 or 0.4 g/kg Trp to the basal diet (0.17% Trp) improved egg shell strength quadratically in Babcock Brown layers, but had no effect on egg internal quality. Conversely, in the current study, we found that dietary Trp from 0.17 to 0.19% increased egg internal quality (albumen height and haugh unit) rather than external quality. The mechanism of Trp regulating egg quality is not well understood due to limited references. Thus, further studies are needed to conduct to verify the role of Trp on egg quality.

It is known that serotonin, with Trp as its precursor, has many functions in the central nervous system to inhibit aggression and modulates stress response [6], suggesting that dietary Trp may sever as a free radical scavenger, and hence have beneficial effects on the antioxidant capacity of animals. Almost all the phenomena of life and pathological processes are related to the perspectives of free radicals that can induce body damage when they are presented in excessive levels [18]. MDA can generally be used as a biomarker for free radical induced damage and can endogenously reflect lipid peroxidation, which is the consequence of diminished antioxidant protection as concentrations of reactive oxygen species increase [19]. SOD and GSH-Px are the main parameters used to assess oxidative status in the enzymatic system, while T-AOC represents enzymatic and non-enzymatic antioxidant defence systems [10, 18]. In the current study, the higher T-AOC level, T-SOD and GSH-px activities and lower MDA concentrations due to supplemental Trp in laying hens reflect a greater antioxidant defence. In accordance with our results, previous researchers also found that Trp increased the serum SOD activity of laying hens [7] and elevated the GSH content in the hepatic tissues of rats [8]. The present study suggested that appropriate Trp levels in the diet may have a positive effect on both the enzymatic and non-enzymatic antioxidant capacity function of laying hens.

Recent studies have also proved that Trp may play an important role in immune function. Dietary Trp deficiency has been demonstrated to reduce the levels of nutrition and to depress immune function to cause a significant increase in the susceptibility to disease infection, morbidity, and mortality in animals [20]. The addition of Trp to the control diet resulted in increasing the levels of serum IgA and IgM in the current study. Substantiating the findings on serum immunoglobulins, a previous study showed that a deficiency of Trp decreased antibody production in rats [5]. In addition, a recent study in laying hens also found that the addition of Trp at 0.4 g/kg to the basal diet (0.17% Trp) resulted in quadratically increasing the levels of serum IgM in Babcock Brown layers [7]. From the nutritional viewpoint, we speculated that amino acids affect the synthesis of effector molecules (immunoglobulins, nitric oxide, lysozyme, and complement). It is worth noting that tryptophan is not the only amino acid that affects the immune function. Researchers observed that a deficiency of phenylalanine decreased antibody production in rats [5], while threonine supplementation increased IgG concentrations in the serum of laying hens [21]. In contrast, excess methionine [5] and leucine [22] suppressed humoral immune function in the rat. Our results indicated that Trp plays important roles in the regulation of poultry humoral immune through regulation of the generation of immunoglobulins; however, our understanding of the function of Trp in the regulation of immune response is far from complete, and its involved mechanisms require further study.

5. Conclusion

In conclusion, supplemental Trp to the control diet can improve laying performance, egg quality, antioxidant capacity, and the immune functions in Xinyang green-shell laying hens. We suggest that the optimum level range of dietary Trp is from 0.19 to 0.21% for Xinyang green-shell laying hens under the current study conditions.

Acknowledgements

The research was supported by the earmarked fun for modern agro-industry technology research system (No. CARS-41-K17).

Author details

Xinyang Dong and Xiaoting Zou*

*Address all correspondence to: xtzou@zju.edu.cn

Animal Science College, Zhejiang University, Hangzhou, China

References

[1] Ruan Z, Yang YH, Wen YM, Zhou Y, Fu XF, Ding S, Liu G, Yao K, Wu X, Deng ZY, Wu GY, Yin YL. Metabolomic analysis of amino acid and fat metabolism in rats with tryptophan supplementation. Amino Acids. 2014;**46**:2681-2691. DOI: 10.1007/s00726-014-1823-y

[2] Woodger TL, Sirek A, Anderson GH. Diabetes, dietary tryptophan, and protein-intake regulation in weanling rats. The American Journal of Physiology. 1979;**236**:R307-R311

[3] Cortamira NO, Seve B, Lebreton Y, Ganier P. Effect of dietary tryptophan on muscle, liver and whole-body protein-synthesis in weaned piglets-relationship to plasma-insulin. British Journal of Nutrition. 1991;**66**:423-435. DOI: 10.1079/BJN19910045

[4] Le Floc'h N, Melchior D, Seve B. Dietary tryptophan helps to preserve tryptophan homeostasis in pigs suffering from lung inflammation. Journal of Animal Science. 2008;**86**:3473-3479. DOI: 10.2527/jas.2008-0999

[5] Gershoff SN, Gill TJ, Simonian SJ, Steinberg AI. Some effects of amino acid deficiencies on antibody formation in the rat. Journal of Nutrition. 1968;**95**:184-189

[6] Martin CL, Duclos M, Aguerre S, Morede P, Manier G, Chaouloff F. Corticotropic and serotonergic responses to acute stress with/without prior exercise training in different rat strains. ActaPhysiologicaScandinavica. 2000;**168**:421-430. DOI: 10.1046/j.1365-201x.2000.00683.x

[7] Dong X, Azzam M, Rao W, Yu D, Zou X. Evaluating the impact of excess dietary tryptophan on laying performance and immune function of laying hens reared under hot and humid summer conditions. British Poultry Science. 2012;**53**:491-496. DOI: 10.1080/00071668.2012.719149

[8] Raju T, Kanth VR, Reddy PUM, Srinivas J, Sobhanaditya J. Influence of kynurenines in pathogenesis of cataract formation in tryptophan-deficient regimen in Wistar rats. Indian Journal of Experimental Biology. 2007;**45**:543-548

[9] Wen H, Feng L, Jiang W, Liu Y, Jiang J, Li S, Tang L, Zhang Y, Kuang S, Zou X. Dietary tryptophan modulates intestinal immune response, barrier function, antioxidant status and gene expression of TOR and Nrf2 in young grass carp (Ctenopharyngodonidella). Fish & Shellfish Immunology, 2014;**40**:275-287. DOI: 10.1016/j.fsi.2014.07.004

[10] Sainio EL, Pulkki K, Young SN. L-Tryptophan: Biochemical, nutritional and pharmacological aspects. Amino Acids. 1996;**10**:21-47. DOI: 10.1007/BF00806091

[11] Corzo A, Kidd MT, Thaxton JP, Kerr BJ. Dietary tryptophan effects on growth and stress responses of male broiler chicks. British Poultry Science. 2005;**46**:478-484. DOI: 10.1080/00071660500157974

[12] Russell GB, Harms RH. Tryptophan requirement of the commercial hen. Poultry Science. 1999;78:1283-1285. DOI: 10.1093/ps/78.9.1283

[13] NRC. Nutrient Requirements of Poultry. 9th rev. ed. Washington, DC: National Acad. Press; 1994

[14] Ohtani H, Saitoh S, Ohkawara H, Akiba Y, Takahashi K, Horiguchi M, Goto K. Research note: Production performance of laying hens fed l-tryptophan. Poultry Science. 1989;**68**: 323-326. DOI: 10.3382/ps.0680323

[15] Pandey NK, Mahapatra CM, Verma SS, Johari DC. Effect of strain on physical egg quality characteristics in White Leghorn chickens. Indian Journal of Poultry Science. 1986;**21**:304-307

[16] Silversides FG, Korver DR, Budgell KL. Effect of strain of layer and age at photostimulation on egg production, egg quality, and bone strength. Poultry Science. 2006;**85**:1136-1144. DOI: 10.1093/ps/85.7.1136

[17] Roberts JR. Factors affecting egg internal quality and egg shell quality of laying hens. Journal of Poultry Science. 2004;**41**:161-177

[18] Zhao X, Yang ZB, Yang WR, Wang Y, Jiang SZ, Zhang GG. Effects of ginger root (Zingiberofficinale) on laying performance and antioxidant status of laying hens and on dietary oxidation stability. Poultry Science. 2011;**90**:1720-1727. DOI: 10.3382/ps.2010-01280

[19] Wang YZ, Xu CL, An ZH, Liu JX, Feng J. Effect of dietary bovine lactoferrin on performance and antioxidant status of piglets. Animal Feed Science and Technology. 2008;**140**:326-336. DOI: 10.1016/j.anifeedsci.2007.02.006

[20] Harden JL, Lewis SM, Lish SR, Suárez-Fariñas M, Gareau D, Lentini T, Johnson-Huang LM, Krueger JG, Lowes MA. The tryptophan metabolism enzyme l-kynureninase is a novel inflammatory factor in psoriasis and other inflammatory diseases. Journal of Allergy & Clinical Immunology. 2015;**137**:1830-1840. DOI: 10.1016/j. jaci.2015.09.055

[21] Azzam MMM, Dong XY, Xie P, Wang C, Zou XT. The effect of supplemental l-threonine on laying performance, serum free amino acids, and immune function of laying hens under high-temperature and high-humidity environmental climates. Journal of Applied Poultry Research. 2011;**20**:361-370. DOI: 10.3382/japr.2010-00308

[22] Aschkenasy A. Prevention of the immunodepressive effects of excess dietary leucine by isoleucine and valine in the rat. Journal of Nutrition. 1979;**109**:1214-1222

Chapter 13

Effects of Dietary Lysine Levels on the Plasma Concentrations of Growth-Related Hormones in Late-Stage Finishing Pigs

Taiji Wang, Mark A. Crenshaw,
Md Shamimul Hasan, Guoyao Wu and
Shengfa F. Liao

Additional information is available at the end of the chapter

http://dx.doi.org/10.5772/intechopen.68545

Abstract

This study was undertaken to investigate the effects of dietary lysine on the plasma concentrations of three growth-related hormones in pigs. Nine late-stage finishing barrows were assigned to three dietary treatments according to a completely randomized experimental design (3 pigs/treatment). Three corn and soybean meal-based diets were formulated to contain three levels of lysine, which were 0.43, 0.71, and 0.98% for Diets 1 (lysine deficient), 2 (lysine adequate), and 3 (lysine excess), respectively. The feeding trial lasted 4 weeks, during which the pigs were allowed ad libitum access to the diets and water. After the 4 weeks, blood was collected and plasma samples were obtained. Then, the plasma concentrations of insulin, growth hormone (GH), and insulin-like growth factor 1 (IGF-1) were measured. No difference in the plasma concentration of insulin or GH among the three treatments was found ($P > 0.10$). However, the plasma IGF-1 concentration was lower ($P < 0.05$) in the pigs fed Diet 1 or 3 than fed Diet 2, suggesting that either dietary lysine deficiency or excess can lead to a lower concentration of plasma IGF-1. It was concluded that IGF-1, instead of insulin or GH, in the blood may be a key controlling growth factor in response to dietary lysine supply for regulating muscle growth in late-stage finishing pigs.

Keywords: lysine, hormone, blood plasma, finishing pig

1. Introduction

The biochemical process of protein turnover in the skeletal muscle of pigs is of great importance for the production of food protein for human consumption [1]. Some nutrients, such as glucose and fatty acids, play important roles not only as energetic substrates but also as cell signaling molecules to regulate the protein turnover in animal body [2–4]. Similarly, besides their function as building blocks for body protein biosynthesis, some amino acids (AAs) can also function as cell signaling molecules regulating those metabolic pathways that are necessary for muscle protein accretion [5, 6]. The regulation of key signaling and metabolic pathways of muscle protein turnover by AAs (Wang et al., personal communication) is closely associated with the concomitant responses of some growth-related hormones [1, 7, 8].

The plasma concentrations of growth-related polypeptide hormones, such as insulin, growth hormone (GH; a.k.a. somatotropin), and insulin-like growth factor 1 (IGF-1), can be affected by animal nutritional status and, in turn, regulate cell and tissue growth and development in animal body [9–11]. Because these hormones are not fat-soluble, they cannot penetrate cell membranes into cytosol. Therefore, they exert cell signaling effects through binding to their corresponding receptors on the cell membranes, where they further activate cell signaling cascades to regulate gene expression and protein turnover [7, 12].

Lysine is the first limiting AA in typical grain-based swine diets [1, 13], and sufficient dietary lysine supply is critical for pig growth performance, especially the growth of skeletal muscle, the largest AA reservoir in the body [14–16]. According to some previous studies on growing pigs [14, 17], dietary lysine supplementation stimulated the insulin secretion, increased the plasma insulin concentration, but not the plasma concentrations of GH and IGF-1, in a dose-dependent manner. On nursery pigs, it was reported that the plasma IGF-1 concentration was reduced when animals were fed a diet lower in lysine level [18]. However, whether or not the effect of dietary lysine on the growth performance of finishing pigs is mediated via these growth-related hormones is still unknown. Therefore, the objective of this study was to investigate the effect of dietary lysine at different levels on the plasma concentrations of three key growth-related hormones, which were insulin, GH, and IGF-1, in late-stage finishing pigs.

2. Materials and methods

2.1. Animal trial and sample collection

All the experimental protocols involving caring, handling, and treatment of pigs were approved by Mississippi State University Institutional Animal Care and Use Committee. A total of nine crossbred (Large White × Landrace) barrows with an average initial body weight (BW) 94.4 ± 6.7 kg were housed in an environment-controlled swine barn at the Leveck Animal Research Center of Mississippi State University. The pigs were randomly assigned to nine individual feeding pens, and then were assigned to three dietary treatment groups according to a completely randomized experimental design. Each treatment consisted of 3 pen replicates ($n = 3$) with one pig per pen.

A corn and soybean meal-based diet (a lysine-deficient diet; defined as Diet 1) was formulated to meet or exceed the NRC [13] recommended requirements of various nutrients, including crude protein (CP) and essential AAs, but not of lysine. Diet 2 (a lysine-adequate diet) and Diet 3 (a lysine-excess diet) were formulated by adding L-lysine monohydrochloride (Archer

Daniels Midland Co., Quincy, IL, USA) to Diet 1 at the expense of corn at the rates of 0.35% and 0.70%, respectively (**Table 1**). The total lysine contents (calculated, as-fed basis) in Diets 1, 2, and 3 were 0.43, 0.71, and 0.98%, respectively. To confirm the contents of major nutrients, samples of the three diets were submitted to the Essig Animal Nutrition Laboratory at Mississippi State University for proximate analysis, and the results are shown in **Table 2**. The AA contents of the three diets were analyzed by using the high-performance liquid chromatography methods [19] at Texas A&M University (College Station, TX, USA), and the results are also shown in **Table 2**.

	Diet 1	**Diet 2**	**Diet 3**
Ingredients (%)			
Corn	90.844	90.494	90.144
Soybean meal	6.400	6.400	6.400
Canola oil	0.800	0.800	0.800
L-Lysine-HCl (98.5%)	0.000	0.350	0.700
DL-Methionine (99%)	0.040	0.040	0.040
L-Threonine (98.5%)	0.090	0.090	0.090
L-Tryptophan (99%)	0.035	0.035	0.035
Limestone	0.650	0.650	0.650
Dicalcium phosphate	0.900	0.900	0.900
Salt	0.200	0.200	0.200
Mineral premix[2]	0.033	0.033	0.033
Vitamin premix[3]	0.008	0.008	0.008
Total	100.000	100.000	100.000
Composition[4]			
Metabolizable energy (kcal/kg)	3319	3323	3326
Crude Protein (%)	10.45	10.75	11.05
Total Lysine (%)	0.43	0.71	0.98
Total Methionine (%)	0.24	0.24	0.24
Total Threonine (%)	0.50	0.50	0.50
Total Tryptophan (%)	0.14	0.14	0.14
Total Ca (%)	0.46	0.46	0.46
Total P (%)	0.43	0.43	0.43

[1]Diets 1, 2, and 3 were formulated to contain dietary lysine at 0.43%, 0.71%, and 0.98% (as-fed basis), respectively, of which Diets 2 and 3 were formulated by adding 0.35% and 0.70% L-lysine-HCl (Archer Daniels Midland Co., Quincy, IL) to Diet 1 at the expense of corn.

[2]The mineral premix contained 13.2% Ca, 1.0% Cu, 8.0% Fe, 5.0% Mn, 10.0% Zn, 500 ppm I, and 300 ppm Se.

[3]Each kilogram of vitamin premix contained the following: 22.05 million IU vitamin A, 3.31 million IU vitamin D3, 66,138 IU vitamin E, 88 mg vitamin B12, 220 mg biotin, 8,818 mg menadione, 15,432 mg riboflavin, 61,728 mg d-pantothenic acid, and 88,183 mg niacin.

[4]Calculated major nutrients.

Table 1. Ingredient and nutrient compositions (as-fed basis) of the three experimental diets fed to the late-stage finishing pigs[1].

Nutrient and Energy	Experimental Diet		
	Diet 1	Diet 2	Diet 3
Dry matter, %	87.10	87.10	87.10
Gross energy, kcal/kg	3,663	3,608	3,559
Crude protein, %	9.77	10.60	10.86
Individual Amino Acids, %			
Lysine	0.42	0.70	1.01
Aspartate	0.98	0.97	0.98
Asparagine	0.74	0.75	0.75
Glutamate	1.01	1.03	1.03
Glutamine	1.42	1.44	1.43
Serine	0.53	0.52	0.54
Histidine	0.34	0.33	0.34
Glycine	0.61	0.62	0.62
Threonine	0.50	0.51	0.50
Arginine	0.73	0.74	0.74
Alanine	0.81	0.83	0.82
Tyrosine	0.52	0.52	0.53
Tryptophan	0.14	0.13	0.14
Methionine	0.25	0.25	0.26
Valine	0.65	0.66	0.65
Phenylalanine	0.66	0.65	0.66
Isoleucine	0.52	0.52	0.53
Leucine	1.42	1.44	1.45
Cysteine	0.26	0.26	0.30
Proline	1.08	1.07	1.09

[1]The energy, crude protein, and dry matter contents were analyzed at the Essig Animal Nutrition Laboratory, Mississippi State University (Starkville, MS, USA). The amino acid contents were analyzed at Texas A&M University (College Station, TX, USA).

Table 2. The analyzed nutrient compositions (as-fed basis) of three experimental diets fed to the late-stage finishing pigs[1].

The feeding trial lasted a total of 4 weeks, during which the pigs were allowed ad libitum access to the experimental diets and fresh water. All the pigs, feeders, waterers, and room temperature were checked two to three times on a daily basis (6:00 am to 7:00 pm). After the 4-week trial period, blood samples were collected by venipuncture of the jugular veins of individual pigs (10 mL/pig) in the early morning (6:00–8:00 am). Immediately after collection, the blood samples were kept onto ice until the plasma was separated by centrifugation (at 800 × g) of the vacutainer tubes at 4°C for 16 min. The plasma samples were then stored in 200 μL aliquots at −80°C until the laboratory analysis of hormones was conducted.

2.2. Laboratory analyses of the growth-related hormones

2.2.1. Analysis of plasma insulin

The plasma concentration of insulin was measured in duplicate using a porcine insulin ELISA kit (ENZO Life Sciences, Farmingdale, NY, USA) according to the manufacturer's instructions. Briefly, the plasma samples were first diluted (1:3) to remove matrix interference. The diluted samples were then added to a pre-coated insulin ELISA plate (100 μL/well/sample) and incubated on an orbital shaker (Forma 420; Thermo Fisher Scientific Inc., USA) at room temperature for 1 h, allowing the liquid to be thoroughly mixed. The plate was then rinsed four times with wash buffer (200 μL/well) and blotted on lint free paper towels after each rinse. After the final blot, the primary insulin antibody solution was added to the plate (100 μL/well) and incubated on the shaker at room temperature for 1 h. Residual primary antibody on the plate was then rinsed and blotted as described above. One hundred microliter of blue solution of horseradish peroxidase (HRP) conjugate was added to each well, and the plate was incubated on the shaker at room temperature for 30 min. After incubation, the plate was rinsed and blotted again to remove the residual blue conjugate if any. Then the HRP substrate solution was added to the plate (100 μL/well) and the plate was incubated on the shaker at room temperature for 30 min again. Finally, the reaction stop solution was added to the plate (100 μL/well), and the optical density (OD) value of the solution was measured at 450 nm using a microplater reader (SpectraMax Plus 384; Molecular Devices, San Francisco, CA, USA). The insulin concentration of each sample was calculated based on the OD values of the standard curve of known concentrations using Curve Expert 1.4 computer program (http://www.curveexpert.net).

2.2.2. Analysis of plasma growth hormone

The plasma concentration of GH was measured in duplicate using a porcine GH ELISA kit (Cloud-Clone Corp., Wuhan, China) according to the manufacturer's instructions. Briefly, the plasma samples were added to a pre-coated GH ELISA plate (100 μL/well/sample) and incubated at 37°C for 2 h. Then, the liquid was removed from each well, the "Detection Reagent A" working solution was added to each well (100 μL/well), and the plate was incubated at 37°C for 1 h. The plate was then rinsed three times with wash buffer (350 μL/well) and blotted on lint free paper towels after each rinse. Following the final blot, the "Detection Reagent B" working solution was added to the plate (100 μL/well), which was incubated at 37°C for 30 min. The plate was rinsed and blotted again for five times as described above. The substrate solution was then added to the plate (90 μL/well), and the plate was incubated at 37°C for 15 min with protection from light. Finally, the reaction stop solution was added to the plate (50 μL/well), and the OD value of the solution in each well was measured at 450 nm using a microplater reader (SpectraMax Plus 384). The GH concentration of each sample was calculated based on its OD values against the standard curve of known concentrations using Curve Expert 1.4 computer program (http://www.curveexpert.net).

2.2.3. *Analysis of plasma insulin-like growth factor 1*

The plasma concentration of IGF-1 was measured in duplicate using a human IGF-1 ELISA kit (ENZO Life Sciences) according to the manufacturer's instructions. Briefly, each plasma sample was mixed in a solution of 100% ethanol (1:5) and 2 N hydrochloric acid (7:1) and incubated at room temperature for 30 min to dissociate IGF-1 from IGF binding proteins. The binding proteins were then pelleted by centrifugation at 9900 × g for 5 min at room temperature. The supernatant was removed and neutralized with an equal volume of the neutralizing reagent. The supernatant sample was then diluted to a final concentration of 1:35 with the assay buffer.

The diluted samples in the assay buffer were added to a pre-coated IGF-1 ELISA plate (100 μL/well/sample) and incubated on the orbital shaker (Forma 420, Thermo Fisher Scientific Inc.) at room temperature for 1 h, allowing the liquid to be thoroughly mixed. The ELISA plate was rinsed five times with the wash buffer (200 μL/well) and blotted on lint free paper towels after each rinse. After the final blot, the primary IGF-1 antibody solution was added to the plate (100 μL/well) and the plate was incubated on the shaker at room temperature for 2 h. Residual primary antibody on the plate was then rinsed and blotted again as described earlier. Then, the blue solution of HRP conjugate was added to the plate (100 μL/well) and the plate was incubated on the shaker again at room temperature for 30 min. After incubation, the plate was rinsed and blotted to remove any residual blue conjugate. The HRP substrate solution was then added to the plate (100 μL/well), which was incubated on the shaker at room temperature for 30 min. Finally, the reaction stop solution was added to the plate (100 μL/well), and the OD value of the solution in each well was measured at 450 nm using a microplater reader (SpectraMax Plus 384). The plasma IGF-1 concentration of each sample was calculated based on OD values against the standard curve of known concentrations using Curve Expert 1.4 computer program (http://www.curveexpert.net).

2.3. Statistical analysis

The plasma concentration of each hormone of each sample was averaged from 2 values of the duplicate measurements and then subjected to analysis of variance (ANOVA) for a completely randomized experimental design using the general linear model (GLM) procedure of SAS 9.4 (SAS Institute Inc., Cary, NC, USA) with dietary lysine level being the main effect and individual pigs being the experiment units. Three means of the treatments were separated by the protected *t*-test using the LSMEANS/PDIFF option in the GLM procedure. Probability values (P) less than 0.05 were considered as significant differences and the P values between 0.05 and 0.10 were considered as tendencies to be different.

3. Results and discussions

3.1. Lysine effect on the plasma insulin concentration

As shown in **Figure 1A**, there are no differences in the plasma insulin concentrations (P = 0.25) among the three dietary treatments, which suggests that the plasma insulin

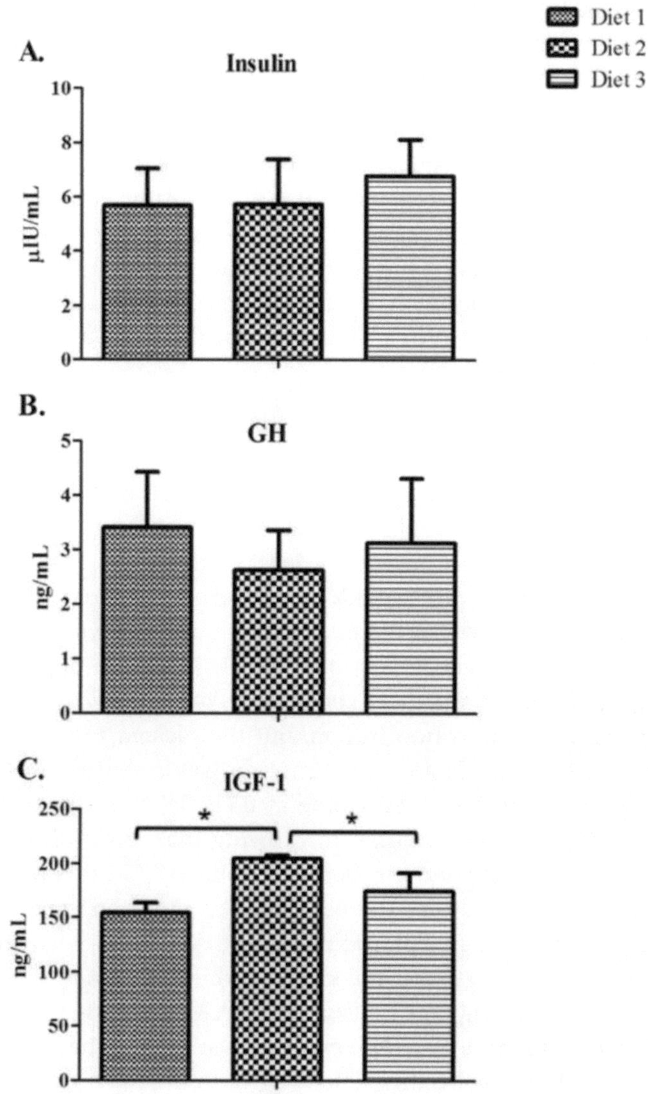

Figure 1. ELISA analyses of the plasma concentrations of (**A**) insulin, (**B**) growth hormone (GH), and (**C**) insulin-like growth factor 1 (IGF-1) in the late-stage finishing pigs fed a lysine-deficient diet (Diet 1), a lysine-adequate diet (Diet 2), or a lysine-excess diet (Diet 3). Shown on the y-axis are the hormone concentrations (μIU/mL or ng/mL) Bars denote means ±SD. All measurements were carried out in duplicate The *signs denote differential concentrations between two treatment groups ($P < 0.05$).

level of the late-stage finishing pigs was not affected by the dietary lysine concentration, at least at the range from 0.43 to 0.98% (**Table 1**). As it is known, insulin plays a critical role in the metabolism of nutrients such as carbohydrates, lipids and proteins, and is a primary acute anabolic coordinator of nutrient partitioning [20]. As a signaling molecule, insulin can activate the insulin signaling transduction pathway, leading to an increase in phosphatidylinositol 3-kinase (PI3K) activity followed by an increase in protein kinase B

(PKB/Akt) activity. The PKB/Akt activity is associated with the phosphorylation and inhibition of the tuberous sclerosis complex (TSC2) and further the mTOR (mechanistic target of rapamycin) kinase activity to regulate protein turnover in skeletal muscle and other tissues [21, 22].

Amino acids and insulin can independently stimulate protein synthesis in skeletal muscle [23, 24]. However, dietary AAs or protein can also affect insulin secretion, although insulin is secreted primarily in response to the elevated blood glucose concentration [20, 25]. The reduction in dietary CP concentration has been shown to decrease plasma insulin concentration in growing pigs [10, 26]. In addition, it was observed that a leucine-induced stimulation of protein synthesis in the skeletal muscle of rats was facilitated by a transient increase in the blood concentration of insulin [27, 28].

Different AAs may have different capacities in stimulating insulin secretion. An intravenous administration of 30 g of AA mixtures or of certain individual AAs to healthy human subjects (19–27 years old) induced prompt and large increases in the level of plasma insulin [25]. While a mixture of 10 essential AAs, or the cationic lysine or arginine alone, appeared to be the most potent, cationic histidine was the least potent, and no obvious common physicochemical property or configuration could characterize the more potent or less potent AAs [25]. In pigs, it was reported that dietary administration of lysine also had a stimulating effect on insulin secretion, and this stimulation is in a dose-dependent manner. For example, the growing barrows fed diets containing 0.45 and 0.75% total lysine showed no difference in plasma insulin concentration, but the plasma insulin concentration was increased by 39% when dietary lysine concentration was raised to 0.98% [14]. Similarly, while the dietary concentration of total lysine at 0.71, 0.95, or 1.20% did not show any influence on plasma insulin concentration in growing pigs, 1.45% lysine in the diet significantly increased the plasma concentration of insulin [17]. In this present study on the late-stage finishing pigs, dietary lysine did not show any effect on the plasma insulin level, and there might be a few reasons responsible for the discrepancy between our results and those reported in the literature: (1) the late-stage finishing pigs may not be as sensitive as young humans or young growing pigs in response to AA stimulation, (2) our dietary lysine concentrations (**Table 1**) might not be high enough to stimulate the release of insulin, and (3) the different blood sample collection time relative to the time of AA administration or feed intake might cause the differences.

3.2. Lysine effects on the plasma concentrations of growth hormone and insulin-like growth factor 1

As shown in **Figure 1B**, there are no differences in the plasma GH concentrations ($P = 0.18$) among the three dietary treatments, which suggests that the plasma GH level of the late-stage finishing pigs was not affected by the dietary lysine concentration, at least at the range from 0.43 to 0.98% (**Table 1**). As shown in **Figure 1C**, however, the plasma IGF-1 concentration of the pigs fed either Diet 1 or Diet 3 was lower ($P < 0.05$) than that of the pigs fed Diet 2, which suggests that either dietary lysine deficiency or dietary lysine excess can lead to a lower level of plasma IGF-1 concentration in the late-stage finishing pigs.

In humans, it has been shown that the plasma level of GH can change in response to a number of physiologic stimuli, including plasma AAs [29]. Different single AAs intravenously administrated into blood vary in their capacities to evoke a release of GH into the blood circulation [29]. Oral administration of lysine (1200 mg) plus arginine (1200 mg) to some healthy volunteers (male, aged 15–20 years) also provoked a release of GH and insulin into the blood, and this effect appeared to be specific to the combination of these two AAs because neither one demonstrated appreciable stimulating activity when administered alone [30].

Primarily consisting of GH, IGF-1, and their associated carrier proteins and receptors, somatotropin axis (a.k.a., GH-IGF-1 axis) is a very critical regulatory pathway for mammalian muscle growth and development, as well as the protein and lipid metabolism in various tissues [31–33]. The stimulating effect of GH on the growth and development of pig muscle has been reported by many previous in vivo studies with daily injection (i.m.) of exogenous porcine GH [34, 35], and it was hypothesized that the actions of GH on muscle growth are mediated by the insulin-like growth factors (IGFs). A great deal of evidence supports the view that the IGFs (especially, IGF-1) are important myogenic agents that could mediate the actions of GH, but this did not demonstrate that GH has no direct effect on the skeletal muscle to stimulate its growth [32].

Essential AAs, especially lysine, can promote swine muscle growth when sufficiently supplied, but it is unknown how or if its promotion effect on swine muscle growth is mediated via the GH-IGF-1 axis. This present study showed that the plasma concentration of IGF-1 in the late-stage finishing pigs fed either a lysine-deficient or lysine-excess diet was reduced ($P < 0.05$), although the plasma concentration of GH was not affected. These results support some previous studies conducted by Roy et al. [14] and Ren et al. [17], who reported that the dietary lysine level had no effect on the plasma GH concentration in growing pigs.

That the dietary lysine deficiency decreased the plasma IGF-1 concentration in the finishing pigs of this study supports the research conducted by Takenaka et al. [36] and Katsumata et al. [18], who showed that the plasma IGF-1 concentration was reduced when young rats (6 weeks of age) and nursery pigs were fed the diets lower in lysine. The previous studies on growing pigs, however, did not show that dietary lysine levels influence the plasma concentration of IGF-1 [14, 17].

The conflicting results regarding the effects of dietary lysine levels on the plasma IGF-1 concentration from different studies may be due to various factors, such as animal species used, growing stages of the animals, and the amounts of lysine provided in the diets. It has been known that a principal hormonal stimulus for IGF-1 production is GH [37]. However, it is interesting to find that the plasma IGF-1 concentration of the late-stage finishing pigs fed either a lysine-deficient or a lysine-excess diet was reduced when compared to the pigs fed a lysine-adequate diet, whereas the dietary lysine level had no effect on the plasma GH concentration. While the animal nutritional status is a key factor in regulating the activity and function of GH-IGF-1 axis [38], the regulation of IGF-1 release may be a key control point for animal muscle growth and nitrogen retention [9, 17]. In terms of animal growth performance, we have reported that there was no further improvement in the average daily gain when excess dietary lysine was provided to the late-stage finishing pigs [16]. Therefore, the lysine promotion effect on pig average daily gain might not be associated with the plasma GH, and the plasma IGF-1 may be one of the limiting growth factors that restricts the pigs from further increasing in average daily gain or further increasing in muscle protein accretion in response to the dietary lysine provision above the NRC [13] recommended requirement.

4. Conclusions

The results generated from this study in late-stage finishing pigs suggest that dietary deficiency or excess of lysine did not affect the plasma concentrations of insulin and GH. However, the plasma concentration of IGF-1 was affected by the dietary lysine levels. In particular, either dietary lysine deficiency or excess led to a reduction in the plasma concentration of IGF-1 in the pigs relative to the pigs fed a lysine-adequate diet. Thus, it can be concluded that IGF-1, instead of insulin or GH, in the blood circulation may be a key controlling growth factor in response to dietary provision of lysine for regulating muscle growth in late-stage finishing pigs.

Acknowledgements

This material is based upon the work supported by a Hatch/Multistate Project (under No. 1007691) funded from the National Institute of Food and Agriculture, United States Department of Agriculture. Authors wish to thank Dr. Jean Feugang and Dr. Derris Burnett in the Department of Animal and Dairy Sciences, Mississippi State University, for their guidance and training of TW on their laboratory work for this study. Donations of various feed ingredients from several swine operations and feed manufacturers, such as the Prestage Farms, Inc. (West Point, MS), Archer Daniels Midland Co., (Quincy, IL), and Ajinomoto Heartland, Inc. (Chicago, IL), are greatly appreciated.

Author details

Taiji Wang[1], Mark A. Crenshaw[1], Md Shamimul Hasan[1], Guoyao Wu[2] and Shengfa F. Liao[1]*

*Address all correspondence to: s.liao@msstate.edu

1 Department of Animal and Dairy Sciences, Mississippi State University, Mississippi, USA

2 Department of Animal Science, Texas A&M University, Texas, USA

References

[1] Liao SF, Wang T, Regmi N. Lysine nutrition in swine and the related monogastric animals: Muscle protein biosynthesis and beyond. SpringerPlus. 2015;**4**:147 (1-12)

[2] Sohal PS, Baracos VE, Clandinin MT. Dietary omega 3 fatty acid alters prostaglandin synthesis, glucose transport and protein turnover in skeletal muscle of healthy and diabetic rats. Biochemical Journal. 1992;**286**:405-411

[3] Fujita S, Dreyer HC, Drummond MJ, Glynn EL, Cadenas JG, Yoshizawa F, Volpi E, Rasmussen BB. Nutrient signalling in the regulation of human muscle protein synthesis. The Journal of Physiology. 2007;**582**:813-823

[4] Blad CC, Ahmed K, Ijzerman AP, Offermanns S. Biological and pharmacological roles of HCA receptors. Advances in Pharmacology. 2011;**62**:219-250

[5] Wu G. Amino acids: Metabolism, functions, and nutrition. Amino Acids. 2009;**37**:1-17

[6] Wu G. Functional amino acids in growth, reproduction, and health. Advances in Nutrition. 2010;**1**:31-37

[7] Schiaffino S, Mammucari C. Regulation of skeletal muscle growth by the IGF1-Akt/PKB pathway: Insights from genetic models. Skeletal Muscles. 2011;**1**:1-14

[8] Rhoads RP, Baumgard LH, El-Kadi SW, Zhao LD. Physiology and endocrinology symposium: Roles for insulin-supported skeletal muscle growth. Journal of Animal Science. 2016;**94**:1791-1802

[9] Straus DS. Nutritional regulation of hormones and growth factors that control mammalian growth. FASEB Journal. 1994;**8**:6-12

[10] Guay F, Trottier NL. Muscle growth and plasma concentrations of amino acids, insulin-like growth factor-I, and insulin in growing pigs fed reduced-protein diets. Journal of Animal Science. 2006;**84**:3010-3019

[11] Davis TA, Suryawan A, Orellana RA, Fiorotto ML, Burrin DG. Amino acids and insulin are regulators of muscle protein synthesis in neonatal pigs. Animal. 2010;**4**:1790-1796

[12] Saltiel AR, Kahn CR. Insulin signalling and the regulation of glucose and lipid metabolism. Nature. 2001;**414**:799-806

[13] NRC. Nutrient Requirements of Swine. 11th rev ed. Washington, DC: The National Academies Press; 2012. p. 420

[14] Roy N, Lapierre H, Bernier JF. Whole-body protein metabolism and plasma profiles of amino acids and hormones in growing barrows fed diets adequate or deficient in lysine. Canadian Journal of Animal Science. 2000;**80**:585-595

[15] Shelton NW, Tokach MD, Dritz SS, Goodband RD, Nelssen JL, DeRouchey JM. Effects of increasing dietary standardized ileal digestible lysine for gilts grown in a commercial finishing environment. Journal of Animal Science. 2011;**89**:3587-3595

[16] Wang T, Crenshaw MA, Regmi N, Armstrong T, Blanton JR, Liao SF. Effect of dietary lysine fed to pigs at late finishing stage on market-value associated carcass characteristics. Journal of Animal and Veterinary Advances. 2015;**14**:232-236

[17] Ren J, Zhao G, Li Y, Meng Q. Influence of dietary lysine level on whole-body protein turnover, plasma IGF-I, GH and insulin concentration in growing pigs. Livestock Science. 2007;**110**:126-132

[18] Katsumata M, Kawakami S, Kaji Y, Takada R, Dauncey MJ. Differential regulation of porcine hepatic IGF-I mRNA expression and plasma IGF-I concentration by a low lysine diet. Journal of Nutrition. 2002;**132**:688-692

[19] Dai ZL, Wu ZL, Jia SC, Wu G. Analysis of amino acid composition in proteins of animal tissues and foods as pre-column o-phthaldialdehyde derivatives by HPLC with fluorescence detection. Journal of Chromatography B. 2014;**964**:116-127

[20] Baumgard LH, Hausman GJ, Sanz Fernandez MV. Insulin: Pancreatic secretion and adipocyte regulation. Domestic Animal Endocrinology. 2016;**54**:76-84

[21] Avruch J, Hara K, Lin Y, Liu M, Long X, Ortiz-Vega S, Yonezawa K. Insulin and amino-acid regulation of mTOR signaling and kinase activity through the Rheb GTPase. Oncogene. 2006;**25**:6361-6372

[22] Proud CG. Regulation of protein synthesis by insulin. Biochemical Society Transactions. 2006;**34**:213-216

[23] O'Connor PMJ, Bush JA, Suryawan A, Nguyen HV, Davis TA. Insulin and amino acids independently stimulate skeletal muscle protein synthesis in neonatal pigs. American Journal Physiology. Endocrinology and Metabolism. 2003;**284**:E110-E119

[24] Suryawan A, O'Connor PMJ, Bush JA, Nguyen HV, Davis TA. Differential regulation of protein synthesis by amino acids and insulin in peripheral and visceral tissues of neonatal pigs. Amino Acids. 2009;**37**:97-104

[25] Floyd JC Jr, Fajans SS, Conn JW, Knopf RF, Rull J. Stimulation of insulin secretion by amino acids. The Journal of Clinical Investigation. 1966;**45**:1487-1502

[26] Caperna TJ, Steele NC, Komarek DR, McMurtry JP, Rosebrough RW, Solomon MB, Mitchell AD. Influence of dietary protein and recombinant porcine somatotropin administration in young pigs: Growth, body composition and hormone status. Journal of Animal Science. 1990;**68**:4243-4252

[27] Buse MG, Atwell R, Mancusi V. *In vitro* effect of branched chain amino acids on the ribosomal cycle in muscles of fasted rats. Hormone and Metabolic Research. 1979;**11**:289-292

[28] Anthony JC, Lang CH, Crozier SJ, Anthony TG, MacLean DA, Kimball SR, Jefferson LS. Contribution of insulin to the translational control of protein synthesis in skeletal muscle by leucine. American Journal of Physiology. Endocrinology and Metabolism. 2002;**282**:E1092-E1101

[29] Knopf RF, Conn JW, Fajans SS, Floyd JC, Guntsche EM, Rull JA. Plasma growth hormone response to intravenous administration of amino acids. The Journal of Clinical Endocrinology & Metabolism. 1965;**25**:1140-1144

[30] Isidori A, Lo Monaco A, Cappa M. A study of growth hormone release in man after oral administration of amino acids. Current Medical Research and Opinion. 1981;**7**:475-481

[31] Tomas FM, Campbell RG, King RH, Johnson RJ, Chandler CS, Taverner MR. Growth hormone increases whole-body protein turnover in growing pigs. Journal of Animal Science. 1992;**70**:3138-3143

[32] Florini JR, Ewton DZ, Coolican SA. Growth hormone and the insulin-like growth factor system in myogenesis. Endocrine Reviews. 1996;**17**:481-517

[33] Etherton TD. ASAS centennial paper: Animal growth and development research: Historical perspectives. Journal of Animal Science. 2009;**87**:3060-3064

[34] Evock CM, Caperna TJ, Steele NC, McMurtry JP, Rosebrough RW. Influence of time of injection of recombinant porcine somatotropin (rpST) relative to time of feeding on growth performance, hormone and metabolite status, and muscle RNA, DNA, and protein in pigs. Journal of Animal Science. 1991;**69**:2443-2451

[35] Thiel LF, Beermann DH, Krick BJ, Boyd RD. Dose-dependent effects of exogenous porcine somatotropin on the yield, distribution, and proximate composition of carcass tissues in growing pigs. Journal of Animal Science. 1993;**71**:827-835

[36] Takenaka A, Oki N, Takahashi SI, Noguchi T. Dietary restriction of single essential amino acids reduces plasma insulin-like growth factor-I (IGF-I) but does not affect plasma IGF-binding protein-1 in rats. Journal of Nutrition. 2000;**130**:2910-2914

[37] Thissen JP, Ketelslegers JM, Underwood LE. Nutritional regulation of the insulin-like growth factors. Endocrine Reviews. 1994;**15**:80-101

[38] Breier BH. Regulation of protein and energy metabolism by the somatotropic axis. Domestic Animal Endocrinology. 1999;**17**:209-218

Chapter 14

Identification and Differential Activity of Glutathione S-Transferase Mu in Strains of *Fasciola hepatica* Susceptible and Resistant to Triclabendazole

Vanesa Fernández

Additional information is available at the end of the chapter

http://dx.doi.org/10.5772/intechopen.69189

Abstract

Fasciola hepatica is a helminth parasite that causes fascioliasis in domestic ruminants and humans. Economic losses due to its infection are estimated in US$ 2000–3000 million yearly [1]. The anthelmintics are at present the only weapon against these parasitic helminths [2]. The parasite resistance to different anthelmintics including that of *F. hepatica* to triclabendazole (TCBZ) is growing worldwide. Glutathione S-transferases (GSTs) are enzymes involved in the detoxification of a wide range of substrates through chemical conjugation with glutathione, so that the product becomes more soluble in water, less toxic and easier to excrete. Eight GST isoenzymes are present in *F. hepatica* [3]. Since the different isoenzymes do not necessarily have the same metabolic activity, in the present work, we evaluated the metabolic activity of total cytosolic GST and GST mu and GST pi isoenzymes in adult strains of *F. hepatica* susceptible (*Cullompton*) and resistant (*Sligo* and *Oberon*) to TCBZ of the highest metabolic activity of total GST. The genetic sequence database at the National Center for Biotechnical Information (NCBI) (GenBank ID: KF680281–KF680282) corresponding to the GST mu gene isolated from *Cullompton strain* (TCBZ-susceptible) and (GenBank ID: KF680283–KF680284) corresponding to the GST mu gene isolated from *Sligo* strain (TCBZ-resistant) in *F. hepatica*. Comparative analysis of both strains, *Cullompton* and *Sligo*, showed two nucleotide changes and change of one amino acid in the GST mu isoenzyme of the TCBZ-resistant strain. These results together with the higher enzymatic activity of GST have a potential relevance as it contribute to the understanding the mechanisms that generate resistance to anthelmintics and the activity, metabolism, and disposition of these drugs in the parasite.

Keywords: *Fasciola hepatica*, triclabendazole, isoenzymes, glutathione S-transferases

1. Introduction

Fasciola hepatica is a helminth parasite that causes fascioliasis in domestic ruminants and humans. Economic losses due to its infection are estimated in U$S 2000–3000 million yearly [1]. The anthelmintics are at present the only weapon against these parasitic helminths [2]. The parasite resistance to different anthelmintics including that of *F. hepatica* to triclabendazole (TCBZ) is growing worldwide [3]. In the last few years, a rise in cattle fasciolosis cases has been reported, probably due to weather changes determining a different distribution of the snail *Galba truncatula,* which is a required intermediate host [4]. Helminth parasites possess different biochemical mechanisms for detoxification. Overall, parasites may evade drug antiparasitic effects by: (i) mutation of target receptors, (ii) overexpression of efflux transport pumps, and/or (iii) overexpression of metabolic enzymatic systems [5].

In fasciolosis, anthelmintic control is based mainly on the use of TCBZ, a halogenated benzimidazole thiol derivative that shows excellent efficacy against both juvenile (immature) and adult stages. The nematodicidal action of benzimidazoles (BZDs) is based on their binding to beta-Tubulin, which produces subsequent disruption of the tubulin-microtubule dynamic equilibrium [6]. When the *in vivo* effect of TCBZ on the distribution of alpha and beta-Tubulin was evaluated in the testis tubules of *F. hepatica* obtained from bovines exposed to this drug, the results obtained confirmed that TCBZ alters the distribution of the microtubules and reaffirmed that this is one of its main mechanisms of action [7, 8]. TCBZ is metabolized into its anthelmintically active metabolite sulphoxide (TCBZ-SO) by the host liver [9] but also by the parasite's subcellular fractions [10]. It has also been reported that *F. hepatica* has significantly higher sulphoxidation activity than nematode and cestode parasites [10]. The resistance to (BZDs) detected in other helminths such as *Haemonchus contortus* [11]. Although the flukicidal activity of TCBZ remains to be fully understood, there are data to support a microtubule-based action of this compound. However, it has been shown that the TCBZ-resistant *F. hepatica* is not associated with residue changes in the primary amino acid sequence of beta-Tubulin [11]. This suggests that there may be an alternative mechanism of TCBZ resistance in *F. hepatica* [5].

The development of drug resistance can be facilitated by the action of xenobiotic metabolizing enzymes (XMEs) of phase I and phase II of detoxification [12]. In all organisms, XMEs serve as an efficient defense against the potential negative action of xenobiotics. Several phase I enzymes are expressed in mammalians, where they introduce or unmask new functionalities on xenobiotic compounds. Examples of these enzymes include cytochrome P450s (Cyt P450), flavin-containing monooxygenases (FMO), alcohol and aldehyde dehydrogenases, and esterases. To eliminate a large array of chemicals, living organisms have developed, in virtually all tissue enzyme systems, XMEs that transform exogenous and endogenous compounds into more hydrophilic derivatives through reactions collectively known as biotransformation. At present, much less is known about the activity of certain phase II enzymes and relatively less attention has been paid to hydrolytic and conjugative pathways. Many phase II reactions in mammals involve conjugating potentially toxic substances to glutathione. These reactions are mediated by the enzyme glutathione S-transferase (GST) enzymes [13]. Recent research has highlighted the importance of these transferases in the establishment of chronic helminth infections. These proteins appear to be the main phase II detoxification system present in parasitic worms. General biological roles of helminth GSTs include xenobiotic detoxification and ligand binding/transport functions [14].

As many as eight GST isoforms have been shown to be present in *F. hepatica* [15]. The genes encoding the mu class of enzymes are organized in a gene cluster on 45 chromosome 1p13.3 and are known to be highly polymorphic. The ability of helminth GSTs to effectively neutralize known cytotoxic products arising from the attack of reactive oxygen species on cell membranes provides evidence that GSTs have the potential to protect the parasite against different xenobiotics. In *F. hepatica*, GSTs are found in the tegument, muscular tissues, parenchymal cells, and the intestine [16]. GSTs account for as much as 4% of the total soluble protein [17] and are major detoxification enzymes in adult helminths, as these organisms appear to lack the important Cyt P450-dependent detoxification reactions [18]. Results obtained in our laboratory confirmed that Cyt P450 are not only involved in detoxification mechanisms but also actively participate in the development of resistance to TCBZ by the trematode [19]. GSTs have been investigated in parasitic worms with respect to their biochemistry and have also been identified as potential vaccine candidates in digenean parasite. This property has been exploited with cGSTs [20–22]. Most studies concerning the metabolic response of *F. hepatica* against the anthelmintic TCBZ have used *in vitro* or *ex vivo* test models [23].

Whereas the background about the interactions with such enzymatic systems may drastically affect the disposition kinetics of different drugs [24], the aim of the present work was to evaluate *in vitro* of total cGST and cGSTmu and cGSTpi isoenzymes in the susceptible (*Cullompton*) and resistant (*Sligo* and *Oberon*) strains of *F. hepatica* and identified and characterized the gene GST mu isolated isolate from *Cullompton strain* (TCBZ-susceptible) and *Sligo* (TCBZ-resistant) in *F. hepatica.*

2. Materials and methods

2.1. Collection of parasite material

Nine (9) parasite-free Corriedale weaned lambs were orally infected each with 200 metacercariae of *F. hepatica.*

2.2. Collection and processing of adult flukes

Adult flukes were collected from bile ducts and liver and processed. The collection of the flukes, their processing to obtain the cytosolic [25].

2.3. Preparation of cytosolic fractions

Parasite specimens (10–15 g) of the TCBZ-susceptible or TCBZ-resistant isolates of *F. hepatica* were rinsed with cold KCl (1.15%) and then transported to the laboratory in flasks filled with phosphate buffer (PB) (0.1 M, pH 7.4) at 4°C. All subsequent operations were performed between 0 and 4°C. Each sample was cut into small pieces and washed several times with PB. Samples were homogenized

(1:1) in PB, pH 7.4 with an Ultra-Turrax homogenizer (IKA Works Inc., Wilmington, USA), centrifuged at 10,000 g for 20 min and the resulting supernatant centrifuged at 100,000 g for 60 min.

The supernatant was collected and stored at −80°C until the analysis. Protein content from the supernatant fractions was determined using bovine serum albumin as a standard [26].

Total cGST activity using 1-chloro, 2,4-dinitrobenzene as substrate (CDNB), GST-pi activity using ethacrynic acid as substrate and GST mu activity using 3,4-dichloronitrobenzene (DCNB) as substrate were monitored by a continuous spectrophotometric method [27].

2.4. Reverse transcription polymerase chain reaction (RT-PCR)

Total RNA was isolated from each strain of adult trematodes ($n = 15$) using Trizol, and reverse transcribed using superscript III RNAase® (Applied Biosystems, Brunn am Gebirge, Austria) following the protocol recommended by the manufacturers [28].

2.5. PCR amplification of cDNA

The PCR product was analyzed by electrophoresis in 1% agarose gel.

3. Results

Total GST activity ($n = 13$) in each strains was different in all strains tested. In the Sligo (1277 ± 32 nmol/min/mg protein) and Oberon (1216 ± 16 nmol/min/mg protein), strains were 59 and 52%, respectively, higher ($P < 0.001$) than in the susceptible *Cullompton strain* (800 ± 60 nmol/min/mg protein) (**Figure 1**). GST mu activity in Oberon (1.37 nmol/min/mg protein)

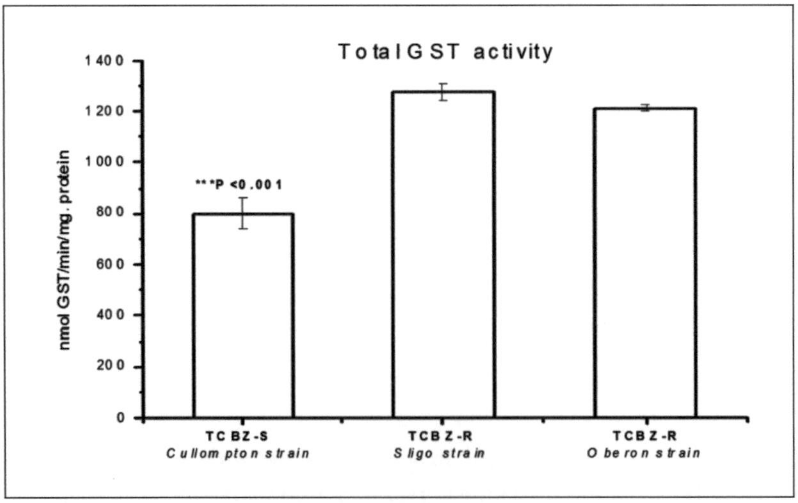

Figure 1. Quantification of total activity of glutathione S-transferase (GST) in *Fasciola hepatica* susceptible (*Cullompton strain*) and resistant (*Sligo* and *Oberon* strains) to triclabendazole.

and Sligo (1.28 nmol/min/mg protein) resistant strains was 17 and 26%, respectively, higher than in the susceptible *Cullompton strain* (0.8 nmol/min/mg protein) (**Figure 2**) while GST-pi activity did not differ between the different strains tested (**Figure 3**). RT-PCR (**Figure 4**). The genetic sequence database at the National Center for Biotechnical Information (NCBI) (GenBank ID: KF680281–KF680282) corresponding to the GST mu gene isolated from the *Cullompton strain* (TCBZ-susceptible) and (GenBank ID:KF680283–KF680284) to the GST

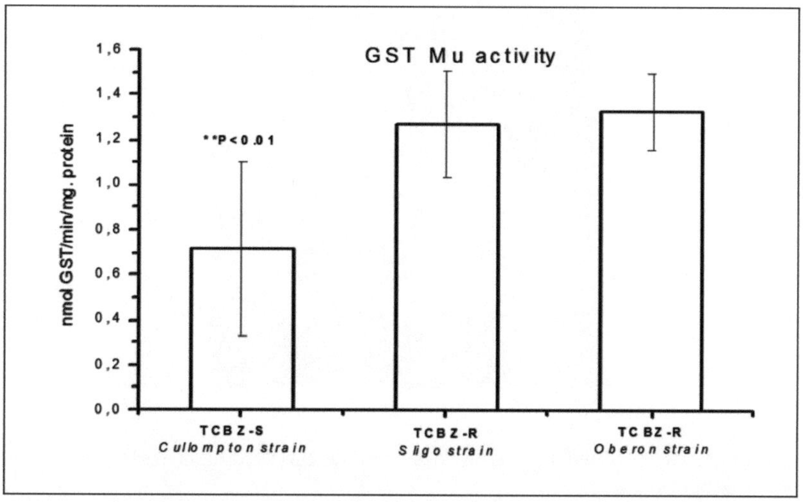

Figure 2. Quantification of activity of mu glutathione S-transferase (GST) in *Fasciola hepatica* susceptible (*Cullompton strain*) and resistant (*Sligo* and *Oberon* strains) to triclabendazole.

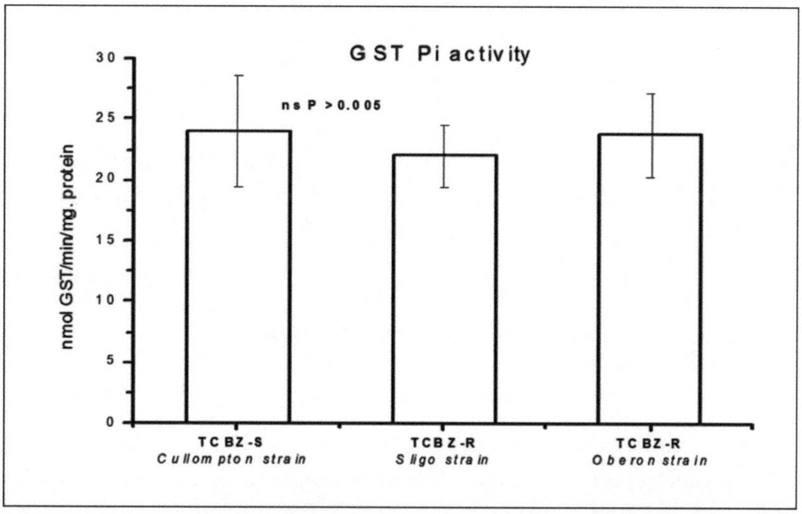

Figure 3. Quantification of activity of pi glutathione S-transferase (GST) in *Fasciola hepatica* susceptible (*Cullompton strain*) and resistant (*Sligo* and *Oberon* strains) to triclabendazole.

Figure 4. A 657-bp band compatible with the expected size in the 100-bp DNA ladder marker (lane 1), *Cullompton strain* (lane 2), and *Sligo* (lane 3) was obtained.

mu gene isolated from the Sligo strain (TCBZ-resistant) in *F. hepatica* (**Figures 5** and **6**) can be substituted by other polar or small amino acids in particular threonine which differs only in that it has a methyl group in place of a hydrogen group found in serine. Serines are quite common in protein functional centers.

4. Discussion

In the absence of an efficacious vaccine, chemotherapy remains the main tool in treating fasciolosis. Although other alternatives exist, current measures to control fasciolosis are based on the use of drugs such as triclabendazole (TCBZ) [29, 30].

In the anthelmintics, has been adults [9, 29, 31].

> *Parasite defense mechanisms include detoxifying and anti-oxidant enzymes that would suppress its oxidative killing [23]. Therefore, it is necessary to know the mechanisms of detoxification and mechanism of anthelmintic resistance of F. hepatica.*

Other possibilities include enhanced substrate affinity of the enzymes brought about by mutations within their encoding genes. Residue changes may also influence the substrate specificity of the enzymes and could explain why TCBZ-resistant flukes remain susceptible to ABZ [14, 32–35].

http://dx.doi.org/10.5772/intechopen.69189

```
Gst Cullompton strain TATGCCAGCCCAAACTCGGATACTGGAAAATAAGAGGGCTCCAACAACCCGTTCGACTCT 60
Gst Sligo strain      TATGCCAGCCCAAACTCGGATACTGGAAAATAAGAGGGCTCCAACAACCCGTTCGACTCT 60
Gst Cullompton strain --------------------------------AGAGGGCTCCAACAACCCGTTCGACTCT 28
Gst Sligo strain      ------------------------------------GGCTCCAACAACCCGTTCGACTCT 24
                                                          ************************

Gst Cullompton strain TGCTCGAATACCTGGGTGAAGAGTATGAAGAACATCTGTACGGTCGTGATGATAGGGAGA 120
Gst Sligo strain      TGCTCGAATACCTGGGTGAAGAGTATGAAGAACATCTGTACGGTCGTGATGATAGGGAGA 120
Gst Cullompton strain TGCTCGAATACCTGGGTGAAGAGTATGAAGAACATCTGTACGGTCGTGATGATAGGGAGA 88
Gst Sligo strain      TGCTCGAATACCTGGGTGAAGAGTATGAAGAACATCTGTACGGTCGTGATGATAGGGAGA 84
                      ************************************************************

Gst Cullompton strain AATGGTTTGGCGATAAATTCAACATGGGATTGGATTTGCCAAATTTACCATACTACATTG 180
Gst Sligo strain      AATGGTTTGGCGATAAATTCAACATGGGATTGGATTTGCCAAATTTACCATACTACATTG 180
Gst Cullompton strain AATGGTTTGGCGATAAATTCAACATGGGATTGGATTTGCCAAATTTACCATACTACATTG 148
Gst Sligo strain      AATGGTTTGGCGATAAATTCAACATGGGATTGGATTTGCCAAATTTACCATACTACATTG 144
                      ************************************************************

Gst Cullompton strain ACGATAAGTGCAAACTGACTCAGTCGGTGGCCATAATGCGGTACATTGCGGATAAGCATG 240
Gst Sligo strain      ACGATAAGTGCAAACTGACTCAGTCGGTGGCCATAATGCGGTACATTGCGGATAAGCATG 240
Gst Cullompton strain ACGATAAGTGCAAACTGACTCAGTCGGTGGCCATAATGCGGTACATTGCGGATAAGCATG 208
Gst Sligo strain      ACGATAAGTGCAAACTGACTCAGTCGGTGGCCATAATGCGGTACATTGCGGATAAGCATG 204
                      ************************************************************

Gst Cullompton strain GAATGCTTGGTTCGACACCCGAGGAACGAGCTCGAATTTCGATGATCGAAGGAGCTGCAA 300
Gst Sligo strain      GAATGCTTGGTTCTACACCCGAGGAACGAGCTCGAATTTCGATGATCGAAGGAGCTGCAA 300
Gst Cullompton strain GAATGCTTGGTTCGACACCCGAGGAACGAGCTCGAATTTCGATGATCGAAGGAGCTGCAA 268
Gst Sligo strain      GAATGCTTGGTTCTACACCCGAGGAACGAGCTCGAATTTCGATGATCGAAGGAGCTGCAA 264
                      ************* **********************************************

Gst Cullompton strain TGGATCTTCGGATGGGTTTTGTTCGTGTTTGTTACAACCCAAAATTTGAAGAAGTGAAAG 360
Gst Sligo strain      TGGATCTTCGGATGGGTTTTGTTCGTGTTTGTTACAACCCAAAATTTGAAGAAGTGAAAG 360
Gst Cullompton strain TGGATCTTCGGATGGGTTTTGTTCGTGTTTGTTACAACCCAAAATTTGAAGAAGTGAAAG 328
Gst Sligo strain      TGGATCTTCGGATGGGTTTTGTTCGTGTTTGTTACAACCCAAAATTTGAAGAAGTGAAAG 324
                      ************************************************************

Gst Cullompton strain GAGATTATCTGAAAGAACTGCCAACAACGTTGAAGATGTGGTCCGATTTTCTTGGAGATC 420
Gst Sligo strain      GAGATTATCTGAAAGAACTGCCAACAACGTTGAAGATGTGGTCCGATTTTCTTGGAGATC 420
Gst Cullompton strain GAGATTATCTGAAAGAACTGCCAACAACGTTGAAGATGTGGTCCGATTTTCTTGGAGATC 388
Gst Sligo strain      GAGATTATCTGAAAGAACTGCCAACAACGTTGAAGATGTGGTCCGATTTTCTTGGAGATC 384
                      ************************************************************

Gst Cullompton strain GTCACTATTTGACAGGTTCTTCAGTTAGCCATGTGGACTTTATGGTTTACGAAGCATTGG 480
Gst Sligo strain      GTCACTATTTGACAGGTTCTACAGTTAGCCATGTGGACTTTATGGTTTACGAAGCATTGG 480
Gst Cullompton strain GTCACTATTTGACAGGTTCTTCAGTTAGCCATGTGGACTTTATGGTTTACGAAGCATTGG 448
Gst Sligo strain      GTCACTATTTGACAGGTTCTACAGTTAGCCATGTGGACTTTATGGTTTACGAAGCATTGG 444
                      ******************** ***************************************

Gst Cullompton strain ACTGTATTCGTTATTTGGCACCACAGTGTCTGGAGGACTTTCCCAAATTGAAGGAATTCA 540
Gst Sligo strain      ACTGTATTCGTTATTTGGCACCACAGTGTCTGGAGGACTTTCCCAAATTGAAGGAATTCA 540
Gst Cullompton strain ACTGTATTCGTTATTTGGCACCACAGTGTCTGGAGGACTTTCCCAAATTGAAGGAATTCA 508
Gst Sligo strain      ACTGTATTCGTTATTTGGCACCACAGTGTCTGGAGGACTTTCCCAAATTGAAGGAATTCA 504
                      ************************************************************

Gst Cullompton strain AGAGTCGTATTGAAGATCTTCCAAAAATCAAGGCATACATGGAATCAGAGAAGTTCATCA 600
Gst Sligo strain      AGAGTCGTATTGAAGATCTTCCAAAAATCAAGGCATACATGGAATCAGAGAAGTTCATCA 600
Gst Cullompton strain AGAGTCGTATTGAAGATCTTCCAAAAATCAAGGCATACATGGAATCAGAGAAGTTCATCA 568
Gst Sligo strain      AGAGTCGTATTGAAGATCTTCCAAAAATCAAGGCATACATGGAATCAGAGAAGTTCATCA 564
                      ************************************************************

Gst Cullompton strain AGTGGCCTTTGAACTCGTGGATTGCTTC-------------------------------- 628
Gst Sligo strain      AGTGGCCTTTAAACTCGTGGATT------------------------------------- 623
Gst Cullompton strain AGTGGCCTTTGAACTCGTGGATTGCTTCTTTCGGTGGTGGAGACGCTGACGCTGGCCTGC 628
Gst Sligo strain      AGTGGCCTTTGAACTCGTGGATTGCTTCTTTCGGTGGTGGAGACGCTGACGCTGGCCTGC 624
                      ********** ************

Gst Cullompton strain ----
Gst Sligo strain      ----
Gst Cullompton strain NAN- 631
Gst Sligo strain      NNNA 628
```

Figure 5. CLUSTAL 2.1 multiple sequence alignment.

```
> Gst Cullompton strain
LGYWKIRGLQQPVRLLLEYLGEEYEEHLYGRDDREKWFGDKFNMGLDLPNLPYYIDDKCKLTQSVAIMRY
IADKHGMLGSTPEERARISMIEGAAMDLRMGFVRVCYNPKFEEVKGDYLKELPTTLKMWSDFLGDRHYLT
GSTVSHVDFMVYEALDCIRYLAPQCLEDFPKLKEFKSRIEDLPKIKAYMESEKFIKWPLNSWIASFGGGD
A

> Gst Sligo strain
CQPKLGYWKIRGLQQPVRLLLEYLGEEYEEHLYGRDDREKWFGDKFNMGLDLPNLPYYIDDKCKLTQSVA
IMRYIADKHGMLGSTPEERARISMIEGAAMDLRMGFVRVCYNPKFEEVKGDYLKELPTTLKMWSDFLGDR
HYLTGSSVSHVDFMVYEALDCIRYLAPQCLEDFPKLKEFKSRIEDLPKIKAYMESEKFIKWPLNSWIASF
GGGDAD

Gst Cullompton strain ----LGYWKIRGLQQPVRLLLEYLGEEYEEHLYGRDDREKWFGDKFNMGLDLPNLPYYID 56
Gst Sligo strain      CQPKLGYWKIRGLQQPVRLLLEYLGEEYEEHLYGRDDREKWFGDKFNMGLDLPNLPYYID 60
                          ********************************************************

Gst Cullompton strain DKCKLTQSVAIMRYIADKHGMLGSTPEERARISMIEGAAMDLRMGFVRVCYNPKFEEVKG 116
Gst Sligo strain      DKCKLTQSVAIMRYIADKHGMLGSTPEERARISMIEGAAMDLRMGFVRVCYNPKFEEVKG 120
                      ************************************************************

Gst Cullompton strain DYLKELPTTLKMWSDFLGDRHYLTGSTVSHVDFMVYEALDCIRYLAPQCLEDFPKLKEFK 176
Gst Sligo strain      DYLKELPTTLKMWSDFLGDRHYLTGSSVSHVDFMVYEALDCIRYLAPQCLEDFPKLKEFK 180
                      ************************** *********************************

Gst Cullompton strain SRIEDLPKIKAYMESEKFIKWPLNSWIASFGGGDA- 211
Gst Sligo strain      SRIEDLPKIKAYMESEKFIKWPLNSWIASFGGGDAD 216
                      ***********************************
```

Figure 6. Amino acid alignments of GST mu isoenzyme of *Fasciola hepatica* susceptible and resistant to TCBZ. Light shading indicates change of one amino acid at position 143 in the TCBZ resistant strain.

GSTs are regulated by a structurally diverse range of xenobiotics, and at least 100 chemicals have been identified to induce GSTs. Many of the compounds that induce GSTs are themselves substrates for these enzymes or are metabolized (by CytP450 or FMO) to compounds that can serve as GST substrates, suggesting that GST induction represents part of an adaptive response mechanism to chemical stress caused by electrophiles [36].

In the present work, cGST was analyzed in three strains of *F. hepatica* (*Cullompton*, a TCBZ-susceptible strain, and *Sligo* and *Oberon*, TCBZ-resistant strains).

The activity of *Sligo* and *Oberon* strains expressed significantly higher metabolic activity than that measured in the cytosolic fractions obtained from the susceptible strain. In this work identified and characterized the GST mu gene isolated from TCBZ-susceptible and TCBZ-resistant *F. hepatica* strains, and comparative analysis of both strains *Cullompton* and *Sligo* showed change two nucleotide and changes GST mu protein: Threonine in the TCBZ-susceptible strain by Serine in the TCBZ-resistant strain can be substituted by other polar or small amino acids in particular threonine which differs only in that it has a methyl group in place of a hydrogen group found in serine [37]. These genetic variations can change an individual's susceptibility to carcinogens and toxins as well as affect the toxicity and efficacy of certain drugs but GST mu protein in active sites not present the classical Asp-His-Ser therefore not would find affected by this change of amino acid mu GST biological activity.

5. Conclusion

GST activity has this great potential importance as it might contribute to generating the phenomenon of resistance to TCBZ. These results contribute to the understanding not only of this metabolic pathway but also of the mechanism of resistance to TCBZ in *F. hepatica*. The results also add information to the knowledge of the response that the parasites have exposure to different xenobiotics.

Author details

Vanesa Fernández

Address all correspondence to: vanesaf@vet.unicen.edu.ar

Laboratory of Immunology, Center of Veterinary Research of Tandil (CIVETAN-CONICET), Faculty of Cs. Veterinary (UNCPBA), University Campus, Tandil, Argentina

References

[1] Boray JC. Disease of Domestic Animals Caused by Flukes; 1994. p. 49

[2] Mas-Coma S, Valero MA, Bargues MD. Fasciola, lymnaeids and human fascioliasis, with a global overview on disease transmission, epidemiology, evolutionary genetics, molecular epidemiology and control. Advances in Parasitology. 2009;**69**:41-147

[3] World Health Organization. Report of the WHO Informal Meeting on Use of Triclabendazole in Fascioliasis Control, Held at WHO Headquarters, October 2006, Geneva, Switzerland; 2007

[4] Mas-Coma S, Valero MA, Bargues MD. Climate change effects on trematodiases, with emphasis on zoonótica fascioliasis and schistosomiasis. Veterinary Parasitology. 2009;**163**:264-280

[5] Alvarez LI, Solana HD, Mottier ML, Virkel GL, Fairweather I, Lanusse CE. Altered drug influx/efflux and enhanced metabolic activity in triclabendazole-resistant liver flukes. Parasitology. 2005;**131**:501-510

[6] Lacey E. The role of the cytoskeletal protein, tubulin, in the mode of action and mechanism of drug resistance to benzimidazoles. International Journal for Parasitology. 1988;**18**: 885-936

[7] Solana H, Scarcella S, Gentile ML, Martínez A, Alzola R. Inmunolocalización de Tubulina en Túbulos testiculares de *Fasciola hepatica* expuesta a triclabendazole "in vivo". In Vet. 2009;**11**(2):105-115

[8] Scarcella S, Fiel C, Guzman M, Alzola R, Felipe A, Hanna R, Fairweather I, McConnell S, Solana H. Reproductive disruption in *Fasciola hepatica* associated with incomplete efficacy of a new experimental formulation of triclabendazole. Veterinary Parasitology. 2011;**176**:157-164

[9] Virkel G, Lifschitz A, Sallovitz J, Pis A, Lanusse C. Assessment of the main metabolism pathways for the flukicidal compound triclabendazole in sheep. Journal of Veterinary Pharmacology and Therapeutics. 2006;**29**(3):213-223

[10] Mottier L, Virkel G, Solana H, Alvarez L, Salles J, Lanusse C. Triclabendazole biotransformation and comparative diffusion of the parent drug and its oxidized metabolites into *Fasciola hepatica*. Xenobiotica. 2004;**34**(11-12):1043-1057

[11] Robinson MW, Trudgett A, Hoey EM, Fairweather I. Triclabendazole-resistant *Fasciola hepatica*: β-tubulin and response to in vitro treatment with triclabendazole. Parasitology. 2002;**124**:325-338

[12] Cvilink V, Lamka J, Skálová L. Xenobiotic metabolizing enzymes and metabolism of anthelmintics in helminths. Drug Metabolism Reviews. 2009;**41**(1):8-26

[13] Robinson MW, Lawson J, Trudgett A, Hoey EM, Fairweather I. The comparative metabolism of triclabendazole sulphoxide by triclabendazole-susceptible and triclabendazole resistant *Fasciola hepatica*. Parasitology Research. 2004;**92**:205-210

[14] Farahnak A, Barrett J. Comparative glutathione S-transferases (GSTs) inhibition assay in the whole extract of *Fasciola hepatica* and sheep liver tissue by hexachlorophene. Iranian Journal of Public Health. 2001;**30**(3-4):125-128

[15] Wijffels GL, Sexton JL, Salvatore L, Pettitt JM, Humphris DC, Panaccio M, Spithill TW. Primary sequence heterogeneity and tissue expression of glutathione-S-transferases of *Fasciola hepatica*. Experimental Parasitology. 1992;**74**:87-99

[16] Panaccio M, Wilson LR, Crameri SL, Wijffels GL, Spithill TW. Molecular characterisation of cDNA sequences encoding glutathione S-transferases of *Fasciola hepatica*. Experimental Parasitology. 1992;**74**:232-237

[17] Brophy PM, Crowley P, Barrett J. Detoxification reactions of *Fasciola hepatica* cytosolic glutathione transferases. Molecular and Biochemical Parasitology. 1990;**39**(2):155-161

[18] Precious WY, Barrett J. The possible absence of cytochrome P-450 linked xenobiotic metabolism in helminths. Biochimica et Biophysica Acta. 1989;**992**:215-222

[19] Lamenza P, Ortiz P, Ceriani C, Solana H. Identification and characterization of phase I detoxification enzymes in isolates of *Fasciola hepatica*to triclabendazole susceptible and resistant. In: Proceedings of the 24th International Conference of the World Association for the Advancement of Veterinary Parasitology; Perth, Australia; 2012

[20] Mannervik B, Widersten M. Human glutathione transferases: Classification, tissue distribution, structure, and functional properties. In: Pacifici GM, Fracchia GN, editors. Advances in Drug Metabolism in Man. Luxembourg: Commission of the European Communities; 1995

[21] Morgenstern R, Guthenberg C, DePierre JW. Microsomal glutathione S-transferase: Purification, initial characterization and demonstration that it is not identical to the cytosolic glutathione S-transferases A, B and C. European Journal of Biochemistry. 1982;**128**:243-248

[22] Andersson C, Mosialou E, Weinander R, Morgenstern R. Enzymology of microsomal glutathione S-transferase. Advances in Pharmacology. 1994;**27**:19-35

[23] Shehab Amel Y, Ebeid Samia M, El-Samak Mohamed Y, Hussein Neveen M. Detoxifying and anti-oxidant enzymes of *Fasciola gigantica* worms under triclabendazole sulphoxide (TCBZ-SX): An in vitro study. Journal of the Egyptian Society of Parasitology. 2009;**39**:73-83

[24] AVMA. Report of the AVMA panel on euthanasia. Journal of the American Veterinary Medical Association. 2001;**218**:669-696

[25] Solana H, Scarcella S, Virkel G, Ceriani C, Rodríguez J, Lanusse C. Albendazole enantiomeric metabolism and binding to cytosolic proteins in the liver fluke *Fasciola hepatica*. Veterinary Research Communications. 2009;**33**(2):163-173

[26] Lowry O, Rosebrough N, Farr A, Randall R. Protein measurement with the Folin phenol reagent. Journal of Biological Chemistry. 1951;**193**:265-275

[27] Habig WH, Pabst MJ, Jakoby WB. Glutathione S-transferases. The first enzymatic step in mercapturic acid formation. Journal of Biological Chemistry. 1974;**249**(22):7130-7139

[28] Jedeppa A, Raina OK, Samanta S, Nagar G, Kumar N, Varghese A, Gupta SC, Banerjee PS. Molecular cloning and characterization of a glutathione S-transferase in the tropical liver fluke, *Fasciola gigantica*. Journal of Helminthology. 2010;**84**:55-60

[29] Fairweather I. Triclabendazole progress report, 2005-2009: An advancement of learning. Journal of Helminthology. 2009;**83**:139-150

[30] Brennan GP, Fairweather I, Trudgett A, Hoey E, McCoy, McConville M, Meaney M, Robinson M, McFerran N, Ryan L, Lanusse C, Mottier L, Alvarez L, Solana H, Virkel G, Brophy PM. Understanding triclabendazole resistance. Experimental and Molecular Pathology. 2007;**82**:104-109

[31] Gusson F, Carletti M, Giuliano Albo A, Dacasto M, Nebbia C. Comparison of hydrolytic and conjugative biotransformation pathways in horse, cattle, pig, broiler, chick, rabbit and rat liver subcellular fractions. Veterinary Research Communications. 2006;**30**:271-283

[32] Devine C, Brennan GP, Lanusse CE, Alvarez LI, Trudgett A, Hoey E, Fairweather I. Effect of the metabolic inhibitor, methimazole on the drug susceptibility of a triclabendazole-resistant isolate of *Fasciola hepatica*. Parasitology. 2009;**136**(2):183-192

[33] Coles GC, Stafford KA. Activity of oxyclozanide, nitroxynil, clorsulon and albendazole against adult triclabendazole-resistant *Fasciola hepatica*. Veterinary Record. 2001;**148**(23): 723-724

[34] Creaney J, Wijffels GL, Sexton JL, Sandeman RM, Spithill TW, Parson JC. *Fasciola hepatica*: Localisation of glutathione S-transferase isoenzymes in adult and juvenile liver fluke. Experimental Parasitology. 1995;**81**(1):106-116

[35] Solana H, Scarcella S, Alzola R, Lanusse C. P-glycoprotein immunolocalization and glutathione-S-transferase activity in *Fasciola hepatica* recovered from triclabendazole treated sheep. Journal of Veterinary Pharmacology and Therapeutics. 2009;**32**(Suppl 1):129-265

[36] Kerboeuf D, Aycardi J. Unexpected increased thiabendazole tolerance in *Haemonchus contortus* resistant to anthelmintics by modulation of glutathione activity. Parasitology Research. 1999;**85**:713-718

[37] Betts MJ, Russell RB. Amino acid properties and consequences of substitutions. In: Bioinformatics for Geneticists. Chapter 14; 2003. pp. 290-316